# ECONOMICS FOR AGRICULTURE

# Food, Farming and the Rural Economy

**Berkeley Hill**
**and**
**Derek Ray**

First published 1987

Published by
MACMILLAN EDUCATION LTD
Houndmills, Basingstoke, Hampshire RG21 2XS
and London
Companies and representatives
throughout the world

Printed in Hong Kong

British Library Cataloguing in Publication Data
Hill, Berkeley
Economics for agriculture: food, farming
and the rural economy.
1. Agriculture—Economic aspects—Great
Britain
I. Title   II. Ray, Derek
338.1′0941   HD1925
ISBN 0–333–35224–6 (hardcover)
ISBN 0–333–35225–4 (paperback)

To Hilarie, Debbie, Tim and Emily
Julia, Douglas, Duncan and Colin

# Contents

# List of Figures

# List of Tables

# Foreword

This book emerged from a teaching course taken by a wide spectrum of second-year undergraduate students at Wye College. Mainly they were studying for degrees in agriculture and had already covered a course in basic economics. Mixed in with these were others following studies in environmental subjects and some specialist agricultural economists. From this original audience the nature of our text should be clear; it is aimed at any student at university or college level who requires a broad picture of the agricultural industry in the UK. In the 1980s it is no longer appropriate to assume that all students are young. There are many post-experience or part-time courses to cater for the wide spectrum of older people interested in rural issues. These also were in our minds and helped shape our approach. The language used is non-technical. Though a knowledge of economic principles is an advantage, a reader completely fresh to this area of study should find no difficulty in understanding our material.

In the past there has been a tendency to regard farming as separate from the consumer, understandable in view of the isolation from the free market that has typified this industry for the last half-century. In response to the increasing awareness that this isolationism is misguided and, for reasons described in this book, cannot continue, we have deliberately stressed the linkage of farming with the rest of the food chain. And as farming is only one user of rural resources, we attempt to look at its activities and use of the countryside within the context of the whole rural economy. Our focus is primarily on farming because this is the point from which most of our readers will start. However, by taking a broad view we hope that we have something to say not only to students involved with agriculture, narrowly defined, but to geographers, economists, natural scientists, sociologists, bankers, businessmen and people in other fields of interest on which agriculture impinges.

As a deliberate policy, when in this text changes are considered (and much of the reason for studying agriculture at all involves its changing nature) attention is given primarily to the medium and longer term. We concentrate on the more permanent trend characteristics that lie behind everyday issues of farming and the market for food in Britain. These trends help explain the problems faced by those actively engaged in farming and those who service the industry in a variety of ways.

In a text covering such a wide range of material it is inevitable that teachers will find that our degree of detail is not uniform. To some extent the balance here reflects our personal interests and estimations of what aspects of agriculture a reasonably well informed student should be aware. Nevertheless we hope that overall the text will be helpful and that we provoke the reader to explore the references.

Detectives of style may wish to know that Derek Ray is responsible for Chapters 2 to 4, Berkeley Hill for Chapters 5 to 9, 11, 12, 14 and 15. The remainder (Chapters 1, 10 and 13) have contributions from both.

BERKELEY HILL
DEREK RAY

# Acknowledgements

The author and publishers wish to thank the following who have kindly given permission for the use of copyright material;

The Controller of Her Majesty's Stationery Office for tables and figures from the *Annual Review of Agriculture* (1986); Croom Helm Ltd for material from *The Changing Countryside* by B. Hill (1985); *The Daily Telegraph* for the graph 'The cost of farmland', August 1973; Department of Trade and Industry for material compiled by the Business Statistics Office, *Business Monitor*; *European Review of Agricultural Economics* for a figure from 'Inhibitors to change in agriculture: is a pluridisciplinary approach needed?' by S. L. Louwes, E.R.A.E. 4 (3); Faber & Faber Ltd for figure from *Agricultural Resources* by A. Edwards and A. Rogers; Food and Agriculture Organization of the United Nations for table from *FAO Agricultural Commodities Projections for 1970* (1962); Food for Britain for figure from 'Food Distribution: its Impact on Marketing in the 1980's' by J. J. Tanburn (1981); The Open University Press for table 'Characteristics of farmers in five categories . . .' from D203 *Decision Making in Britain: Agriculture*, copyright © The Open University Press (1972); The Statistical Office of the European Community for material from EC publications Green Europe, National Accounts 1978 as Agricultural Situation in the European Community.

Every effort has been made to trace all the copyright-holders, but if any have been inadvertently overlooked the publishers will be pleased to make the necessary arrangement at the first opportunity.

# Abbreviations and Glossary

| | |
|---|---|
| AAS | Annual Abstract of Statistics (London: HMSO) |
| ACP | African, Caribbean and Pacific countries (linked to the EC through the Lomé Agreement) |
| ADAS | Agricultural Development and Advisory Service |
| *Am. J. Ag. Econ.* | *American Journal of Agricultural Economics* |
| ATB | Agricultural Training Board |
| BSU | British Size Unit |
| *Can. J. Ag. Econ.* | *Canadian Journal of Agricultural Economics* |
| CAP | Common Agricultural Policy of the EC |
| CAS | Centre for Agricultural Strategy, University of Reading |
| CCAHC | Central Council for Agricultural and Horticultural Cooperatives (incorporated into FFB, 1984) |
| CCD | Common Customs Duty (Also called Common External Tariff. For all EC/non-EC trade. The level of tariff varies from product to product) |
| CEAS | Centre for European Agricultural Studies, Wye College, University of London |
| CGT | Capital Gains Tax |
| CLA | Country Landowners Association |
| COMA | Committee on the Medical Aspects of Food Policy (which reported to the DHSS) |
| CSO | Central Statistical Office |
| CTT | Capital Transfer Tax |
| DAFS | Department of Agriculture and Fisheries for Scotland |
| DANI | Department of Agriculture for Northern Ireland |
| DHSS | Department of Health and Social Security |
| EAGGF | European Agricultural Guidance and Guarantee Fund (also called FEOGA, its French acronym); |

|   |   |
|---|---|
|   | that part of the European Fund of the EC allocated to CAP expenditure |
| EC | European Communities (the official name for the European Economic Community, European Coal and Steel Community, and the European Atomic Energy Commission, which merged in 1967 to form the EC) |
| ECU (or ecu) | European Currency Unit |
| EDC | Economic Development Committee for Agriculture, part of the National Economic Development Office |
| EEC | European Economic Community |
| ESU | European Size Unit |
| EUR (9) | European Community with 9 member states; the addition of Greece raises the size to 10, and Spain and Portugal to 12 |
| *Eur. Rev. Ag. Econ.* | *European Review of Agricultural Economics* |
| FADN | Farm Accountancy Data Network (also called RICA), based on national surveys of farm accounts in the EC |
| FAS | see FMS |
| FDI | Food and Drink Industries |
| FFB | Food From Britain |
| FMS | Farm Management Survey (in Scotland, the Farm Accounts Survey) Part of the survey results are sent to the EC for FADN |
| GATT | General Agreement on Tariffs and Trade |
| GB | Great Britain (England, Wales and Scotland) |
| GDP | Gross Domestic Product |
| GNP | Gross National Product |
| ha | hectares |
| HMSO | Her Majesty's Stationery Office |
| IDS | Institute of Development Studies |
| IGD | Institute for Grocery Distribution |
| IR | Inland Revenue |
| JACNE | Joint Advisory Committee on Nutrition Education (formerly NACNE), reporting to the DHSS |
| *J. Ag. Econ.* | *Journal of Agricultural Economics* |
| LDC | Less Developed Country |
| LFA | Less Favoured Area (renamed Severely Disadvantaged Area) |
| LIC | Lands Improvement Company |
| MAFF | Ministry of Agriculture, Fisheries and Food |

| | |
|---|---|
| MLC | Meat and Livestock Commission |
| MMB | Milk Marketing Board |
| NACNE | see JACNE |
| NEDO | National Economic Development Office |
| NFS | National Food Survey |
| NFU | National Farmers Union |
| OECD | Organisation for Economic Cooperation and Development |
| PMB | Potato Marketing Board |
| SGM | Standard Gross Margin |
| *Scot. J. Ag. Econ.* | *Scottish Journal of Agricultural Economics* |
| SGM | Standard Gross Margin |
| SLA | Scottish Landowners Association |
| SMD | Standard Man-days |
| SPI | Survey of Personal Incomes (by the IR) |
| UAA | Utilised Agricultural Area |
| UHT | Ultra Heat Treated |
| UK | United Kingdom (England, Wales, Scotland and Northern Ireland) |
| USDA | United States Department of Agriculture |

# Introduction and Overview 1

## 1.1 Changing Perspectives of Agriculture

Over the last decade agriculture in the United Kingdom has experienced a traumatic change, almost a reversal, in the public esteem it for long had enjoyed. Until the late 1970s farmers were generally regarded favourably by the public, given the occasional carp about feather-bedding, and their activities were assumed to be contributing to the general welfare of the country. Agriculture was admired for its efficiency, and the way that its productivity had increased was held up as an example for other sectors of the economy. It was seen as a triumph for the virtues of the small family business. Farming was an industry which had moved with the times; compared with the agricultures of other countries it was widely thought the most efficient. In the countryside farming was perceived as the most important user of land and, if there was a conflict with other users such as road builders or housing developers, there was an in-built feeling that the agricultural use should be heavily weighted in the deliberation of planning committees or public enquiries. Spending on support for farmers was accepted as necessary and desirable. Few questioned that the best way of assisting prosperity in the countryside was to channel public funds into farming; higher farming incomes would lead to higher rural incomes and by a multiplier process more business activity in rural areas. The owners of land and its farming users were regarded as the most sympathetic custodians of the countryside; after all, it was they who were largely responsible for giving its present shape.

By the mid-1980s much of this had changed. For a combination of reasons rather than a single cause farmers have become a favourite media subject for severe criticism. A stream of books appeared after 1979[1] detailing the case against the agricultural community and the policies which

1

support them. Farmers are more likely to be regarded by the public now as a group seeking to maximise profits even at the cost to society of spoiling the countryside, denuding it of traditional features such as hedges and buildings made of local materials, destroying wildlife habitats, polluting water courses and poisoning produce with chemicals. This they do in pursuit of ever greater output encouraged by inflated prices paid from the Common Agricultural Policy's budget, which actually offers financial incentives to carry out practices which are against the wider national welfare, such as draining ancient marshes. This is done despite the fact that the Community is over-supplied with farm products; a sizeable proportion of output can only be disposed of by subsidising exports, with international repercussions for both countries rich and poor. Support has meant that farms have grown and adopted more intensive practices; the much-praised growth in labour productivity, achieved since the mid-1950s largely through state-aided capital investment, is now viewed as reducing employment opportunities at a time when employment in other sectors of the economy is difficult to find. Hired agricultural workers are very poorly paid and are often trapped in agriculture by housing linked to their job. In sharp contrast, farmers, far from being a disadvantaged sector of society in need of assistance, are mostly wealthy and enjoy privileged lifestyles. Furthermore, farming is increasingly restricted to the families of existing farmers as the wealth necessary to become a farmer rises with time.

Neither view of agriculture is a fair representation. Those who blindly supported the virtues of the industry in the past were generally interested parties (farmers and their representatives) and aware that the prosperity of farming depended in no small measure on continued government spending, and the more the virtues could be expounded the greater the likelihood that support would continue. In large measure they were continuing a line of policy developed during the Second World War and the postwar period of recovery during which a strong case was made for the expansion of farm output for very basic reasons – to feed the nation and to relieve an acutely difficult trade position. Even in the 1950s and 1960s a special place was given to farming's role in assisting the UK's balance of payments position by an expansion of output. Once such a train of policy is set up, and farmers, their unions, government ministries, politicians and the general public have all adopted a way of thinking about the role of agriculture in the economy, there is a momentum established that is self-sustaining, even in the face of substantially changed conditions in the country and of altered requirements for its food-providing sector. The negative attitude towards the agricultural sector displayed in Britain does not as yet seem to be replicated in other Member States of the European Community, even though they share the same basic agricultural policy. This reflects not only the structure of farming but also differences in the attitudes to the social role played by farmers, particularly small farmers, in the national fabric.

Modern critics of agriculture, while concentrating on the negative aspects of current farming practice, are in danger of underrating the contribution the industry plays to the economic system. Public concern is frequently about the periphery rather than the core of agricultural production, at least in the view of farmers. Producing the wrong dietary nutrients, denying the public access to footpaths and polluting water courses with nitrates are not the issues which farmers themselves focus upon. Yet they are increasingly required to take into account complaints over these peripherals. There is also the danger of blaming farmers for actions which are predictable responses to signals received from policy-makers, the national government or the CAP. It is at these that the invective might more properly be levelled.

Because farming uses land, and land provides simultaneously many services to society other than just a space for farm production (a habitat for wildlife, a means of recreation for town dwellers and so on) real conflicts occur between modern farming and non-agricultural users of the countryside. With this multiple-function characteristic of land it is unlikely that a free market will result in the best pattern of use from society's standpoint. Government intervention in the markets for farm products and for the inputs that it uses are not, however, usually concerned with corrections of the price system to resolve such conflicting interests. Rather it is directed at achieving other aims, such as supporting the incomes of farmers. Farmers generally have freedom to operate within the market system, and there is a danger that, so long as policy is used to modify the prices and costs of farming for narrow agricultural ends, the outcome will conflict with the other objectives society now has for the countryside.

The cases of both the pro- and anti-farming lobbies are argued on the basis of myths, that is, beliefs about the way in which the real world works; we take up the nature of myths in Chapter 14. Examples encountered in the course of this book are the assumed link between greater self-sufficiency in food and a more reliable supply for consumers (Chapter 4), that the incomes of farmers would be unsatisfactorily low without state aid (Chapter 12), and that more prosperous agriculture is a prerequisite for farmers to spend time and money on environment protection (Chapter 15). Others are that food manufacturers are solely concerned with short-term profits and do not care about the long term consequencies of food additives (Chapter 3), that lowering cereal prices would stop the encroachment of pasture land (Chapter 10), and the myth that subsidies to capital investment has been responsible for the shedding of labour, to the cost of small villages (Chapter 11). Many of these myths contain an element of truth, but for the protagonists it is sufficient that they believe them to be true. In this text we are concerned more with the evidence by which they may be tested.

## 1.2 Basic Trends and the State

Two themes pervade this book. The first is that agriculture and the food industry are characterised by some long-term trends, driven by fundamental economic forces which are part of the capitalist market system and which are inexorable; the direction of trend inevitable. The most important is that the supply of farm products in the UK, indeed virtually all developed countries, is expanding faster than is the demand for them. A great deal of this text is concerned with the reasons behind this trend and the implications for agriculture and the rural economy. Chapters 2 to 4 look at the demand side and Chapters 5 to 9 the supply side; within each there are further trends.

The second theme is about the involvement of the state with farming and the food industry. For nearly half a century agriculture has been protected, regulated and insulated from the direct influence of free market forces. Though farmers operate as independent resource owners and decision-makers, and for the majority their ownership extends to the land they occupy, the economic environment in which these decisions are made is an artificial one, dictated by policy-makers who influence heavily the prices received, the costs paid, the types of technical changes that go on, the marketing mechanisms employed and even the level of farmers' incomes. The ways in which the state intervenes in agricultural markets are described principally in Chapter 13, although when considering each of the factors of production separately (Land, Chapter 6; Labour, Chapter 6; Capital, Chapter 7) the shadow of government is clearly visible. To a large extent government (or state) intervention is defensive in the sense that it attempts to protect agriculture against the long-term trends. The objectives of public policy towards this industry are dealt with in Chapter 14. In many cases the state can only slow down the inevitable process of change, yet in doing so it frequently builds up costs which are impossible to accept. Rather like damming a river, if no water is let through eventually the dam has to be raised to contain the water behind it. The higher the dam the more the pressure builds up. But the dam cannot suddenly be removed without causing a catastrophy for those who live in its protection. Similarly, agricultural policy could not be suddenly abandoned; the costs (economic, social, environmental and not least political) of making a sudden large adjustment if farming were again exposed rapidly to full market forces would be unacceptably high. When government farm policy is found no longer acceptable the problem is how to make adjustments with tolerable costs, in terms of the dam example letting a little more water through at a time.

What seems to have worsened the change in attitude towards agriculture is the failure of government institutions to grasp how far farming has progressed technically and how these changes have altered its accustomed environmental role in the country. Many, if not most, individuals working for government or farming bodies are well aware of the need to alter policies, but unless the political will exists they are impotent. The way that farming presently relates to the rest of the rural economy is described in Chapter 15, although some of the important issues are touched on in earlier chapters. In Chapter 15 we see that the role of farming in the UK rural economy is by no means as important as is commonly supposed and that, although agriculture has a part to play in the process of rural development, it is only a minor one.

## 1.3 The Integration of Agriculture

In this book we emphasise the integration of farming into the national economy. The linkages between farming and other sectors are increasingly recognised as important to the changing pattern of farming, and frequently these linkages are becoming formalised. The market for food is of obvious importance to the prosperity of farming, yet the industry has traditionally neglected its vertical connections in the food chain, the result of a long period stretching from at least the mid-nineteenth century to the 1970s where there were imports which could always be displaced by any extra food produced at home. With farm production now in excess of domestic demand in many major products there is a pressing need for producers to pay attention to the market; also farmers are keen to capture for themselves some of the consumer spending that now goes on processing and packaging. Already a third of UK farmers have some other economic activity they run in parallel with their farm, and this proportion is increasing. The amount of inputs farming buys in from other sectors has risen, making it more sensitive to developments in the rest of the economy. (Space does not permit a separate treatment of input industries.)[2] To some extent agriculture has lost its uniqueness, becoming more like other forms of production in its use of resources and approach to supplying markets. The importance of some of these linkages are shown by the input/output analysis described in Chapter 4.

This reduction in uniqueness and greater integration with the rest of the economy means that any attempt to describe the condition of agriculture must embrace more than just the narrow activity of farming. For this reason we start with a description of the demand for food and nature of the food industry beyond the farm gate. This is a logical starting point when farming is seen not as an autonomous activity but as one stage in the process of turning the country's resources of land, labour, capital and management skills into consumer goods. We end by looking at agriculture within the wider rural economy, because it is increasingly realised that farming is only one way by which many objectives of society can be approached. Some observers argue that rural areas could be made richer and more jobs created if the support currently given to farming were channeled into alternative activities, such as promoting small factories and tourism.

Notwithstanding the importance of the links between agriculture and its economic, social, political and natural environment, the bulk of this book is concerned with farming, the resources it employs and the attempts by the state to ensure that the farming population, as well as seeking their own personal objectives, also serve those of society at large. This is justified on the grounds, firstly, that many of the problems of agriculture are rooted in its technical relationships between inputs and outputs. Secondly, for most of our readers farming and agricultural production are the starting point for understanding the food chain and rural issues. There is more than enough material for one book in this method. Essentially our approach is one first of description, followed up by the use of theory to try to explain what has been happening in food, farming and the rural economy. This is not to deny the usefulness of sometimes starting from a theory and deducing the likely outcomes, which can then be tested empirically, but a straightforward establishment of the patterns of consumption, production, resource use and policy intentions are more appropriate to our end.

## Notes

1. For example, see the following:
   Newby, H. (1979) *Green and Pleasant Land?* (Harmondsworth: Penguin; reissued 1985, London: Temple Smith).
   Shoard, M. (1980) *The Theft of the Countryside* (London: Temple Smith).
   Body, R. (1982) *Agriculture: the Triumph and the Shame* (London: Temple Smith).
   Bowers, J. K. and Cheshire, P. (1983) *Agriculture, the Countryside and Land Use* (London: Methuen).
   Body, R. (1984) *Farming in the Clouds* (London: Temple Smith).
   Pie-Smith, C. and Rose, C. (1984) *Crisis and Conservation: Conflict in the British Countryside* (Harmondsworth: Penguin).
   Howarth, R. (1985) *Farming for Farmers?* (London: Institute of Economic Affairs).
   Blunden, J. and Curry, N. (eds) (1985) *The Changing Countryside* (London: Croom Helm/Open University).
2. See Chapters 10 and 13 especially. Also, papers by Dawson, P. J. and Lingard, J. and McCorriston, S. and Sheldon, I. presented in December 1986 to the Agricultural Economics Society conference on the input industries.

# The Consumer End of the Food Chain 2

## 2.1 Introduction

In the next three chapters we deal with the concept of a food chain, which reaches from the farmer at one end to the consumer at the other. The next chapter deals with the farmer and with those links in the chain between the farmer and the consumer which we refer to as the food industry, but in this chapter we deal with the last link in the chain – the consumer.

This Chapter is divided into two sections. In the first section we describe the levels of food actually consumed in the UK, and in the second we deal with the economic concept of the demand for food.

## 2.2 Food Consumption in the UK

In 1984, consumers in the UK spent £28.4 billion on household food (14.6 per cent of total consumer expenditure), £11.0 billion on catering and eating away from home (5.6 per cent of total consumer expenditure), and £14.4 billion on alcoholic drink (7.3 per cent of total consumer expenditure)[1] Since the early 1970s household expenditure on food has increased slowly in real terms at a rate of about 0.2 per cent per annum, but it has gradually declined as a proportion of the total. In the 1950s, 25–30 per cent of household expenditure was on food but in 1972 it was 18.4 and in 1982 15.6. This long-run relative decline in the share of total expenditure, and very slow real increase, reflects the fact that the wants of UK consumers in this direction are already largely satisfied. Consequently, as incomes rise the extra spent on food is proportionately less than that on non-essentials and luxury items. The size of the market is limited by the dietary and nutritional requirements of people and the fact that consumers are limited in how much (and how little) food they can eat. These product characteristics of food also lead to consumer expenditure on food showing more stability than consumer expenditure in total. During the economic recessions of 1975–6 and 1980–2, consumers' spending on food remained stable in real terms. We return to this relationship between income and the demand for food later when we consider the income elasticity of demand.

A similar pattern is found in other industrial countries. Furthermore, there is a negative association between incomes and the proportion being spent on food which crosses national boundaries as is shown in Table 2.1. The richer the country, the lower the proportion of income devoted to purchasing food.

Table 2.2 illustrates the size of the markets in the UK for broad food groups. Ninety per cent of food expenditure is accounted for by six distinct groups of foodstuffs: (1) meat, (2) dairy products, (3) oils and fats, (4) sweetened foods (sugar, confectionary, preserves and soft drinks), (5) bread and cereals and (6) fruit, potatoes and other vegetables. The distribution of expenditure among these broad food groups has remained surprisingly stable from as far back as 1938,[2] reflecting the fact that most changes in the food market have been within food groups, substituting one product for another.

Within broad food groups, consumers' expenditure allocated to food will alter as a result of both changes in physical consumption and changes in prices. If the price of one foodstuff declines over time, the quantity bought will increase, but this increase in volume in proportional terms is usually less than the decrease in price. Demand for food is considered later in this chapter, and we only introduce it at this stage to point out that increases in food consumption and decreases in consumers' expenditure on food are likely to result from falling food prices, other things being equal. Similarly, higher food prices will often be associated with higher consumers' expenditure on food and lower consumption of food.

### 2.2.1 *Measuring food consumption*

The consumption of food means the physical quantity that people eat from whatever source. Thus there is a distinction between the consumption and the quantity purchased, and for a product grown in back gardens such as apples the purchased quantity may only account for 85–90 per cent of consumption, varying with yields. This should be borne in mind in the following discussion, although we will ignore the distinction at this stage.

Estimates of the level of consumption of different foods may be obtained in two ways, and for Britain both are available. These estimates began during the Second World War in response to wartime food supply planning

**Table 2.1**
**Expenditure on Household Food, 1983**

| | Per capita GDP[1] | Per capita household consumption FDT[2] as % of total | | Absolute per capita expenditure on FDT[2] |
|---|---|---|---|---|
| | (ECU)[3] | (%) | (food %) | UK=100 |
| USA | 14 743 | 15.7 | (11.2) | 128 |
| Canada | 14 425 | 20.4 | | 131 |
| Denmark | 12 053 | 24.5 | (16.7) | – |
| Japan | 11 339 | 24.2 | | 118 |
| W. Germany | 11 977 | 19.4 | (14.6) | 131 |
| Luxembourg | 11 833 | 21.3 | (16.3) | 115 |
| France | 11 776 | 21.2 | (17.5) | 128 |
| Belgium | 11 176 | 22.3 | (18.3) | 126 |
| Netherlands | 10 702 | 19.6 | (15.1) | 111 |
| Eur (10) | 10 593 | 31.2 | (26.7) | – |
| UK | 10 238 | 20.2 | (14.7) | 100 |
| Eur (12) | 10 064 | – | – | – |
| Italy | 9 102 | 29.8 | (25.6) | 124 |
| Spain (1982) | 7 616 | 31.7 | | 100 |
| Ireland | 7 040 | 41.2 | (23.1) | 128 |
| Greece (1982) | 5 759 | 41.5 | (35.6) | 105 |
| Portugal (1982) | 5 001 | 37.0 | | 52 |

NOTES   1. Gross domestic product, adjusted to reflect purchasing power.
        2. Food, drink and tobacco.
        3. In 1983, 0.587 ECU: £1.
SOURCE  EC (1986), *The Agricultural Situation in the EC* (Brussels).
        Eurostat (1985), *Basic Statistics of the EC*, 23rd edition.

requirements. Britain had to plan for an appropriate level of imports of food from North America in order to feed the population adequately.

The first method of estimating, known as the **supply approach**, is to measure consumption indirectly by assessing the supply of food. Supplies coming from domestic farms are added to imports, a deduction made for exports and adjustments made for changes in stocks and waste incurred in the marketing process. This shows the level of supplies available for

**Table 2.2**
**Consumers' Expenditure on Food, 1983 and 1973–83 Percentage Change**

|  | 1983 expenditure (£ million) | As % total food | 1973–83 % change, constant 1980 prices |
|---|---|---|---|
| Bread, cereals | 3 649 | 13.4 | 0 |
| Meat, bacon | 7 346 | 27.0 | +6 |
| Fish | 907 | 3.3 | +10 |
| Milk, cheese, eggs | 3 975 | 14.6 | −9 |
| Oils, fats (incl. butter) | 951 | 3.5 | −1 |
| Fruit | 1 564 | 5.8 | +4 |
| Potatoes | 1 191 | 4.4 | +14 |
| Vegetables | 2 032 | 7.5 | +13 |
| Sugar | 368 | 1.4 | −3 |
| Sugar preserves, confectionery | 2 447 | 9.0 | −3 |
| Coffee, tea, cocoa | 884 | 3.3 | +11 |
| Soft drinks | 968 | 3.6 | +41 |
| Other manufactured food | 866 | 3.2 | −17 |
| Total food | 27 148 | 100 | +1.6 |
| Alcohol | 13 372 |  | +6 |
| Tobacco | 6 208 |  | −23 |
| Total | 46 728 |  | −3 |
| Catering (food & accom.) | 9 967 |  | +2 |
| Total Consumers' Expenditure | 182 427 |  | +13 |

SOURCE   CSO (1985) *Annual Abstract of Statistics*, Table 14.10.

consumption (sometimes called 'apparent consumption') and is the measure most readily obtained from aggregate data. The Food and Agriculture Organisation (FAO) of the United Nations uses this approach in the preparation of Food Balance Sheets to assess the availability of food supplies in countries, and uses this assessment in conjunction with an estimate of domestic utilisation as an indicator of the possibility of food shortages.[3] One disadvantage of this approach is that it largely ignores the distribution of food among families in the country concerned (some families may have too little while others have much more than they can eat) and may therefore give an over-optimistic view of the general availability of food. Nevertheless, in countries with reliable production and marketing estimates the supply approach to assessing food availability can yield a useful set of data, particularly when supplemented with retail and household surveys. The Economic Research Service of the United States Department of Agriculture adopts a supply approach to estimating food consumption and the method has been used in, for example, a survey of consumption patterns for twenty-one OECD countries.[4]

In Britain the supply approach estimates take the form of the Consumption Level Enquiry (CLE) by the Ministry of Agriculture, Fisheries and Food (MAFF).[5] This is particularly useful in revealing information about raw foodstuffs, and the level of self-sufficiency in different agricultural products can be measured in this way. Also, food supply can be expressed as changes in nutrient supply to provide a long-term estimate of changes in nutrition over a century or more.[6]

The second approach is to estimate directly food purchased for final consumption, usually by surveys of households. The estimates obtained are in terms of foodstuffs as sold in their often processed and packaged form rather than of agricultural products and are thus generally of more interest to those engaged in food manufacturing and distribution and to consumer organisations than the supply approach estimates. This **direct consumption approach** also offers much greater detail and allows the relationships between prices, incomes and purchases of individual foods to be quantified. The main drawback is that coverage is limited to household food consumption and it is not easy to allow for meals and snacks away from home. This can be overcome by weighing and analysing food intake but this is a costly and time-consuming process.

In Britain the National Food Survey, conducted annually by the MAFF, takes the direct approach. A sample of about 7000 housewives, drawn from the electoral role, keep a record for a week of all food purchased (or otherwise obtained) and consumed in the home. Results are published in detail[7] and are not confidential to clients as is the case for most other market research survey results. The information has proved of great value to anyone interested in the British food market, despite criticisms that the results are probably biased towards housewives more willing and able to keep a close record of their purchases. The Food Survey has gradually expanded its coverage and now reports on special items such as soft drinks and the effects of freezer ownership on food purchases. Consumption data are available relating to nutrients consumed as well as to types of food. Estimates based on the National Food Survey of the total market size for particular farm products are difficult to make because such a high proportion of food is consumed in a processed form. The estimates are closest for foods consumed mainly in a raw or semi-processed state with little waste, and household survey estimates of market size are around 80 per cent of supply-based estimates for products such as poultry and fruit and vegetables. The percentages are much lower for carcase meats (about 50 per cent) and sugar (33 per cent). In these cases, changes in household consumption patterns of the semi-processed product alone may not be closely related to changes in the total supplies actually coming onto the market.

### 2.2.2 The pattern of food consumption in Britain

Table 2.3 illustrates the pattern of food consumption in Britain in 1982 and how this changed over the preceeding forty-four years. The supply approach to measurement is adopted in order to take account of all food consumed, not just household food consumption. In the early part of the period, war-time shortages changed consumption patterns. Consumption of meat, fats, sugar and fish were curtailed, whilst consumption of bread and potatoes increased. During the 1950s as shortages were overcome, consumption levels of many types of food recovered and rose to levels well above pre-war levels.

**Table 2.3**
**Supplies Moving into Consumption, UK, kg. per person per year**

| Individual foods | Pre-war | 1943 | 1955 | 1965 | 1975 | 1980 | 1982 |
|---|---|---|---|---|---|---|---|
| Liquid milk (litres) | 95.8 | 131.7 | 143.6 | 141.7 | 142.3 | 128.9 | 124.9 |
| Cheese | 4.0 | 5.2 | 4.1 | 4.6 | 6.3 | 6.0 | 6.2 |
| Beef[a] | 24.9 | 15.0 | 21.2 | 20.1 | 23.6 | 20.6 | 18.3 |
| Mutton and lamb | 11.4 | 11.9 | 11.1 | 10.5 | 8.3 | 7.5 | 7.3 |
| Pork | 5.6 | 3.4 | 8.1 | 11.7 | 10.3 | 12.5 | 13.0 |
| Bacon and ham | 12.0 | 8.4 | 11.2 | 11.7 | 8.7 | 9.1 | 8.4 |
| Poultry meat[b] | 4.0 | 2.1 | 3.1 | 7.9 | 11.4 | 13.3 | 14.4 |
| Fish[c] | 9.9 | 6.4 | 8.6 | 7.8 | 6.1 | 5.4 | 5.0 |
| Eggs in shell[d] | 201 | 106 | 210 | 250 | 232 | 227 | 216 |
| Butter | 11.2 | 3.4 | 6.7 | 8.8 | 8.4 | 6.3 | 5.8 |
| Margarine (net) | 3.9 | 7.7 | 8.0 | 4.4 | 5.0 | 6.9 | 7.3 |

| | | | | | | | |
|---|---|---|---|---|---|---|---|
| Lard, etc. | 4.2 | 5.4 | 4.8 | 6.1 | 5.9 | 5.4 | 5.4 |
| Other oils and fats | 4.6 | 3.0 | 4.7 | 5.3 | 5.5 | 6.0 | 7.6 |
| Fresh citrus | 12.9 | 0.9 | 8.4 | 8.7 | 9.3 | 10.2 | 9.2 |
| Other fresh fruit | 22.7 | 12.1 | 21.7 | 24.7 | 22.3 | 27.7 | 26.2 |
| Potatoes | 86.2 | 112.9 | 106.2 | 100.9 | 101.9 | 105.0 | 105.6 |
| Fresh tomatoes | 4.7 | 4.0 | 7.1 | 5.9 | 5.9 | 5.8 | 6.0 |
| Beer (litres) | 108.6 | 112.6 | 79.7 | 91.4 | 117.3 | 116.4 | 108.2 |
| Wine (litres) | 2.1 | 0.5 | 1.5 | 2.9 | 6.3 | 8.1 | 9.0 |
| ***Broad food groups*** | | | | | | | |
| Milk solids | 17.4 | 22.7 | 23.9 | 25.1 | 26.4 | – | – |
| Meat[e] | 53.9 | 41.1 | 53.4 | 58.6 | 57.0 | – | – |
| Oils & fats[f] | 21.4 | 17.7 | 21.8 | 22.4 | 21.8 | – | – |
| Sugars | 43.5 | 30.3 | 48.9 | 49.2 | 45.9 | – | – |
| | | | | | 42.8[g] | 39.9 | 40.2 |
| Cereal products | 95.3 | 112.9 | 88.9 | 77.3 | 72.3 | 69.8 | 68.1 |
| Fruit[h] | 56.2 | 30.6 | 51.3 | 55.4 | 53.6 | – | – |
| Vegetables[h] | 54.6 | 57.3 | 55.2 | 60.9 | 60.5 | – | – |

NOTES  a. bone-in equivalent
        b. includes rabbit and game
        c. fresh, frozen and cured; fresh equivalent
        d. number of eggs
        e. edible weight
        f. total fat content
        g. total sugar content. New basis, 1975 onwards.
        h. fresh equivalent

SOURCES  Angel, L. J. and Hurdle, G. E. (1978) 'The nation's food – 40 years of change',
          *Economic Trends* (April) 97–105.
          CSO (1985) *Annual Abstract of Statistics* (London: HMSO) p. 190.

In the 1960s and 1970s, several of these consumption trends were reversed. First the consumption of sugar began to decline, then that of milk and fish, bacon, beef, lamb, eggs and butter. One reason was that people were doing less physical work and required less energy in their diets. Another was that prices, incomes and population characteristics were changing in Britain. We consider the separate impact of such factors later but require at this stage a summary of the more important changes, contained in Table 2.3, and confirmed by National Food Survey results for household food consumption.

### 1  Dairy products

Butter is included in Table 2.3 under 'oils and fats' and consumption has been in decline, whilst cream and other dairy products have increased in consumption, though not sufficiently to offset the decline in butter. In the 1980s, a trend has also begun towards a rapid increase in semi-skimmed and skimmed milk consumption at the expense of full cream milk. Liquid milk consumption in total has declined, partly following the reduction in the distribution of subsidised milk to schools and other outlets and partly as a consequence of a change in tastes towards other drinks. Soft drink and fruit juice consumption has increased rapidly, and the range of other drink mixtures available to consumers has widened. Nevertheless, liquid milk consumption in Britain remains relatively high in comparison with many European countries. One factor involved is the system of Milk Marketing Boards which together have managed the supply of milk in Britain for over fifty years (Chapter 3). Another is continued doorstep deliveries.

## 2   Meat and fish

Total consumption of meat took a long time to recover from wartime shortages and then grew steadily until 1970, when a plateau was reached. Trends in individual meat consumption show a declining consumption of lamb and bacon and a rising consumption of poultry and pork. Technical improvements in poultry and pork production have led to lower prices and encouraged higher levels of consumption, whilst the decline in bacon consumption reflects the demise of the traditional British cooked breakfast. Beef consumption was maintained until 1981, with high prices a feature of the market during the 1970s.

The household consumption of processed and frozen meats has increased markedly over recent years and probably the same has been true in the case of food eaten away from home. Some supermarkets see a small but expanding market in the 1980s for speciality products (smoked and cured meats and meat products). Fish consumption is largely as fresh, frozen or cured fish but the overall level of consumption has never recovered since the war. More recently, the depletion of fishing grounds has led to increased prices and lower consumption and also the taste for fresh fish seems to have declined, whilst that for frozen fish has increased. Fish farming may perhaps raise total consumption sometime in the future by increasing supplies of cheaper and more preferred types of fish .

## 3   Oils and fats

As with meat, the consumption of oils and fats did not recover from the war until the 1950s and consumption subsequently rose to a peak in the 1960s. Since then, three trends have been discernable. Firstly, there has been a decline in 'yellow fats' (margarine and butter) associated with a reduction in the consumption of bread. Secondly, margarine has increased at the expense of butter, particularly as a result of an increase in butter prices after entry to the EC in 1973. Thirdly, those margarines and spreads rich in polyunsaturated fats have increased as consumers have responded to the recommendations concerning diet and heart disease and as new products have become available. Other changes in the fats and oils market are the increase in consumption of cooking and salad oils and decline in the consumption of lard.

## 4  Sugar and syrup

Supplies of sugar rose sufficiently in the 1950s for consumption to exceed the prewar level but since then household consumption of bagged sugar has declined. There has been a strong trend in the United States towards replacing sugar with corn syrup (and other non-sugar sweeteners), the sale of which is effectively restrained within the European Community by strict controls on manufacturers and importers. The sugar industry has attempted to counter the image of sugar as an unnecessary and fattening food but with little impact so far on the level of household consumption.

## 5  Cereal products

Cereals are mostly consumed as bread, though the gradual decline of bread consumption since the 1950s and increase in breakfast cereal consumption has reduced that dominance. Cake and biscuit consumption has also gradually declined. A notable trend has been the rapid increase in wholemeal and brown bread consumption at the expense of white bread since 1978. Tastes have clearly changed, perhaps linked to the ownership of freezers (which reduce the need to have the good keeping qualities of sliced, wrapped white bread) and encouraged by the 'healthy eating' emphasis on dietary fibre.

These changes, however, are of only limited significance to the supply of cereals from Britain's farms, partly because only a small proportion of home-grown wheat is of bread-making quality. The decline in the consumption of milk and butter, and hence in the market for dairy cattle feedstuffs, and a switch from intensive, grain-fed beef to grass-fed beef, are likely to have more of a negative impact on grain producers in the long run.

## 6  Fruit and vegetables

Fruit consumption took a long time to recover to its pre-war level and citrus fruits have never re-established their dominance of the fresh fruit market. Expectations of lower prices, greater supplies and increased fruit consumption in Britain, resulting from entry to the EC, may finally be fulfilled during the 1980s, although consumers appear increasingly interested in the quality and novelty value of their fruit purchases. Fruit juice consumption has increased rapidly since the 1970s, reflecting better packaging and a changed attitude to the product on the part of British consumers who used to regard fruit juice as a luxury breakfast drink.

Vegetable consumption is still dominated by potatoes, where the steady decline in total potato consumption since the war has been halted by the increased consumption of processed potato products, such as crisps, frozen chips and canned potato products (the operations of the Potato Marketing Board are discussed in Chapter 3). Other vegetables have increased in consumption, particularly salad vegetables such as fresh tomatoes (see Table 2.3) and also onions and mushrooms. The consumption of cabbages and several other outdoor vegetables has declined, reflecting, in the main, changes in tastes and eating habits. Finally, there has been a growth in the consumption of several processed vegetable products, notably frozen peas and canned beans.

## 7   *Alcoholic drink*

The total consumption of alcoholic drink increased during the 1960s and 1970s. The consumption of spirits, wine and cider went up, whilst beer consumption rose and then stabilised. An increasing proportion of the beer has been of the lager type and Table 2.3 suggests that overall beer consumption may be set to decline.

### *2.2.3   Regional variations in food consumption*

Regional differences in food consumption still exist in Britain, despite improvements in communications and shifts in population. Current regional diet characteristics have even been traced back to the different cooking cultures of, on the one hand, Anglo-Saxon ovens in the South and East (breads, buns, puddings and roast meat, reaching up as far as Yorkshire pudding) and, on the other hand, Celtic pots warming over hearths in the North and West (stews, soups, porridge and vegetables).[8] Where the two cultures meet, odd dietary habits seem to persist, such as the exceptional tastes for sugar, vinegar and pickle in the Midlands. Welsh consumption seems noteworthy – high in white bread, butter, total fats and sugar; low in fish, fruit and brown bread. As we see later, this is not a diet of which nutritionists would approve on health grounds. The Scottish diet is also different in a number of ways, with the consumption of beef and oatmeal relatively more important compared to the rest of Britain. Consumption of fresh fruit and vegetables is particularly low, and this and other Scottish patterns of diet have traditionally affected the North of England (especially the North-East).

Consumption patterns in East Anglia and the South-East of England and in Greater London reflect the relatively greater affluence of households in these areas as much as any regional dietary patterns. Foods which are consumed more than elsewhere include fresh fruit and fresh green vegetables, poultry, pork and coffee, whilst households buy less bread, potatoes, margarine, lard and bacon. In the South-West, fresh green vegetables, pork and flour are favoured, margarine and fish less so. Again, the average household disposable income in the region is above the level for Britain as a whole.

In an attempt to isolate factors such as regional preferences for food from income levels and the availability of foods locally (reflected in price differences) a multivariate analysis was conducted by MAFF using data for 1982.[9] This confirms that regional preferences are important even after allowing for differences in incomes, prices and other factors (size of household, the age of the housewife and so on).

Turning to international comparisons, Britain is one of twelve member countries in the European Community and food consumption in Britain in part reflects a pattern common to other northern European countries such as the Netherlands, West Germany and Denmark. In all these countries, potatoes, sugar, butter and pigmeat are important items in consumption. In contrast, Southern and Mediterranean countries of the European Community consume more cereals, fruits and vegetables, vegetable fats, wine, mutton and goatmeat.[10] There are signs of a slow narrowing of national consumption differences for some products, such as sugar, eggs, rice, and some fruits and vegetables, but for others, such as citrus fruit, consumption remains widely different. For certain foods such as butter, national consumption differences appear to be widening. It must be remembered that these differences are usually expressed as national averages and therefore often conceal large variations within countries, associated with factors such as the level of incomes and prices and the food traditions of a particular region within the country. These are certainly important influences on the pattern of food consumption within Britain.

## 2.3 The Demand for Food

So far we have described the pattern of food *consumption* and have rarely refered to the *demand* for food. The two are not synonymous, and the quantity of food consumed is not necessarily the same as the quantity demanded or supplied, although they may be. To explain how they relate it is necessary to clarify what is meant by the term 'demand'.

It is not the intention here to give a full account of the economic theory of the price mechanism of demand and supply. This can be found in any elementary economics textbook. Nevertheless, it is perhaps worthwhile reminding ourselves that the *quantity demanded* refers to the amount buyers are willing to take from a market at given prices and over a specified period and the *quantity supplied* refers to the amount suppliers who sell goods are willing to supply to that market, again at given prices and over a specified period.

On the buyers' (demand) side, the amount of a food they will wish and are able to buy will depend on a number of variables:

1. the price of the food
2. the prices of other goods, such as substitute foods or items which complement the food
3. the incomes of the consumers
4. tastes and preferences of consumers
5. the size of the population and the structure of the population (in terms of age and sex)
6. the income distribution within the population.

On the supply side, the amount suppliers will wish to sell will depend on a similar list, including the prices they can get for the food, the prices for other goods they can supply, costs of production, the state of technology, and the aims of the suppliers, usually assumed to be the maximisation of profits. (See Chapter 10 for a further discussion of supply.)

Figure 2.1 contains the familiar demand curve, showing the amounts food buyers are willing to purchase at a range of prices. The lower the price the greater the quantity buyers want. Similarly, the supply curve shows that at higher prices suppliers are willing to sell more. In drawing these curves, it is assumed that factors other than price which can affect demand or supply do not change. In a free market a price would result at which the quantity which buyers wished to buy exactly equals the quantity which suppliers wish to supply, the point in the diagram where the two curves intersect. The market is said to be in equilibrium at this point and factors which move either the demand or the supply curves left or right will change the equilibrium price at which the curves intersect. Using this simple diagram, it is possible to explain why consumption (the quantity actually bought) need not equate demand (the quantity that consumers are willing to buy) or supply.

When the market is in equilibrium, the relationship between quantity demanded at the market price and consumption is one of equality; people actually consume the equilibrium quantity. When the market is *out* of equilibrium, the relationship depends partly on whether price is above or below the equilibrium level. As can be seen from Figure 2.1, when price is for some reason stuck above the equilibrium at ($P_1$), say by the government fixing prices at a high level, then there will be excess supply on the market and consumption will equal the quantity *demanded* ($Q_1$) at this price. Conversely, when the price is stuck below the equilibrium, perhaps by the government setting prices below market levels, at ($P_2$), there will be excess demand; people will be forced to buy less than they want and

**FIGURE 2.1**
**The Relationship between Consumption and Supply, Demand Curves**

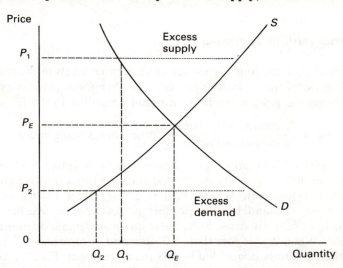

consumption will equal the quantity *supplied* at this price. In the case of the latter, some means of rationing out the limited supply will be required and a black market may develop. In Britain, supply determined consumption for several foods in the 1940–53 period when the government set many food prices at low levels, but since then prices have generally been free to rise to equilibrium levels and consumption has been the same as quantity demanded at the market price. The supply constraint does not normally occur in peace time where there is access to the world market, unless the weather is exceptionally bad over many regions of the globe (as occurred with grain in 1972, sugar in 1974 and coffee in 1976).

Trends in consumption are not necessarily an accurate guide to changes in the demand curve over time and are difficult to use even in conjunction with prices unless there is some knowledge of the behaviour of supply. Consumption will equal the quantity demanded at the market price when a competitive market is either in equilibrium or when a price is above the equilibrium level. However, over time, consumption and expenditure will alter to reflect the interaction of supply *and* demand.

The demand for food will be affected by the factors listed above and it is logical to examine first how the demand for a foodstuff varies with the price of the food. From Figure 2.1 it can be seen that changes in price result in movements along the demand curve as quantity demanded responds to the new level of price. The first question concerning the demand for food is how responsive is demand to price? The second question is, what factors cause the demand curve to shift?

### 2.3.1 Price elasticity of demand

The demand for some foods is less sensitive to price levels than others. The responsiveness of the quantity demanded to changes in price is estimated by calculating the **price elasticity of demand** according to the formula.

$$E_{dp} \text{ equals } \frac{\% \text{ change in } QD \text{ of } A,}{\% \text{ change in price of } A} \text{ other things being equal} \qquad (1)$$

Price elasticity of demand $(E_{dp})$ will normally be negative, reflecting the negative response in quantity demanded (i.e. less) to a positive change in price (i.e. a price rise). When $(E_{dp})$ is greater than 1.0 (ignoring the negative sign), demand for the product is described as *elastic*; a price change of 1 per cent will cause a change in quantity demanded of *more than* 1 per cent. When $E_{dp}$ is less than 1.0, demand is *inelastic*; a 1 per cent change alters quantity demanded by less than 1 per cent. Finally, when $E_{dp}$ equals 1.0, demand has unitary elasticity and the percentage change in price and quantity are the same.

Table 2.4 shows price elasticities calculated for households in Britain derived from the National Food Survey, using data for the period 1978–83. There are two important aspects of the response to price of foodstuffs illustrated by Table 2.4. First, nearly all the food products have inelastic demand ($E_{dp}$ is less than 1.0), a reflection of the absence of substitutes

---

**Table 2.4**
**Estimates of Price Elasticity of Demand for Selected Household Foods, 1978–83**

| | | |
|---|---|---|
| Liquid milk | −0.30 | a |
| + Carcase meat | −1.49 | |
| beef and veal | −1.89 | a |
| steaks (less expensive) | −0.74 | |
| steaks (more expensive) | −1.84 | |
| pork | −2.00 | a |
| pork chops | −0.66 | |
| joints | −2.37 | |
| mutton and lamb | −1.48 | a |
| + Fats | −0.14 | |

| | | |
|---|---:|---|
| butter | −0.02 | a |
| New Zealand butter | −2.34 | |
| Danish butter | −0.39 | |
| UK butter | −0.16 | |
| margarine | −0.51 | a |
| + Sugar and preserves | −0.03 | |
| sugar | −0.61 | a |
| marmalade | −1.42 | c |
| honey | −0.79 | c |
| + Fresh vegetables (excl. fresh potatoes) | −0.68 | |
| cauliflower | −2.08 | a |
| leafy salads | −0.85 | a |
| cabbage | −0.10 | b |
| all processed potatoes | −0.83 | a |
| canned beans | −0.50 | b |
| Potatoes, fresh | −0.18 | a |
| + Fresh fruit | −0.30 | |
| oranges | −0.97 | a |
| bananas | −0.42 | b |
| fruit juices | −1.23 | a |
| + Bread | −0.62 | |
| standard white | −0.79 | a |
| brown | −1.66 | b |
| wholemeal and wholewheat | −2.72 | a |
| + Beverages | −0.45 | |
| tea | −0.19 | c |
| coffee (bean and ground) | −1.07 | b |

NOTES   + Broad foodgroups, taking account of cross-price elasticities. Proportion of variation in monthly average purchases explained by the price elasticity and any significant seasonal or annual shifts in demand: 0–33 per cent (a), 34–66 per cent (b), 67–100 per cent (c).

SOURCE   MAFF (1983) *National Food Survey* (London: HMSO), Tables 3 and 7.

available for several types of foods. Secondly, and related to this point, price elasticities for narrowly defined products are invariably higher than for broad food groups. This reflects the existence of closer substitutes for, say, margarine (butter, cooking oil, lard) than for fats and oils taken together. The greater the number and the greater the closeness of substitutes, the more willing consumers are to respond to price by switching to or from the substitute.

The price elasticity of demand also varies according to how close to the farmgate level is the demand response that is being measured. This is illustrated and explained in Figure 2.2. Price elasticity of demand will be low when the product is only a small proportion of the consumer's budget,

## FIGURE 2.2
## Demand at Retail and Farm Level

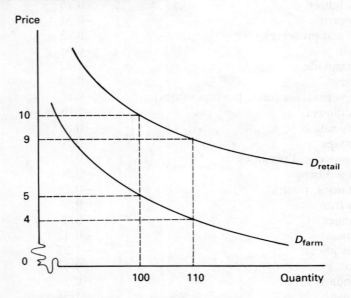

Let us assume that the retail price is 10 and the farmgate price is 5, and that the difference is being accounted for by the costs and profit margins of wholesalers, manufacturers and distributors. A cut in the retail price from 10 to 9 leads to an increase in quantity demanded from 100 to 110. Price elasticity of demand at the retail level can be estimated for small changes in price using the formula in equation (1a), derived from equation (1)

$$\text{Retail } E_{dp} = \frac{Q_2 - Q_1}{Q_1 + Q_2} \cdot \frac{P_1 + P_2}{P_2 - P_1} \qquad (1a)$$
$$= \frac{10}{210} \cdot \frac{19}{-1} = -0.905$$

This gives the price elasticity of demand over the range of the demand curve between quantity demanded of 100 and of 110 (it is termed the average or arc elasticity).

Now, assume that the margin between shop and farm is fixed at 5, so that the farmgate offer price falls to 4, maintaining a fixed margin between the new shop and farmgate prices. At the farmgate level,

$$\text{Farmgate } E_{dp} = \frac{10}{210} \cdot \frac{9}{-1} = -0.429$$

(fixed margin)

Thus, the retail price elasticities shown in Table 2.3 will be higher than the farmgate price elasticities in circumstances where fixed margins operate along the food chain. The use of a percentage margin is more common for fruit and vegetables whilst fixed margins are more common for meat. It can be seen that where a percentage margin operates, the farmgate elasticity is the same as the retail elasticity.

$$\text{Farmgate } E_{dp} = \frac{10}{210} \cdot \frac{9.5}{-0.5} = -0.905$$

(percentage margin)

since there is little to be gained financially by the consumer responding to price changes. However, the presence of substitutes among the many small food items which are purchased makes it difficult to isolate this influence on price elasticity. Where the market for a product is traditional, the elasticity will also be very low – the near-zero price elasticity of demand for sugar and preserves in Table 2.4 is an example of this. The inelastic nature of the demand for food at the retail level (and the even lower price elasticity for foods traded with fixed margins at the farmgate level) has several implications for the agricultural industry which are described later.

The demand curve shows how the quantity demanded varies with price so long as other factors which could affect demand are held constant. We must now turn from considering movements along the demand curve to shifts in the demand curve when these other factors are no longer constant.

### 2.3.2 Factors that shift the demand curve for food

#### 1 Prices of substitutes and complements, and cross-price elasticity

Two products are **substitutes** if an increase in the price of one, other things being equal, causes an increase in the quantity demanded of the other. Butter and margarine are good examples. The demand curve for butter will be moved to the right if margarine prices rise. Two products are **complements** if an increase in the price of one, other things being equal, leads to a decrease in the quantity demanded of the other. Beef and horseradish sauce are a favourite example; a rise in beef prices will shift the demand curve for sauce to the left. Two products are unconnected if a price change in one product has absolutely no effect on the quantity demanded of the other. This relationship can be quantified by calculating the cross-price elasticity of demand using the following formula.

$$E_{dcp} = \frac{\text{percentage change in } QD \text{ of } A,}{\text{percentage change in price of } B} \quad \text{other things being equal}$$

(2)

If the products are close substitutes (say peas and beans, or, even closer, different brands of frozen peas) the cross price elasticity will be positive and high.

Table 2.5 shows cross price elasticities calculated for households in the National Food Survey for three types of fresh fruit although it must be emphasised that the figures are averages and only estimates. The co-efficients are interpreted as follows. A 1 per cent increase in the price of oranges results in a 0.89 per cent decline in the quantity of oranges demanded, a 0.03 per cent increase in the quantity of apples demanded, and a 0.35 per cent increase in the quantity of pears demanded, other things remaining the same. It is evident from column 1 in Table 2.5 that pears are probably a closer substitute for oranges than are apples. The second column shows that pears are also a closer substitute for apples than oranges are for apples, whilst column 3 shows a 1 per cent increase in the price of pears results in a greater rise in the quantity demanded of oranges (0.11) than of apples (0.03). We may conclude that consumers regard pears as substitutes for oranges and apples *more than* they regard oranges and apples as substitutes for pears. In fact, they hardly regard apples as a substitute for pears at all. When the price of apples increases, consumers buy more pears (+0.23), but when the price of pears increases, consumers do not buy more apples (+0.03) and the response in the orange market is also weak (+0.11).

The reason why the demand for pears responds to the orange and apple markets but not *vice versa* lies in the relative size of the three markets. Table 2.5 shows that the relatively small size of the market for pears results in this market demand being influenced by other markets, but not influencing these other markets in turn very much. This is an important consideration when interpreting cross-price elasticities.

Another example is the relationship between meat and cheese. Using cross-price elasticities calculated for the same period as in Table 2.5, a 1 per cent increase in the price of cheese leads to a 0.14 per cent increase in the demand for carcase meat. In contrast, a 1 per cent increase in the price of carcase meat leads to a 0.66 per cent increase in the demand for cheese, other things being equal. As one might expect from such a result the two markets are of different size: household expenditure on carcase meat is currently about five times higher than household expenditure on cheese.

Several foods have complementary relationships, such as between beverages (tea, coffee, cocoa, etc.) and sugar. A 1 per cent increase in the price of beverages reduces quantity demanded by 0.45 per cent and also reduces the quantity of sugar demanded by 0.20 per cent. The relationship is weaker in the other direction, since cheaper sugar is less of an influence on the demand for beverages in Britain, though during the industrial revolution the demand for tea, milk and sugar were probably more closely interrelated.[16] Today, a 1 per cent increase in the sugar price reduces sugar consumption by 0.42 per cent and beverage consumption by only 0.13 per cent.

When the price of an important foodstuff or group of foodstuffs changes, this has an additional effect on the real incomes of consumers. Since 1955, meat has accounted for between 27 per cent and 33 per cent of household expenditure on food and a fall in the average price of meat relative to other foods causes not only substitution effects (less fish and dairy products demanded) but also it has the effect of raising the consumer's real income (more of all foods can be purchased, including fish and dairy products).

**Table 2.5**

**Estimates of Price and Cross-price Elasticities of Demand, 1983**

| | Elasticity with respect to the price of | | | $R^2$ | Per person per week[a] | | (% households)[b] |
| | oranges | apples | pears | | (oz) | (pence) | |
| --- | --- | --- | --- | --- | --- | --- | --- |
| oranges | −0.89 (0.17) | 0.08 (0.09) | 0.11 (0.09) | 0.28 | 2.81 | 4.64 | 24 |
| apples | 0.03 (0.04) | −0.28 (0.08) | 0.03 (0.03) | 0.11 | 6.51 | 11.62 | 51 |
| pears | 0.35 (0.27) | 0.23 (0.23) | −1.41 (0.27) | 0.34 | 0.99 | 1.74 | 12 |
| | | | | | 18.27[c] | 36.28[c] | 72[c] |

NOTES
a. Average household consumption and expenditure.
b. Percentage of households purchasing during the survey week.
c. All fresh fruit.

The figures in brackets are estimates of the standard errors and indicate the confidence with which the coefficients can be accepted as accurate measures of these elasticities. The values of $R^2$ give the proportion of the residual variation in monthly average purchases (after the removal of seasonal and annual shifts) explained by the own- and cross-price elasticities.

SOURCE  MAFF (1985), *National Food Survey 1983*, pp. 30, 35 and 205.

Therefore, the overall effect on the demand for different foods is more complex than the cross-price elasticity alone.

In the market for processed foods, manufacturers and retailers frequently seek to promote customer loyalty by 'branding' products. This may be interpreted as a strategy to weaken the substitution effect and reduce the coefficient of cross-price elasticity for the branded product. Then an increase in price of the branded product leads, via a reduced substitution effect, to a smaller reduction in quantity demanded than was hitherto the case. The price elasticity for meat pies in general may be 0.6, whilst the price elasticity for Smith's meat pies following a successful promotion of the brand may be only 0.4. The cross-price elasticity of Smith's meat pies with respect to the price of other pies will also have fallen, to reflect the fact that other pies are no longer such a close substitute for Smith's pies. In this way, the promotion of Smith's pies may be a profitable exercise; manufacturers and retailers are likely to guard their estimates of such responsiveness to price at the brand level as valuable commercial information.

The relationship between food prices and the general level of prices of retail goods and services is often of interest to governments because food is such an important item in total consumer expenditure. It is often politically unwise to permit food prices to rise too fast and, as Table 2.6 illustrates, food prices in Britain have risen no faster than other retail prices over the last two decades, except during the 1971–6 period. Food prices would have risen relative to other prices even faster (because of difficult weather, high world prices and the effect of British entry to the EC in 1973) but for a programme of food subsidies which operated from 1974 to 1976.

## 2  Income levels

The demand for food is influenced by the incomes of consumers with a rise in incomes likely to shift the demand curve to the right for most (but not all) foods. The response of the demand for food to changes in income is calculated as the income elasticity of demand ($E_{dy}$) for which the general formula is:

$$E_{dyq} = \frac{\% \text{ change in quantity of food}}{\% \text{ in income}}, \text{ other things being equal} \quad (3a)$$

In practice, it is often found that consumers 'trade up', buying more expensive food products, thereby spending more money but not necessarily buying a greater *quantity* of food. This applies to foods such as meats, bread and cakes, and fruit, where there are a wide variety of products at different prices and increases in income permit consumers to switch from cheap to more expensive products within the same food group. Under such

**Table 2.6**
**Inflation and Food Prices, 1961–84**

| Period | Average annual rate of increase (%) | |
|---|---|---|
| | All retail prices[a] | Food prices |
| 1961–66 | 5.8 | 3.2 |
| 1966–71 | 6.0 | 6.1 |
| 1971–76 | 16.0 | 17.4 |
| 1976–80 | 13.3 | 12.5 |
| 1980–82 | 9.4 | 8.2 |
| 1982–84 | 4.6 | 4.4 |

NOTES    a. All items, excluding housing, in retail price index.
The weights given to food in the RPI excluding housing were 30 per cent in 1961 and 24 per cent in 1984.
SOURCE    Derived from CSO (1986) *Social Trends* (London: HMSO) Table 6.4.

circumstances it becomes desirable to be able to measure demand response not only in physical quantity terms but also according to the amount of expenditure. For the latter purpose the usual formula is:

$$E_{dye} = \frac{\% \text{ change in expenditure on food,}}{\% \text{ change in total expenditure}} \text{ other things being equal}$$

$$(3b)$$

It has been observed for a hundred years that as household expenditure increases, the proportion devoted to purchasing basic commodities such as food declines; this is known as Engels Law. As incomes rise, the percentage rise in spending on food is less than the percentage rise in total spending. In these circumstances the expenditure income elasticity coefficient is less than one. In wealthy industrial countries such as Britain, most people do not wish to eat a greater physical volume of food (some of course are actively engaged in trying to eat less). The proportion of total household expenditure devoted to food in the UK was over 60 per cent before 1914 [11] and has gradually declined; as we saw at the beginning of this chapter, the proportion was 14.6 per cent in 1984. There has been an increase in eating meals away from home, though even allowing for that only raises the proportion of consumers' expenditure on food and eating out to 20.2 per cent in 1984.

An estimate of the household expenditure elasticity for food taken overall for Britain is 0.2 and has not changed very much for a decade. Note that Engels Law does not require the elasticity itself to decline; the proportion of total expenditure allocated to food declines so long as $E_{dye}$ is less than one. In fact, though, the average income elasticity of expenditure for food in Britain (calculated from household data) has gradually declined over the years, from 0.3 in the 1960s to 0.2 today and many foodstuffs have become less positively responsive to income changes.

Table 2.7 shows income elasticities (expenditure and quantity) for selected foods for 1983 and illustrates four points. First, virtually all food products are characterised by an inelastic response to income and many have elasticities of 0.5 or less. Exceptions are often high value products which are unimportant in the total of household expenditure on food. Secondly, a number of foodstuffs show hardly any response to income at all (liquid milk, eggs and total fats for instance). Thirdly, some food products

**Table 2.7**
**Estimates of Income Elasticities of Demand for Selected Household Foods, 1983**

|  | expenditure | quantity |
|---|---|---|
| Liquid milk | 0.03 | 0.02 |
| Cheese | 0.46 | 0.40 |
| + Carcase meat | 0.29 | 0.20 |
| beef and veal | 0.28 | 0.14 |
| mutton and lamb | 0.29 | 0.26 |
| pork | 0.31 | 0.23 |
| Other meats (uncooked) |  |  |
| bacon and ham | 0.16 | 0.05 |
| broiler chickens | 0.34 | 0.22 |
| sausages: beef | −0.39 | −0.40 |
| pork | 0.13 | 0.08 |
| + Fats | 0.05 | −0.07 |
| butter | 0.18 | 0.18 |
| margarine | −0.16 | −0.29 |
| + Sugar and preserves | −0.15 | −0.24 |
| sugar | −0.29 | −0.33 |
| jams, etc. | −0.12 | −0.15 |
| honey | 0.34 | 0.34 |

| | | |
|---|---|---|
| + Fresh green vegetables | 0.42 | 0.25 |
| cauliflower | 0.43 | 0.38 |
| leafy salads | 0.57 | 0.51 |
| cabbage | 0.15 | 0 |
| + Processed vegetables | 0.09 | −0.04 |
| frozen peas | 0.67 | 0.68 |
| all processed potatoes: frozen | −0.11 | −0.03 |
| other | 0.18 | 0.16 |
| + Fresh fruit | 0.62 | 0.48 |
| oranges | 0.23 | 0.15 |
| bananas | 0.60 | 0.56 |
| fruit juices | 1.15 | 1.16 |
| + Bread | −0.10 | −0.18 |
| white: sliced | −0.60 | −0.59 |
| unsliced | −0.11 | −0.10 |
| brown: | 0.29 | 0.30 |
| wholemeal and wholewheat | 0.54 | 0.56 |
| Flour | −0.28 | −0.25 |
| Breakfast cereals | 0.31 | 0.29 |

NOTE    + Broad food groups, taking account of cross-price elasticities.
SOURCE  MAFF (1985) *Household Food Consumption and Expenditure: 1983*, Annual Report of the National Food Survey Committee (London: HMSO), Appendix B, Table 2.

have negative income elasticities; these products are described as 'inferior' goods of which less is demanded as incomes rise. The reason for this negative income effect upon bagged sugar, lard, white bread and potatoes is that consumers with higher incomes switch to other food products. Fourthly, quantity elasticities are characteristically lower than those based on expenditure because of the 'trading up' effect mentioned earlier.

In general, an increase in household income has only a small positive effect on the quantity of food demanded, shifting the demand curve to the right at all levels of price. Most of the increase is accounted for by 'trading up' and spending more on higher value food products. Overall, a low income elasticity for food implies a stable market size whether incomes are growing in a period of prosperity or declining in a period of recession. This has interesting implications for investors and entrepreneurs who may consider an increased involvement in the food market when other, more income-elastic sectors, of the consumer goods economy are in decline.

The income elasticities calculated and reproduced in Table 2.7 are for one year and represent a cross-section, from householders with high incomes to householders with low incomes.[12] Figure 2.3 illustrates how

**FIGURE 2.3**
**Potato and Fruit Consumption, 1983, by Income Group**

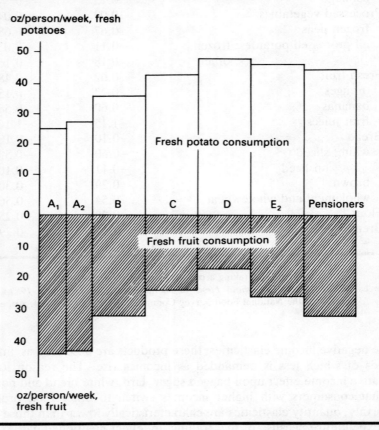

SOURCE    MAFF (1985), *Household Food Consumption and Expenditure Survey*, Report of the National Food Survey, 1983 (London: HMSO).

consumption per person of fresh potatoes and fresh fruit varied in 1983 according to household income in Britain. This is not necessarily the same as indicating what may happen when the incomes of the *same* people rise over time, although it may be an acceptable guide, as many other factors which affect demand may also be at work.

For example, fresh fruit consumption has not changed very much over the last 20 years, during which time disposable real incomes have risen by more than a quarter and income elasticities calculated using cross-sectional data have never been less than 0.5. Purchases of fruit have fluctuated up and down but shown no upward trend. Either other factors, such as household composition and prices, have altered to disguise the income – consumption relationship of fresh fruit, or people do not acquire the fruit-eating tastes of the better-off when they become more prosperous.

Measures taken by the government to change the distribution of income in society affect the demand for food because the rich and the poor have different demands for food, both in terms of types of food and quantities of food. Income redistribution that helps the poor members of society is likely to raise the overall demand for food, though not necessarily for each type of food. There is a greater willingness to buy additional food out of extra income among the poor, arising from their restricted choice of food and poorer nutritional status. Evidence suggests that in the UK, 'There is a major restriction on food choice of those with low income or with large families. The fact that the evidence is too imprecise to indicate whether or not malnutrition still exists should not lead us too readily to assume by inference that because of this the case for no malnutrition has been conclusively proven.'[13]

The government may intervene if it believes poor people are unable to buy sufficient food and this has proved popular in the USA, particularly since the 1964 Food Stamp Act. In the 1981 recession, 22 million Americans were using food stamps. This intervention is unlikely to alter the types of food consumed and, indeed, the primary aim of food stamps in the USA has been to dispose of surplus agricultural products which are typically wheat flour, dairy products and meat.[14] In the UK, governments have avoided rationing, except in wartime and the postwar period of 1945–53, and prefered to treat the poverty problem and inadequate diet by financial transfers rather than food transfers. The exceptions have been free or subsidised school meals for poor children and welfare transfers of orange juice and milk (1939–71). In 1982 welfare foods only constituted 2.7 per cent of state benefits paid to the poor and only 1 per cent of their average disposable incomes.[15]

Income distribution is an important factor for food producers and distributors because it affects the types of products they can sell. When the number of poor households is increasing, basic foodstuffs will be increasingly in demand, whereas when the number of rich households is increasing, it is the demand for speciality foods that will be growing, although in general the better-off are more important in terms of market size than the poor. In 1984, for instance, the 20 per cent of households with the lowest average weekly expenditure on all goods in Great Britain spent an estimated £3.2 billion on food (at home and eating out), whereas the 20 per cent of households with the highest expenditure spent £11.4 billion on food (poor households were on average larger than rich households but this does not affect the point being made here). The better-off allocated only 15 per cent of their total spending to food (including eating out), compared

**Table 2.8**
**Population Changes in Industrial Countries**

|  | Overall % change, 1971–83 | 1983 population (million) | 1983 density (persons/km²) |
|---|---|---|---|
| UK | 1.3 | 56 | 231 |
| France | 6.6 | 46 | 100 |
| West Germany | 0.2 | 61 | 247 |
| Italy | 5.2 | 57 | 189 |
| Spain | 12.0 | 38 | 110 |
| Netherlands | 9.1 | 14 | 352 |
| Belgium | 9.9 | 10 | 323 |
| USSR | 11.2 | 273 | 12 |
| Japan | 12.9 | 119 | 320 |
| USA | 13.3 | 234 | 25 |
| Canada | 15.3 | 25 | 2 |
| Australia | 19.4 | 15 | 2 |
| New Zealand | 14.3 | 3 | 12 |

SOURCE   CSO (1984) *Social Trends* (London: HMSO) p. 23.

with 30 per cent for the poor, but in absolute terms the rich spend nearly four times as much on food as the poor because their average incomes were nearly seven times greater.[16] As a result, the food industry in a free market devotes more effort and resources to satisfying the wants of the rich than it does to satisfying the wants of the poor.[17]

## 3   *Population: its changing size and composition*

The size of the population is the most important determinant of the overall quantity of food demanded, and together with the level of income, is the main determinant of the total expenditure on food in a country. Table 2.8 shows how population has grown at different rates in industrial countries, slowest in West Germany and the UK, fastest in Australia during the 1971–83 period. In Britain, the slow growth of population and of *per capita* disposable incomes, combined with a low income elasticity for food, has led to a very slow growth in the overall size of the market for food. The British population is predicted to grow only marginally faster, at 0.14 per cent per annum from 1981 to 1996 in contrast to 0.11 per cent per annum from 1971 to 1983.[18] Therefore the British market for food is unlikely to expand any faster in the future and changes will continue to involve the substitution of one food product for another.

The population of Britain, and of most of the industrial countries (except some Mediterranean region countries, such as Turkey and Spain) is ageing and the proportion of people in Europe aged 60 or more is forecast to rise to 20 per cent by the year 2000.[19] Old people have an average lower incomes, live in smaller households, consume smaller quantities of food and spend less than average on food and eat out less. Also, the types of food in demand by old people are likely to reflect the diet to which they became accustomed when they were younger, so the future demand for food by old people will in part reflect the present demand for food by the middle aged.

A. *Size of households* The sizes and types of households is a factor which influences the demand for food only marginally on its own, though it is a significant factor taken in conjunction with others. Households have become smaller as couples have married later, have had fewer children and as the number of single parent and pensioner households has increased. One-person households constituted 12 per cent of households in 1961 but 23 per cent in 1983, whilst the average size of household fell from 3.06 persons in 1961 to 2.71 in 1981, a decline of 11 per cent.[20] Smaller households require smaller quantities of food and the greater diversity of household types increases the likelihood of diversity in shopping habits, meal preparation and food consumption. It is, perhaps, in conjunction with changes in work patterns that changes in household structure have their greatest impact in the food market.

B. *Work patterns* There has been a decline in the amount of work done in manual and outdoor occupations. Employment declined between 1971 and 1982 in agriculture, forestry and fishing ($-19$ per cent), energy and water supply ($-11$ per cent) and manufacturing ($-24$ per cent) whilst employment in physically undemanding jobs increased, such as employment in banking and finance ($+27$ per cent) and the public services ($+15$ per cent).[21] Over the same period, there was a decline of 14 per cent in the average consumption of calories derived from household food, from 2530 kcal per person per day in 1971 to 2180 kcal per person per day in 1982.[22] One trend consistent with this decline is the reduction in eating cooked breakfasts (egg and bacon consumption has declined) and growth in the demand for breakfast cereals and fruit juice. Another is the shift from red meats which are perceived to have a relatively high fat content to white meats with a low fat content,[23] whilst a third is the shift from cooked to salad vegetables.

Some nutritionists argue that in fact people have not reduced their consumption of calories sufficiently to compensate for factors such as the decline in physical work and, as a result, the population has become unacceptably obese. One estimate suggests 6–8 per cent of adults are obese (more than a third above the target weight for their height and age) and 32–9 per cent are overweight.[24] The proportion of the total civilian labour force that is economically active (employed or actively seeking paid employment) has increased as more women have joined the labour force, because of the increase in part-time jobs available in service industries, and this may have implications for food demand and consumption as well.

The majority of married women with paid jobs work away from home, although home workers numbered 1 million in 1981 (male and female) and accounted for 4 per cent of the labour force. Because they go out to work, such women and their families can be expected to have an increased demand for convenience foods, semi-prepared meals and frozen foods and a reduced demand for foods associated with a long time spent in food acquisition in shops, preparation and cooking at home (meat, cooked vegetables, potatoes). Data are more scarce which distinguish between households with one income and those with two incomes, nor between housewives who are economically inactive and active, so these relationships remain speculative. The increase in the numbers of working wives has been contemporaneous with a growth in car and freezer ownership (43 per cent and 57 per cent of households respectively in 1983) and the development of one-stop shopping at superstores and out of town hypermarkets. In this way, the supply of and demand for convenience foods have increased together, along with the technology for cool-chain marketing and home-storing of perishable foods. The part-time nature of many new jobs has also perhaps encouraged the practice of taking snacks and smaller, lighter meals, of packed lunches and of buying take-away meals.

Whilst more married women have found paid employment, there has been a growth in male unemployment since 1980 and an increase in the intractable nature of unemployment in industrial countries, especially in Britain. Definitions of unemployment vary but officially 3.16 million people (13.1 per cent of the labour force) were unemployed in 1984,[25] directly affecting perhaps 15–20 per cent of households. In households where a wife earns a wage whilst her husband is unemployed, there may be particular implications for the demand for food, though we do not consider the matter further here.[26]

C. *Ethnic diversity* About 6 per cent of the population of Great Britain is officially classified as non-white, about half born in Britain. The importance for the demand for food of the mix of ethnic origins in Britain lies in the introduction of new foods and tastes and new ways of preparing foods. Those cities with important ethnic communities (Leeds and Bradford, Birmingham and Coventry, West and South London, for instance) have experienced the initial demonstration effect and new ideas are likely to gradually diffuse to other parts of the country. The increasing expenditure by households on tropical fruits, herbs and spices and on eating out also owes something to foreign travel, television and improved availability. Immigration into Britain largely ceased in the 1970s and as an increasing proportion of the ethnic population is now born in Britain, their continuing impact on the demand for food will depend on how far their dietary tastes and preferences continue to reflect their family backgrounds. The continued expansion of the market for ethnic and foreign foods, eaten at home and away suggests that the diversity of tastes and foods is likely to continue to increase.[27]

## 4   Tastes and preferences

Even after we have accounted for the effects on the demand for food of prices, incomes and population variables, there remains a residual part of demand which is unexplained and which can only be related to the tastes and preferences of consumers. As we have already noted, increases in the consumption of fruit, vegetables, dairy products and, up to the 1950s, in sugar and the decline in the consumption of bread and potatoes can be explained to a large extent in terms of higher incomes and increasing prosperity in Britain. Changes in the types of meat consumed can be explained largely by recourse to a combination of changes in relative prices and increases in incomes. The increasing demand for convenience foods (canned, frozen and prepared foods) in large part can also be explained by increased incomes and by new patterns of living, working and shopping in Britain.

Similarly, the increasing preference for foods perceived as 'healthy' which we consider in this section has strong associations with income as well as being an example of a change in tastes. The government has accepted the need to alter dietary habits in Britain and the future is likely to see consumers officially encouraged to alter their tastes and preferences for food. Therefore, it is logical to consider the issue of food and health as a taste change, rather than as one associated with a growth in income. Nevertheless, it remains true that several foods perceived as 'healthy' do have relatively high income elasticities and this suggests the possibility that a government inspired campaign to encourage 'healthy' dietary habits in Britain may be more successful during a period of prosperity and rising incomes than during a period of stagnation and decline.

Food and health have been related in the minds of those in authority from time to time in our history and often special measures have been adopted to protect groups in society which have been viewed as 'at risk' from dietary deficiencies. Examples include the Admiralty supplementing the diets of British sailors with lime juice to prevent scurvy and of steelmasters providing their furnace workers with rations of beer to offset dehydration and give them more energy. In wartime, troops and manual workers in reserved occupations have also been given larger rations than others of high energy foods, and during the 1930s slump measures were introduced to help poor mothers and infants with issues of free or subsidised milk. In contemporary Britain, people variously defined as 'at risk' from poor diets may include pregnant women and nursing mothers and infants, housebound women of Asian extraction and poor families subsisting on low incomes, often with illness and unemployment to contend with besides a low quality inadequate diet.

However, official concern over food and health in recent years, at least demonstrated by the establishment of committees and publication of reports has been not only towards these specific groups of people but also towards society in general. This wider concern relates particularly to heart disease. The origins of this concern can be traced back to the 1960s when medical and nutritional research teams were seeking explanations for the rise in the death rate from what became known as the diseases of affluence – heart disease, cancers and various stress-related diseases. The death rate from coronary heart disease (CHD) in particular rose steadily from 1930 to 1960 and thereafter stubbornly refused to decline in Britain until the 1980s. The position remains unacceptably high according to medical opinion, with the UK ranking high in the world for CHD mortality. CHD is responsible for 38–40 per cent of all deaths and in England and Wales 30 000 men under the age of 65 die from CHD annually.[28]

It was for these reasons that CHD became an issue of public concern and the establishment of the Health Education Council (HEC) under the Department of Health and Social Security (DHSS) in 1968 provided a vehicle for a campaign to reduce the death rate for CHD, though as regards diet, it was to be seventeen years before official recommendations were finally agreed. Dietary habits were viewed and continue to be viewed as only one element in explaining why CHD has increased in Britain. Other elements which have been identified include smoking, a lack of exercise, a hereditary disposition to CHD, soft water, high blood pressure, age and sex (men being more prone to CHD than women).[29] With regard to diet, some researchers identified the consumption of refined foods and particularly sugar as being excessive among the population and most nutritional dietary recommendations now include either a ceiling or a reduction in sugar consumption as an important measure in preventing excessive weight gain.[30] Alcohol consumption is similarly fattening and consumption is usually recommended to be reduced. About forty per cent of the British population of middle age are estimated to be significantly overweight and there has been a growing emphasis on influencing the young to acquire healthier dietary habits than those of their parents.[31]

A controversial aspect of food and health and one that is potentially damaging for many sectors in the farming and food processing industries is the consumption of fat, both the total amount and the balance between saturated and polyunsaturated fats in the diet. Successive governments in

Britain have been reluctant to become involved in nutrition policies with implications for food and diet[32], in contrast to their willingness to devise new policies for agriculture. The eventual acceptance of dietary recommendations by government came after a series of reports and research projects during the 1960s and 1970s, by which time several trends towards a healthier diet were already evident. The demand for skimmed milk, wholemeal bread, breakfast cereals and at least some fruit and vegetable products have increased in Britain since the late 1970s, whilst bagged sugar, butter and white bread have continued to show reductions in demand.

On the other hand, some trends in demand appear to be contrary to the emerging consensus on nutritional advice – more soft drinks, wine, cream, processed foods and continuing high levels of salt intake for example. Also, fresh potatoes and total bread consumption continue to decline. Therefore, it is apparent that convenience, tradition and palatability, together with prices and incomes, continue to influence food demand as well as the growing awareness of the health issue. Further, there is evidence from the market for meat that whilst consumers understand the health message, they do not always choose to act upon it.[33]

The issue of fat consumption is a good illustration of the difficulties which can arise when the government and scientific opinion are brought together in an attempt to shape nutritional guidelines; we have yet to see whether these efforts produce any profound changes in diet in Britain. The Committee on Medical Aspects of Food Policy (COMA) first reported to the DHSS in 1974 on the need to reduce fat consumption[34]. However, little action was taken on this report, or on a report with similar findings by the Royal College of Physicians in 1976[35]. In 1978 the HEC published a booklet encouraging people to eat more fibre and less fat, but with no emphasis on reducing saturated fats as such[36].

In 1978–9, the Centre for Agricultural Strategy at the University of Reading considered the interrelationships of food and diet and especially the saturated fat component in a series of reports and seminars.[37] These reviewed research in several countries besides Britain and reported on the growing body of opinion that western governments should formulate national targets relating to food and health based on nutritional requirements and involving lower levels of a healthy diet of saturated fat consumption. Such a food and nutrition policy should in turn be linked to an agricultural policy in a consistent way, with the implications for meat producers and dairy farmers of lower levels of red meat and butter fat consumption recognised.

In 1980, the DHSS and its National Advisory Committee on Nutrition Education (NACNE) set up a subcommittee to report on the consensus of expert opinion in order to plan a programme of nutrition education for the public. This took three years to report, during which time the Expert

Committee of the World Health Organisation (WHO) recommended a significant cut in the consumption of saturated fats in the diets of people living in Western countries.[38] The problem that NACNE faced was to make general dietary recommendations for individuals to follow which could only be based on the available research findings and evidence. This in turn is drawn largely from samples of small groups of people in experiments in which the participants are controlled only in their dietary intake and not in other aspects of their lives and personal behaviour.

The NACNE recommendations[39] called for a reduction in saturated fat levels to provide no more than 10 per cent of total energy, out of a total fats contribution to energy of 30 per cent (in contrast, the proportions in 1983 were about 20 per cent and 40 per cent respectively). The implications for farming according to one estimate would involve a 50 per cent drop in butterfat requirements and a 30 per cent reduction in the consumption of red meats.

In 1984, a report was published by COMA (updating their 1974 report) and this proved more acceptable than the NACNE report to both the government and the food and farming industries. The emphasis of this report was on CHD rather than on diet and health in general and that led to an emphasis on fat consumption. Targets were set for individuals which were more modest than the NACNE targets (COMA recommends a maximum of 15 per cent of energy from saturated fats and 35 per cent from total fats). MAFF accepted these and the other COMA recommendations as a basis for future food policies (less fats, sugar, alcohol and salt, in the diet and more dietary fibre) and in 1985 took steps to improve food labelling legislation and to reduce the subsidies paid to those selling beef and lamb carcases with a high fat content.[40]

The Joint Advisory Committee on Nutrition Education (JACNE) succeeded NACNE in 1985 and was charged with advising the HEC on the production of a new booklet which could translate the COMA recommendations into practical dietary measures. An initial attempt in October 1985 was criticised for recommending too many processed foods and not enough whole foods, fruits and vegetables. This illustrates the continuing difficulty of officially recommending specific dietary targets in a market for food as competitive and highly organised as that in Britain.

The issues of obesity, saturated fat consumption and CHD have been joined in recent years by other concerns regarding the number and types of food additives used in processed foods and the presence in fresh and processed foods of harmful chemicals and residues. In time, and given a gradual improvement in food labelling, the food industry argues that there will be increasing scope for consumers to choose between those foods which are for instance cheaper, more coloured, store longer and contain additives and those foods which do not. Similarly, farm produce that is wholly organically grown or semi-organically grown (and pesticide free) is already available although at markedly higher prices than conventionally produced food, reflecting higher costs of production and scarcity.

The European Community has banned the use of hormone implants in beef and veal production, to take effect from 1987 (1989 in the UK) and this can be taken as an example of a policy measure that satisfies three interest groups. It pleases consumer associations in the EEC, and those beef producers in the Community particularly those in France and Italy not using hormone implants, and thirdly it pleases those charged with balancing the Community budget since it discourages future increased production of surplus beef, thus reducing future expenditure from the EC budget. In this case scientific opinion that the practice was harmless was ignored in favour of the political and financial advantages of a ban. It will be interesting to see whether this order of priorities is repeated with respect to food additives and pesticide residues in food products in the European Community during the 1980s and 1990s.

Undoubtedly we are in a period of conflicting trends in consumer preferences for food in Britain. What is healthy and regarded as safe is not always the same as what is convenient and palatable, and it remains to be seen to what extent consumers actually do alter their dietary habits in line with recommendations of bodies such as COMA, and how far they are swayed by other considerations.

# Notes

1. CSO (1985) *Annual Abstract of Statistics* (London: HMSO).
2. Angel, L. J. and Hurdle, G. E. (1978) 'The Nation's Food – 40 years of change', *Economic Trends* (London: HMSO) (April).
3. FAO (1984) *Food Balance Sheets, 1979–81 Average* (Rome).
4. USDA (1985) *National Food Review*, *24*: 137 (Washington). Blandford, D. (1984) 'Changes in Food Consumption Patterns in the OECD Area', *Eur. Rev. Ag. Econ.* 11–1, 43–64. Blandford's analysis omits Greece, Turkey and Iceland, leaving 21 countries in Western Europe, Australasia and North America.
5. MAFF published the CLE results as 'Output and Utilisation' reports until the 1980s. Tables are published in a number of CSO reports including the *Annual Abstract of Statistics* (London: HMSO).
6. See for example, Centre for Agricultural Strategy (1979) *National Food Policy in the UK* (As Report No. 5 University of Reading) p. 64.
7. Reports are entitled *Household Food Consumption and Expenditure*, Annual Report of the National Food Survey Committee, MAFF, (London: HMSO). They are generally published two years after the data is collected and MAFF publish quarterly updates of Food Survey statistics in the series 'Food Facts'. For a critique of the survey see, Frank, J. D., Fallows, S. J. and Wheelock, V. J. (1984) 'Britain's National Food Survey', *Food Policy* 53–67.
8. Allen, D. E. (1968) *British Tastes – An Enquiry into the Likes and Dislikes of the Regional Consumer* (London: Hutchinson). Allen, D. E. (1976) 'Regional

variation in food habits', in Oddy, D. J. & Miller D. (eds) *The Making of the Modern British Diet* (London: Croom Helm). Regional variations in average purchases are summarised quinquennially in the National Food Survey: see for instance the reports for 1975 and 1980. For a more accurate statistical estimate of regional preferences, see 9 below.

9. Lund, P. J. and Derry, B. J. (1985), 'Household Food Consumption: the influence of household characteristics', *J. Ag. Econ.* 36:1, 41–58.

10. EC (1985) *The Agricultural Situation in the European Community* (Brussels).

11. See for instance, Burnett, J. (1979) *Plenty and Want: A Social History of Diet in England from 1815 to the Present Day* (Rev. edn, London: Scolar Press).

12. More strictly a cross-section of households with a housewife and in which the earnings of the head of household ranged from high to low. This may be an inaccurate measure of household income now that a growing proportion of households contain more than one income-earner (see Frank *et al.* (1984) note 7 above).

13. McKenzie, J. G. 'Poverty: Food and Nutrition Indices' in Townsend, P. (ed.) (1970) *The Concept of Poverty* (London: Heinemann) p. 85.

14. Food stamps were used by 21 million Americans in 1983–4. USDA (1985) *National Food Review*, 29.

15. Calculated from *Social Trends* (London: HMSO, 1986) Chs 5 and 6.

16. CSO (1985), *Social Trends*, Table 5.19.

17. Scitovsky, T. (1977) *The Joyless Economy. An Inquiry into Human Satisfaction and Consumer Dissatisfaction* (Oxford: Oxford University Press).

18. CSO (1984), *Social Trends*, p. 20.

19. Kirk, M. (1981) *Demographic and Social Change in Europe, 1975–2000* (Liverpool University Press).

20. CSO (1984) *Social Trends*, Table 2.5.

21. CSO (1984) *Annual Abstract of Statistics*.

22. MAFF (1983) *Household Food Consumption and Expenditure*.

23. The extra fat on red meats may be trimmed off before or after cooking and so not be eaten. See Meat and Livestock Commission (1986) *Meat: The health issue in perspective* (2nd edn, Banbury).

24. Taken from NACNE (1983) *Discussion Paper on Proposals for Nutritional Guidelines for Health Education in Britain*, (London: Health Education Council). Prepared for NACNE under an ad hoc working party chaired by Professor W. P. T. James.

25. CSO (1985) *Monthly Digest of Statistics* (November). The unemployment rate in 1980 was 6.8 per cent.

26. The relationship of poverty to diet in the north of England is explored in Lang, T. *et al.* (1984). *Jam Tomorrow?* (Food Policy Unit, Manchester Polytechnic).

27. Hill, S. (1985) *Specialist Foods* (Watford: 19D).

28. COMA (1984) *Diet and Cardiovascular Disease*, report of the Committee on Medical Aspects of Food Policy, DHSS (London: HMSO).

29. Ibid.

30. Sugar is recognised as a major cause of dental caries. See, for instance, British Dental Health Foundation (1984) *The No Sugar Cook Book* (Milton Keynes).

31. Health Education Council (1984) *Food for Thought*. p. 15. Besides a number of Health Education Council booklets and DHSS surveys reported in 1968 and 1975 of the diet of children, the London Food Commission has published guidelines for pre-school children. *Food and Drink for Under-Fives* (June 1985) p. 21.

32. For a criticism of the slowness of government action, see Walker, C. and Cannon, G. (1984) *The Food Scandal* (London: Century).
33. Foxall, G. R. and Haskin, C. G. (1985) 'Naughty but Nice': Meanings of Meat in the 1980s', *Food Marketing 1*:3, 56–69.
34. COMA (1974) *Diet and Coronary Heart Disease*, DHSS (London: HMSO).
35. Royal College of Physicians (1976) 'Prevention of coronary heart disease', *Journal of the RCP, 10*: 213–75.
36. Health Education Council (1978) *Eating for Health* (London).
37. Centre for Agricultural Strategy, University of Reading: Robbins, C. J. (ed.) (1978) *Food, Health and Farming: a Report of Panels on the Implications for UK Agriculture*; Report No.4 (1978) *A Strategy for the UK Dairy Industry*; Report No.5 (1979) *National Food Policy in the UK*. Since 1982, the Food Policy Research Institute at the University of Bradford, under Professor Vernon Wheelock, has produced a number of discussion papers and articles exploring these and related themes further.
38. WHO (1982) *Prevention of Coronary Heart Disease* (Geneva) WHO Technical Report Series, 678: p. 53.
39. NACNE (1983) op. cit. See also Cannon and Walker (1984), op. cit. and Meat and Livestock Commission (1986), op. cit.
40. MAFF (1985) Press Release, No. 162, 24 June 1985. MAFF (1985) Press Release, No. 78, 12 March 1985. It may be possible to accommodate much of the recommended reduction in fat consumption by marginal changes in the fat content of carcase meat and milk. See Jones A. S. (1984) '*Food quality and human health: how can farmers respond?*' pp. 64–73 in RURAL, *The Way Ahead* (Bore Place, Chiddingstone, Kent).

# The Farmer and the Food Industry in the Food Chain  3

## 3.1 The Importance of the Food Industry

Very little of the food consumed in the UK is produced in its final form by the same people who grow it. Vegetables and fruit grown in domestic gardens fall into this category, but little else. Not much goes straight from farmer to consumer either, although farm shops and Pick-Your-Own are cases where this does happen. About 75 per cent of food produced on farms in Britain is processed in some way before it reaches the consumer, 85 per cent if carcase meat is included.[1] Most food passes along some form of **food chain**, a term used to refer to the different stages through which farm products are gradually transformed into food products. Figure 3.1 illustrates a simplified food chain involving domestic farming, imports, food manufacturers, wholesalers, caterers (probably under-estimated in 1979), retailers and consumers. Food manufacturing embraces processes such as bread-making, meat processing, dairy products, confectionery and many other types of product. Other stages can be found: pre-packers, storers, primary and secondary wholesalers (the former operating between farmers and the next stage, the secondary wholesalers operating between markets of one sort or another and retailers). The term **food industry** refers to all these elements in the food chain with the exception of the farmer and the consumer. Some elements are private companies but there are also cooperatives, Marketing Boards and Commissions appointed in part at least by government.

In the UK, the food industry has been an important part of the economy for well over a hundred years. By the mid-nineteenth century, 75 per cent of the population lived in large cities and food processing and distribution

**FIGURE 3.1**
**The Food Chain in Britain £'000 million. Estimates for 1984 (1979 in brackets where available)**

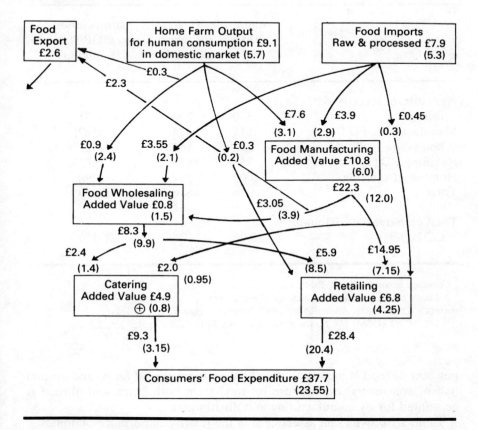

NOTES    Agricultural Input Industries and Resources are excluded in this presentation of the food chain. Also, note that the term 'Added Value' used here differs from the concept of Value-Added that is used in the National Accounts. Estimates are at current 1984 (and 1979) prices.
      ⊕ A new, broader definition was used for the 1984 catering estimate.

SOURCES   Tanburn, J. (1981) *Food Distribution: its Impact on Marketing in the '80s*, (London: Central Council for Agricultural and Horticultural Co-operation). Food from Britain (1986). Meeting the needs of the market. London: Market Tower, New Covent Garden.

(especially by rail) provided regular supplies of cheap food to factory workers. Currently, the food chain as a whole contributes about 7 per cent to Britain's National Income, roughly equal proportions coming from domestic agriculture, food manufacturing and food distribution. About 40

**Table 3.1**
**Agriculture and the Food, Drink and Tobacco (FDT) Industries 1984**

|  | employees (GB) ('000) | (of which food only) | contribution to GDP (UK) (£ million) |
|---|---|---|---|
| Agriculture, forestry & fishing | 330 | 330 | 5971 |
| Manufacturing FDT[1] | 611 | 424 | 9582 |
| Wholesaling FDT | 282 | 166 | 2427[2] |
| Retailing FDT | 755 | 649 | 6562[2] |
| Hotels and catering | 1000 | 307 | 8298 |
| Total | 2978 | 1876 | 32840 |
| Total employees (GB) and GDP (UK) | 20690 |  | 280129 |

[1] Coverage is wider than in Table 3.3.
[2] Author's estimates, based on number of employees.
SOURCES  CSO (1986), *Annual Abstract of Statistics*, Table 6.2.
           CSO (1986), *UK National Accounts*, the Blue Book, Tables 2.1, 2.2.

per cent of food is imported, mainly in an unprocessed form, and around half of consumers' expenditure on food from both home and abroad is accounted for by manufacturing and distribution.

Table 3.1 provides an assessment of the relative importance of domestic agriculture and the food industry by comparing numbers of employees and gross value added; on these counts the food industry is over five times as important as agriculture to the UK economy. In the rest of the EC, agriculture and the food industry are of a more comparable size. Thus in 1982, whilst the UK only accounted for 8 per cent of employment in EC agriculture, it accounted for 19 per cent of employment in the EC food industry.[2]

Employment and output in the food industry are relatively stable over time in comparison with other industries. This stability reflects the overall static nature of consumers' expenditure upon food in Britain that was illustrated in Table 2.2. However, the economic fortunes of agriculture and the food industry do not go hand in hand; farm prices and food prices do not always move in unison and that creates the opportunity for changes in the profits and incomes of agriculture to be different from those in the food industry. For instance, during the 1981–3 period, when net farming income was low (described in Chapter 12), the profit rate in the food industry (adjusted for inflation) rose to a ten-year peak.[3]

One major difference between agriculture and the food industry in Britain lies in the number of businesses and degree of concentration. Technical change has led to an increasing size of business and fewer businesses in all sections of the food chain. In agriculture, production is dominated by the largest 20 000 farms (Chapter 11), whereas in the various branches of the food industry production is often dominated by fewer than five firms, a point we return to later.

## 3.2  Understanding the Food Chain

The question is often asked, why don't farmers receive a greater proportion of consumer spending of food? Behind the question lie two main anxieties: first, that the food industry is wasteful and inefficient, and second, that large food manufacturers and distributors can use their superior bargaining power to exploit farmers. A review of the landmarks in agricultural policy-making (see Chapter 14) reveals that these anxieties have led to legislation in the past, and to various attempts by government to provide an institutional framework for the marketing of the products of the agricultural industry. In the next section we consider this further.

### 3.2.1  Marketing

The marketing of agricultural and food products is generally viewed in one of two ways: either as a commercial activity carried out by farmers and food companies along the food chain or as an area of concern for policy-makers in government.[4] We shall consider both of these in turn, though the reader should refer to basic marketing textbooks for a fuller treatment of the subject.[5]

Marketing can be defined as follows: 'Marketing is the process whereby, in order to fulfil its objectives, an organisation accurately identifies and meets its customers' wants and needs.'[6] From this and other definitions the essential ingredients of marketing can be extracted. They are that:

1. Marketing is all-embracing. 'The concept actively involves in the marketing operation all those concerned with research, development, design, production, finance, distribution, after-sales service, as well as the labour force.'[7] From this, we note that successful marketing requires organisational integration and that where individual farms and food firms are unable to provide sufficient integration, we may expect to see integration imposed by commercial expansion or by government.

2. Marketing is customer-orientated, rather than designed to dispose of what has already been produced. The focus on the customer is relatively modern and has only become a dominant feature of business marketing since the 1930s.

3. Marketing is a behavioural science. Increasingly, consumers are viewed as having different wants and demands according to where they live, what level of income they have, how large their family·is, etc. We have observed in Chapter 2 that such factors have a discernable effect on the demand for different types of food. When concern is with a particular food product, the behavioural factors become very important. (Consider, for instance, factors which might explain the level of the demand for (1) dairy products in general and (2) low-fat, fruit-flavoured yoghurt in 500g. packs sold in superstores.) The motivation behind consumer choice is so important that a subsector of marketing activity has emerged, originally in the United States, called **market research**. This is devoted to providing information which allows firms to organise their activities in line with a **marketing concept**, 'identifying, anticipating and satisfying customer needs and desires – in that order'.[9] Six stimuli to marketing can be identified and these may be viewed together as the **marketing mix**:

1. the product
2. the package
3. price
4. the promotion
5. distributive structure
6. the corporate image.

The marketing concept and marketing mix have become established as an essential ingredient in commercial success wherever competition is active. This is particularly true of markets in the food chain, where the overall limitations of the size of the domestic market means that firms can expand sales only at the expense of competitors. This is sometimes termed a 'zero-sum game'. Since 1973, British farmers and food firms have faced increasing competition within the EC and there have been a number of individual and joint efforts to increase the influence of the marketing concept upon what is produced and therefore upon how resources are allocated. From the early 1980s, an attempt has been made to make farmers more dependent on market returns and this will inevitably continue to increase the relevance of the marketing concept to the food chain.

### 3.2.2 Agricultural marketing

There are special features which make agricultural marketing different from other forms of marketing. There are a large number of small farm businesses each supplying more or less the same product. In 1982 there were, in England and Wales, 29 000 dairy farms, 10 000 horticultural farms and 10 000 pig and poultry farms. In all, there were about 100 000 full-time farms supplying the food chain with a range of food and agricultural products which are only marginally graded according to type and quality (e.g., milk by butterfat content, grain mainly according to whether it is suitable for breadmaking or animal feed). Many of these farms are limited to a few production possibilities and individual farmers are seldom in much of a position to apply the marketing concept. They must instead rely on joint action through cooperation or Marketing Boards. We consider these responses to the marketing problem of farms operating within a small business structure later. We note in passing that in other countries where farms are much smaller, such as in Spain and Italy, the marketing disadvantage becomes greater. It is for this reason that the EC has introduced policies to improve marketing through cooperatives, particularly in Mediterranean countries. Where farms are small but cooperatives are strong and well organised, as in Denmark and the Netherlands, this marketing problem has been substantially overcome.

The pattern of large numbers of farmers supplying indirectly large numbers of consumers leads to a 'dumb-bell' shape in the food chain and there are three stages for consideration. In the first stage, food and agricultural products are collected from farms, concentrated and bulked up into an economic scale. It is in this first stage that government intervention is most likely. Besides the questions of the efficiency and cost along the food chain and of unequal bargaining power to which we have already alluded, there is the tendency for prices to fluctuate. This is a problem of agricultural markets, rather than of agricultural marketing, and one to which we return in Chapter 13. In the second stage of agricultural marketing, many agricultural products are transformed in some way into food products whilst in the third stage the products are broken down into smaller quantities and distributed to different localities and shops for final sale to consumers. We consider each of these stages in turn and show that changes in the relative dominance of manufacturers, wholesalers and retailers in the food chain has accompanied an increasing market orientation amongst farmers.

The traditional starting point for an analysis of markets is the concept of adding **utility** to a commodity and the activities and resources used in the food chain are justified on the grounds that they are useful in transfering the agricultural raw material to the final consumer. This is achieved in four main ways. Firstly, the product is made more useful by a change of **form** (from grain to bread for instance). Secondly, by transporting a product such as grapes from Cyprus to Newcastle, it is put into a more useful location for consumers on Tyneside (the utility of **place** is provided). Thirdly, many agricultural products are most available at certain times of the year and, by storing them, businesses in the food chain provide the utility of **time**. (Many firms will do all three; for instance a meat firm that buys live cattle, butchers them, transports the meat to a city and freezes it in cold store provides two changes in form as well as changes in place and time). The fourth utility is that of **ownership**. A firm that buys from A and sells to B may add nothing to form, place or time (except a few days) and yet it takes the risk of ownership. Foods can perish and become inedible, so this can be the most useful of all additions to a commodity's value on its way along the food chain. Where prices fluctuate unpredictably, cereal growers may prefer a merchant to own the grain and accept the risk of a price fall at the time of sale.

It is implicit in this utility theory of marketing that consumer sovereignty reigns in the market and that utilities are provided according to the wishes of consumers to the extent that they are a reflection of purchasing power. Where the distribution of income is particularly uneven, a large number of poor consumers will have a relatively small impact on total effective demand, whereas a small group of rich consumers and those in the middle will have much more impact (see Chapter 2). In such circumstances, the utilities added to food by the food chain may not be in accordance with the wishes of a majority of consumers but rather reflect the desires of the better off minority.[10]

The simplest utility of form that can be added to an agricultural product is grading and this is often an important issue for farmers. For consumers, grading is the first step in product differentiation and helps them allocate their money more effectively. Grading has the further advantages that it allows buyers to purchase without closely inspecting each consignment, and makes it easier for buyers and sellers to negotiate contracts. It also permits 'futures' trading to develop, and lastly it encourages consumers to discriminate between high and low 'quality'. In fact, high quality in the economic sense simply means a grade of quality that commands a high price from consumers. For instance, consumers have traditionally regarded high quality fruit as being unblemished, regular sized and of a certain colour. It is possible that in the future consumers may come to associate high quality with organically grown and pesticide-free fruit, whatever its shape, size or colour. Similarly with meat, 'high quality' can mean to some consumers low visible fat whilst to others it may mean a strong flavour. For this reason, grading standards should be changed to match changing tastes.

Consumers are willing to pay more for a graded product because they are gaining more satisfaction from it. As long as the extra revenue covers the costs of grading, producers will also benefit. However, markets in the food chain are often dominated by a few firms and these may require a standard of grading so high that there is little net return to producers. Many farmers will find the costs of both grading and finding a market for a high proportion of outgrades not worth the benefit.[11]

Statutory grading exists in the UK for eggs and certain fruits and vegetables, based on EC grades: extra, Class I, II and III for horticultural products. Few producers find it profitable to attempt to grade at the 'extra' level and the EC often forbids the general sale of Class III produce on the open market because of chronic over-supply. Producer associations can also request 'derogation', which means setting a minimum class and size of marketable produce.

The four utilities which are added during the food chain also add to costs and consumers are only prepared to pay higher prices as long as they can recognize, and want, the added usefulness of the product. Promotion and advertising is used to inform and educate the public of added utilities. Sometimes all products of one type are advertised (e.g., all beef or all British beef) and this is termed generic advertising. Such advertising is likely to be most effective when combined with quality control through grading and there have been two attempts in Britain to promote the food products of the whole farming and food industries based on improved quality. We discuss advertising by farmers themselves later and concentrate here on national promotions.

The first National Mark was introduced in 1928 and was a short-lived failure.[12] The Mark was applied in cases where producers could demonstrate a high standard of quality and continuous supply. Thus, an egg producer who was accepted could use the National Egg Mark and draw on public funds to promote and advertise. Few farmers bothered to apply and fewer remained in the scheme. Its failure, according to the Lucas Report, 'Left the market for highly standardised, attractively presented and well-advertised bulk lines of produce, mostly sold under a brand name, to be still further exploited by their overseas competitors'.[13] The Food From Britain (FFB) Quality Mark was introduced in 1985 with a £1 million advertising budget. It introduced a system under which foods (again, mainly processed) could be offered for inspection to obtain the FFB logo. Initial responses have encouraged FFB to continue with the Mark, despite the hurdle that multiple retailers prefer their own names on foods rather than anyone else's. Government funding for FFB tapers off in 1988 and it is then up to farmers and the food industry to decide whether a Quality Mark can increase returns and achieve what the National Mark failed to achieve sixty years earlier. Interestingly, both were launched in a Britain over-supplied with food from home and abroad. Holland, Denmark, France, Spain and West Germany all promote their foodstuffs on the British market with advertising budgets at least equal to that of FFB. Other countries advertise specific products often through an Export Marketing Board (e.g., New Zealand apples).

There are several product-specific promotional bodies in Britain. For instance, the National Dairy Council promotes the consumption of British dairy products and is financed jointly by the producers (through the Milk Marketing Boards) and the Dairy Trades Federation. Meat promotions are organised by similar jointly-financed promotional bodies, with producers represented through the Meat and Livestock Commission. At the other end of the scale, there are a number of small producer associations which raise levies on farmers to promote a foodstuff. Some, such as that for mushrooms, have been notably successful, whilst others have witnessed controversies among members over issues such as which variety should be advertised and at what time of year. An exception to the pattern of national promotional bodies in Britain is the Butter Information Council, which has promoted butter from any source since 1954.[15]

The food chain can be analysed in a number of ways and we adopt a mixture of three of the more popular approaches. One way is to concentrate on the institutions in the market (the firms, Marketing Boards and so on), whilst a second approach is to look at particular commodities and channels of distribution. A third approach is to identify which functions are being performed in the Food Chain, functions such as buying and selling, storing and processing, financing and risk-bearing.

The rest of this chapter considers the food chain at three stages: first at the farmgate level, second at the stage of processing and third at the retail level. In practice, many business firms and other organisations add value to agricultural products at all stages of the food chain, perform several functions and provide all four types of usefulness. Nevertheless, the problems and practices are sufficiently different to make it sensible to consider the food chain divided into three stages.

## 3.3   Stages in the Food Chain: (1) From the Farm Gate

A key aspect of this stage in the food chain is the urge by farmers to attain more power through collective or governmental action and to improve their returns from marketing by integrating either horizontally and/or vertically (we explain these terms later).

### 3.3.1   Individual marketing

In pre-industrial communities farmers market their own produce, taking it to village or town markets and staying until the produce is sold. For marketing in the larger towns, wholesale traders purchase foods and add usefulness through various activities. Such practices are still the normal channels for marketing food over large areas of the globe.[16] Interestingly, there are signs of a resurgence of such direct and informal marketing in North America where small farmers and traders bring fresh produce to 'truck markets' in cities.[17]

The most common forms of direct selling in the UK are farmshops and Pick-Your-Own (PYO) fruit and vegetables (called 'U-Pick' in North America). In the UK there was a rapid growth of PYO during the 1970s. This form of selling is geared to car ownership and the willingness of consumers to provide their own marketing functions. Considerations include the weather, the attractiveness of an area for visiting, and more mundane factors such as the provision of toilets, play areas for children and adequate car parks.[18] Other schemes for direct selling do not yet provide a significant option for farmers in the marketing of food (e.g., fresh cream and dairy products at the farm gate, lamb and pig carcases for freezing and farm-processed foods). The costs of such operations generally require a more reliable outlet than passing trade can provide. However, there are signs of a growing interest in farm-scale food processing in Britain, particularly speciality foods which can be sold to delicatessens, health food stores and, eventually perhaps, the large multiple stores. Small dairy farmers in Wales and Western England, restricted by a milk quota insufficient to provide a reasonable income but with the talent to produce and market speciality foods, are likely to become suppliers of this growing demand in the food market, a market hitherto dominated by imports.[19] At the other end of the scale, some farms are sufficiently large to operate as individuals making contracts with processors or retailers, whilst medium-sized producers are more likely to form associations of some kind. We consider the issue of contract farming and of cooperation later.

The marketing position of the farmer is that of a price-taker with such a small output relative to the size of the market that individually he has no market power. Consequently, he is subject to unstable and unpredictable fluctuations in price, both short- and long-term. Also, both in buying and selling, the farmer deals with industries frequently dominated by a few large firms which are able to exploit a monopolistic position in the market. History has shown that a free market in agricultural products has problems which necessitate some form of regulation by governments and Governments use market intervention for a range of reasons. The main ones are:

1. *Inherent instability* We mentioned in Chapter 2 the problems of weather-induced supply variations that interact with highly inelastic demand. Prices fluctuate very widely to achieve a balance of supply and demand. This may discourage investment by farmers and leave consumers to face food shortages which threaten those too poor to afford high prices. Virtually all government measures which directly or indirectly intervene in

the market have as one objective the stabilising of prices and the encouraging of 'orderly marketing'.

2. *The threat of the market power of the food industry* The dumb-bell shape of agricultural markets suggests that farmers may be subject to exploitation by a few large buyers; consequently they may benefit from joint action. By reducing the oligopolistic powers of food processors and distributors, or by joint action, farmers may obtain higher prices for their products. It has been argued that there is little evidence to support this and that since the technical barriers to entry in the food industry are low and small plant size is often economic, there is no *a priori* strong reason to suggest farmers are exploited.[20] Nevertheless, joint action by farmers through cooperatives or Marketing Boards to increase market power continues to prove popular.

3. *Alleged inefficiency in the food industry* Farmers often observe that the retail prices of the foods they produce are two or three times as much as they receive at the farmgate level. Intervening in the market to encourage adjacent links in the food chain to be closer is seen as a way of reducing costs, cutting out the 'middle man' and making the food chain more efficient. Both consumers and farmers can benefit from this vertical integration (unlike horizontal integration by farmers which is more likely to be in the interest of farmers alone).

4. *Adding value by farmers* The fourth argument relates to the incomes that farmers receive. Much of the value added to agricultural products once they leave the farm (grading, packing, etc.) can be done by groups of farmers even if the relevant stages of the industry are already efficient. By a combination of horizontal and vertical integration farmers can invest their labour and capital profitably in their own Food Industry.

It can be seen that the major beneficiaries of intervention for reasons 2, 3 and 4 are farmers and the losers are commercial organisations already trading in, and processing, foodstuffs. There are two other arguments for intervention, however, which need not be controversial and can satisfy both farmers and the Food Industry, though not necessarily consumers.

5. *Raising product prices by controlling and limiting supply* Controlling supply can result in higher prices and higher returns for both farmers and the food industry so long as demand is inelastic and alternative supplies (e.g., imports) cannot spoil the market. This is simply the process of creating a monopoly in a previously competitive market. This has usually proved beyond the organisational abilities of farmers acting collectively, unless implemented through government-backed schemes such as the Milk Marketing Boards which have been able to limit supplies to a particular market (liquid milk) in order to fix price and maximise revenue. We consider the Milk Boards in more detail later.

6. *Promoting and expanding the size of the market* As we have seen,

promotion at the national level can lead to competitive advertising by promotional bodies. Individual farmers seldom find advertising their products worthwhile because of their individually small output. If one farmer were to promote a particular product other farmers would be the free beneficiaries of his advertising. Collectively farmers may have an impact but many remain unenthusiastic about advertising, partly because of the fixed size of the food market, where spending more on one product means spending less on other products.

Farmers and governments intervene for all six purposes listed above and farmers stand to gain from policies which attempt to achieve them. If these policies are successful, there is scope for farmers to become more self-supporting and this is increasingly a goal of governments with budgetary constraints. Therefore, we must include:

7. *Making farmers as a whole more self-reliant* Actions which increase farm incomes from marketing reduce the need for direct subsidies by government. The essential element is to direct the gains from intervention towards farmers with particular economic handicaps whom government would otherwise feel obliged to subsidise directly. We have already mentioned measures taken in the EC to strengthen cooperatives and these can be viewed in part as an attempt to transfer some of the burden of agricultural support from direct government support to farmer-run marketing organisations.

### 3.3.2 Cooperatives, groups and horizontal integration

Cooperation between farmers in their marketing is often seen as a partial answer to the disadvantaged position of individual farmers. The essential features of agricultural cooperatives in Britain are that membership is voluntary, control is exercised through farmer shareholders and the objective is to provide services to members at cost, rather than to maximise profits. Thus the farmers are both shareholders and also the cooperative's trading partners.[21] Some cooperatives are run to buy requisites (fertiliser, sprays, etc.) and have become national trading concerns. West Cumberland Farmers Ltd had over 17 000 members in 1985 and a turnover in excess of £100 million. Others are smaller. For example, West Coast Growers Ltd of Ormskirk had 8 members in 1985 and a turnover of less than £1 million. Most cooperatives are termed **Primary cooperatives** since they organise production and marketing at the farm level. **Secondary cooperatives** are sometimes set up to coordinate and facilitate marketing or other functions on behalf of a number of primary cooperatives. An example is Home Grown Fruits Ltd, a secondary cooperative that is a

Federation of 8 primary cooperatives and 23 individual fruit growers. Taken as a whole, there were 318 000 members of more than 600 agricultural cooperatives in the UK in 1984–5 (clearly many farmers are members of more than one), with 16 000 employees and a combined turnover of £2794 million[22].

Reasons for cooperation are many, though most relate in some way to economies of size as regards physical operations, administration or particular marketing functions.[23] Cooperatives are an example of **horizontal** integration which bring together farmers at the same stage or level in the food chain. This is sometimes complemented by a measure of **vertical** integration, when a cooperative moves along the food chain, taking over functions performed by traders or processors and increasing the added value accruing to farmer-members. An example of this occurs when a contractual agreement by the cooperative requires a more rigorous standard of grading or processing than hitherto and members invest in new buildings and equipment. The **vertical** movement by cooperatives is considered later under contract farming.

European Community grant aid has encouraged farmers in Britain to combine to obtain grants. Cooperatives have an important impact on farmers marketing in areas of the EC with smaller farms and poorer rural facilities than is the case in Britain. In these continental circumstances, cooperatives often provide ancillary services (banking in particular) besides supplying requisites and handling marketing.[24]

There have been two periods of particular interest in the application of cooperation to agricultural markets in Britain.[25]

One occurred in the 1920s and stemmed from three influential findings of the **Linlithgow Report**.[26] The first was that 'the spread between producers and consumers' prices is unjustifiably high' (p. 11). The second was that 'the cooperative organisation of farmers on the right lines and in the right places should be encouraged' (p. 27). The third was that cooperatives would bring 'educative value and social advantages generally' (p. 33). However, an over-valued pound in the 1920s, which lasted until 1931, meant that imports were cheap. Also world markets were glutted so that cooperatives in Britain had little power to influence prices or contractual arrangements. Processors could find alternative and cheap sources of supply and this, coupled with lukewarm support from farmers and hostility from traders who felt threatened, brought the first major attempt to establish agricultural cooperatives in Britain to an end.

The second period of interest straddles 1967, the year in which the **Central Council for Agricultural and Horticultural Cooperatives** (CCAHC) was set up. This resulted from both the coming to power of the Labour government (1964–70) and the 'Move Towards Europe', which began in earnest in 1967.[27] The CCAHC was set up to encourage and foster such developments and administer new grants for cooperatives. Compulsory grading and hygiene standards were introduced in 1967 and 1973 to harmonise British standards to those in the EC and these served to emphasise quality control and a role for cooperatives. Barker estimates that by the late 1970s cooperatives were responsible for the following market shares: oilseeds and top fruit (38 per cent), eggs (29 per cent), glasshouse vegetables (19 per cent), cereals (15 per cent), potatoes (11 per cent) and livestock (8 per cent).[28]

In 1984, the CCAHC was restructured to the Cooperative Division of **Food From Britain** and commenced a broader, more market-orientated era of cooperative encouragement. Various commodity areas have been identified to form a basis for future cooperative effort. Nevertheless, despite the number of cooperative members in Britain being high relative to the number of farms, the proportions of output marketed by cooperatives quoted above are well below those found in many other EC countries.[29] The reasons are in many ways historic and include the political overtones of cooperatives for some people. There is also a preference among many farmers in Britain for other approaches such as marketing boards and producer groups. The marketing board has the advantage that standards can be legally enforced if necessary. Producer groups can restrict membership more readily than cooperatives to those farmers who are willing and able to maintain high standards and market discipline. This gives the group approach an advantage in the development of contract farming.[30]

### 3.3.3  Contract farming

Contractual arrangements between farmers and either processors or distributors can vary greatly in their conditions. At one extreme a farmer may become almost an employee of the firm, at the other he may simply agree to supply a certain tonnage and grade after harvest. Contractual arrangements in general offer a farmer security in exchange for a loss of freedom; they also usually require cooperative action on the part of a group of farmers.

It became evident in the 1960s that farming under contract to processors and distributors was increasing, particularly for products such as poultry, pigmeat and vegetables. A government-appointed committee of enquiry concluded favourably on this development in 1972, noting that contracting was not new in agriculture and that the situation in Britain was not very different from that in other developed countries. About 11 per cent of total agricultural production was farmed under contract and the proportion ranged from 43 per cent of fat pigs, 29 per cent of poultry and 22 per cent of eggs, to 2 per cent of fat cattle and sheep. In all, about 12 per cent of farmers in Britain were affected by contractual arrangements.[31]

The amount of contract farming was not seen as a threat to the 'family farm', although other commentators are concerned about the social implications.[32] The committee saw scope for extending the practice, so long as market prices could be forecast with some degree of certainty. A problem occurs when there is an increasing volume of lower grades which are below the contract standard but saleable on the open market. This depresses market prices and adversely affects the basis for negotiating future contract prices.

The 1970s also witnessed the growth of 'joint venture' or 'market linked' contracts. Under these, processors pay for crops on the basis of final prices received for the processed products or a joint company is set up with processors to freeze, store and pack produce. These arrangements prove risky when the market becomes over-supplied, such as happened in the case of frozen vegetables in the early 1980s when farmers who had over-expanded sold the excess onto the open market and further weakened prices.[33]

One of the least risky forms of cooperative action is for farmers to market at the first stage of the food chain through a statutory Marketing Board, and we consider these next.

### 3.3.4 The Marketing Boards

As one way of assisting British farming to survive the agricultural depression of the late 1920s onwards, legislation was introduced which permitted the establishment of Marketing Boards.

The Agriculture Act of 1931 permitted farmers to submit to the government a marketing scheme which, if approved, would be made compulsory on all producers. A two-thirds majority of producers was required and every scheme had to have a safeguard protecting the rights of consumers. Such Boards could be set up for primary products (e.g., pigs, milk) or secondary products (e.g., bacon, butter). The Agricultural Act 1933 (part I) allowed a Board to bridge the gap and combine two stages in the food chain (such as the Pigs and Bacon Board). Finally, the Agricultural Act 1933 (part II) gave the Boards the right to request import quotas or supply restriction measures for domestic producers if they felt the market was over-supplied. The powers of the Boards were suspended from 1939 to 1953, although they continued to perform advisory and administrative functions under the direction of the Ministry of Food. Boards were re-established as a result of the 1949 Agricultural Marketing Act, with stronger government control and strengthened protection for consumers.

As well as the five Milk Boards, Egg Board and the Pigs and Bacon Boards, there was a Hops Marketing Board and a Potato Marketing Board set up in 1932–3 (the Hops Board became a limited company in the 1980s) and two more began in 1950, the Wool Board and the Tomato and Cucumber Board. The last named is the only Board to have been voted out of existence by its members (in 1969) and tomato growers have continued to struggle to find a satisfactory way of influencing the market. Pigs and Bacon also proved a difficult area for Marketing Boards and the Pig Industry Development Authority (PIDA) was established as an alternative in 1957 and absorbed ten years later by the Meat and Livestock Commission. As we note above, pigs producers have been in the forefront of contract farming and this provides one solution to their marketing requirements.

It is apparent that many major commodities are missing from the list of Boards and this is because farmers have felt they were more likely to maximise their returns from private and cooperative trading arrangements for products such as wheat and feed grains, malting barley, oilseeds, beef, mutton and lamb, poultrymeat and field vegetables. Sugar has always been governed by contracts between farmers and the monopolised sugar factories. There was pressure from farmers for a Fatstock Board but this has never been set up. We now consider the main features of those Boards still operating for Milk, Potatoes and Wool.

## 1 The Milk Marketing Boards

There are five such Boards in the UK, one in England and Wales, one in Ulster and three in Scotland. The England and Wales Board accounts for 83 per cent of milk produced in the UK (88 per cent of milk destined for liquid consumption) and is organised in eleven regional areas. All milk producers (except those producing sheep and goatsmilk) must register with the appropriate Board and must sell all their milk to the Board. The only exceptions are the small and declining number of 'producer-retailers' who are licensed by their Board to retain milk for direct sale (though the milk is still in theory sold to the Board). There were 120 000 registered producers in 1933 and, after peaking at 196 000 in 1950, numbers have declined to 49 000 in 1985. The number of cows has remained steady but yields have risen and milk production increased from 4.6 million litres in 1933 to 15.2 million litres in 1984–5. These numbers give some indication of the size of the marketing operation involved. 88 per cent of dairy farmers sell all their milk to the MMB. Another 10 per cent of farmers sell some or all of their milk retail and 1 per cent use some or all of their milk for on-farm processing into cheese and other products.[34]

The main objective of the Milk Boards is to maximise the returns farmers obtain for their milk and they do this by dividing the market into two distinct parts. The first is the market for liquid milk and here the Boards have faced little competition from imports because of the difficulties of shipping fresh milk from abroad, and the obligation on local

**Table 3.2**
**Price Structure for Milk, 1984–5, England and Wales**

*pence per litre*

| | |
|---|---|
| 37.98 | maximum average retail price for liquid milk |
| 18.39 | EWMMB average selling price for milk to liquid market |
| 15.93 | "          "          "          "          " all milk |
| 13.77 | "          "          "          "          " milk to manufacturers |
| 13.26 | price charged by EWMMB to cheese-makers |
| 13.08 | "          "          "          "          " butter-makers |
| 14.60 | farmers' average return (average costs, 12.98) |

authorities in Britain to inspect and approve of pasteurising facilities. Only Ultra Heat Treated (UHT) milk is presently imported and consumer preference for other milk has been maintained, in part by promotional campaigns financed by the Boards. The situation will change since the UK must liberalise imports of fresh milk from the rest of the EC by 1989. This follows a ruling in the European Court. Continental dairies tend to sell liquid milk cheaper than the Boards do in the UK. They also produce and sell a wider range of the more expensive dairy products on the Continent. There will therefore be a period of adjustment in the EC market following liberalisation. Until 1985, the UK government regulated the price of fresh milk in association with the Boards, taking care to keep the price at a level that neither overly discouraged consumption nor encouraged UHT imports. Since 1985, milk prices have been agreed in the Joint Committees of the Boards and buyers (originally set up in 1954) and the buyers have played an increased role on these Committees.

The Milk Boards operate as discriminating monopolies, charging a higher price to dairies which sell liquid milk than to buyers who intend to produce butter, cheese, cream, skimmed milk powder and other manufactured products. The Board's price in these latter markets is determined by competition from imports, and in the EC these prices are around the intervention level (see Chapter 13). For instance, Table 3.2 illustrates the price structure adopted by the England and Wales Board in 1984–5 (the government still set a maximum retail price for milk in that year).

By discriminating between the liquid and manufacturing markets, the Milk Boards are able to maximise the returns to their members. Technically, the Boards equate marginal revenue and marginal costs in each market and exploit their position as monopoly suppliers of milk in their areas. The importance of this power relates to the importance of milk to British agriculture; over 20 per cent of British farmers are dairy farmers and milk represents a similar proportion of gross agricultural output (the value of milk and milk products to farmers has been about £2400 million in recent years). The importance of the Boards to farmers was demonstrated in 1978, when over 99 per cent of UK dairy farmers voted for the Boards to continue. The Treaty of Rome forbids producer-monopolies with compulsory membership under its rules of free competition, and Marketing Boards are only exempted so long as they can be shown to be in the interest of consumers (by maintaining a high level of consumption) and acting with the consent of the majority of producers. The Milk Boards also have to demonstrate that they are maintaining a relatively high level of liquid milk consumption in Britain.

In 1979 the Boards purchased 16 creameries owned by Unigate Ltd and became major manufacturers and distributors of milk under the name Dairy Crest, besides selling milk on behalf of farmers. By 1982, over a third of milk sold by the Boards was being handled by their own creameries and dairies.[35] The ownership of a substantial share of milk manufacturing capacity by the Board has proved a controversial matter. Prior to 1986 there was evidence that Dairy Crest was being run primarily in the interest of farmers rather than to maximise profits from manufacturing. Independent creameries argued that this was unfair and that Dairy Crest should be floated on the Stock Exchange. Following a government commissioned study which reported in 1986, the representation of Milk Board members on the Board of Dairy Crest was reduced and the Ministry of Agriculture took up a watching brief. However, the call to extend share ownership in Dairy Crest was resisted.

As, on the one hand, butter consumption in Britain has declined and the demand for skimmed milk has risen (see Chapter 2), and, on the other hand, milk production has increased, Boards have been faced with problems of finding markets for their milk. Self-sufficiency in butter for the UK rose from 9 per cent in 1933 to 73 per cent in 1984; for cheese over the same period, self-sufficiency rose from 27 per cent to 62 per cent. Intervention purchases in Britain (on behalf of the EC) of butter and skimmed milk powder increased as a result, until milk quotas were introduced in 1984. Dairy surpluses are a long-standing problem for the EC and one measure introduced in the 1970s was the co-responsibility levy, paid by milk producers (and consumers) and used to finance ways of increasing demand. The levy has been about 2 per cent of the milk price and has been ineffective in either taxing farmers or raising demand. Milk quotas are likely to have more impact, especially now they are transferable between farmers.

The Boards have also (since 1954) supported the Butter Information Council in its activities (which in recent years have aimed to slow down the decline in butter consumption rather than expand the market as originally intended). It operates on behalf of all butter suppliers to the UK market.

The milk marketing system adopted in Britain means that farmers who produce milk near cities, destined for liquid consumption, are penalised. This is a result of the practice of averaging returns from milk sales and paying all farmers the same, pooled price. The system subsidises farmers in remote areas who produce milk destined for manufacturing (although some fresh milk from Wales, Cumbria and Cornwall is still transported to conurbations). Also, winter milk producers are paid a higher price to reflect winter feed costs. The price to consumers is held constant so winter milk consumers are being subsidised by summer milk consumers (though of course they are largely the same people). However, the extent of the transfer of revenue from one group of dairy farmers to another has not been sufficient to cause any significant dissention within the Boards' membership. The elaborate system of prices operated by the Boards has been the subject of several official enquiries, and has been researched by academics as well, often on behalf of the Boards. The conclusion has generally been that the system has worked well, providing consumers with reasonably priced fresh milk, avoiding excess production and maintaining the incomes of farmers in remote areas.[36]

It is apparent from this section that the future holds a number of challenges for the Milk Boards Dairy Crest and dairy farmers. Four in particular can be identified. First, the COMA recommendations on fat consumption (discussed in Chapter 2) imply a reduction in the consumption of butter and (high fat) cheese. They may not be accepted fully by consumers, but the weight of nutritional evidence may make it difficult to slow down the decline in butter consumption or to raise the consumption of other dairy products sufficiently to compensate. Also, the growth of skimmed milk consumption raises the problem of disposing of more dairy fat. One development might be for the Boards to cease paying farmers a price premium for milk with a high butterfat content. COMA recommendations on fat consumption could be met partly by reducing the fat content of milk from 3.8 per cent to 2.8 per cent.[37] The premium has already been modified to take into account non-fat milk solids.

A second challenge concerns liquid milk. The EC instituted quotas on dairy production in 1984 but at a level of production still above consumption. Milk producers in countries such as Denmark, the Netherlands and France have been successful in demanding easier access to the UK liquid milk market at a time when the consumption of liquid milk in Britain appears to be declining.[38] The twin pressures of market share and market size may mean the Boards obtain a smaller price premium from the liquid market and can only offer lower average prices to farmers in the future. Another factor is the proportion of liquid milk sold through supermarkets. In 1978, Britain, with 92 per cent of liquid milk delivered to the doorstep, was in sharp contrast to Holland with 50 per cent, and to the rest of the EC where this practice has either disappeared or never existed.[39]

A third challenge is the growing diversity of dairy products in demand.

Consumers are seeking new and different products which require a more rapid marketing response than the Boards may be able to provide. It may prove more difficult to market effectively speciality cheeses and desserts than to dispose of liquid milk. On farm processing is, as yet, of negligible importance to the Boards but it is increasing.

A fourth challenge is the impact of milk quotas on dairy farmers themselves as the size of quotas is gradually reduced. If milk prices also rise, those farmers with a large initial quota will continue to prosper and the value of their land will be maintained. Others will not be so lucky and many farms with small quotas will face the choice of growing indebtedness or selling their land and/or their quota.

## 2 The Potato Marketing Board[40]

Potatoes are an important product, worth around £500 million in recent years and grown by 26 500 farmers in Britain. Production has varied from 5.8 to 6.8 million tonnes. All growers with one hectare or more must register with the Potato Marketing Board (PMB) and are allocated a basic area on which a quota percentage is imposed. Basic area can be traded between registered growers and the annual adjustment of the quota percentage provides a broad control over supply. By limiting fluctuations in the area grown, the PMB endeavours to stabilise the supply and the price, with the object of obtaining an average price level that is not high enough to attract imports. The power to import maincrop potatoes freely from countries in the EC has been available in Britain since 1979.

There remains the problem of shortages and gluts appearing on the market as a consequence of wide fluctuations in annual yields. The Board has adopted various methods which attempt to overcome this problem. Until 1974, the Board could change the minimum riddle size permissible for selling potatoes at any point in a season to remove enough potatoes from the market to maintain prices. Since 1974 the minimum riddle size must be defined in advance and is applied for the whole season. The PMB also has a system of advance contracts with farmers to buy potatoes, if the market price falls below a pre-defined floor price for a particular part of the season. Growers can then sell potatoes to the Board for disposal to uses other than human consumption contract. However, the PMB has only had funds to contract for around 5 per cent of production, which in some seasons may limit its effectiveness. Until 1984, the government also funded a deficiency payment scheme payable on total estimated tonnage sold for human consumption if the average price for the whole season fell below the Guarantee Price. The Board has also been active in encouraging the establishment of a Futures Market in London for potatoes. Futures trading offers growers and merchants some degree of individual price security in advance even though hedging opportunities may not be available at profitable levels in some seasons. Furthermore, the market as a whole may not be more stable as a result of future activity.

The PMB faces some of the same difficulties as the Milk Boards, despite the differences in the two product markets. Each is having to adapt to EC rules on competition and face an increasingly diverse set of markets for processed foods. The PMB does not as a rule purchase potatoes, except in times of a glut, but the expansion of contracts between growers and processors is likely to modify the role of the Board in the future. The Board also pioneers marketing innovations of various kinds through its Experimental Storage and Packing plant at Sutton Bridge. A recent example was the attempt by the PMB to launch its own quality mark for potatoes ('Great Brits'). The Board also continues to be active in the inspection of potato grades, provision of market information and the funding of research.

## 3   The British Wool Marketing Board

Anyone in the UK (except the Shetlands) with four or more adult sheep who wishes to sell wool must, by law, register with the Wool Board. There are about 90 000 registered producers, shearing 20 million sheep annually (another 15 million sheep are slaughtered before they can grow a full fleece). The main objective of the Board is to provide farmers with countervailing market power so that their returns are maximised and stabilised through the season and from year to year. The Council of the Board includes representatives of government and users of wool as well as farmers.

The Board operates through 30 local wool merchants. Each farmer's wool clip is allocated to a merchant who arranges transport, storage and grading (according to over 200 grades). The Board pays farmers a guaranteed price on the basis of their graded wool. The wool is then bulked up by grade and sold at fortnightly auctions in Bradford and Edinburgh to the wool trade. Over half of the 40 000 tonnes produced is exported. Once the wool is scoured and carded, manufacturers buy it for spinning. The Wool Board operates its own manufacturing subsidiary as well.

The Board stabilises prices to farmers by using a Special Account. When auction prices exceed the guaranteed price, the surplus accumulates in the Account. When auction prices are below the guaranteed price, the deficit is covered by drawing on the Account. If the Account becomes exhausted, the government makes a deficiency payment to the Board to cover the deficit. However, such payments are treated as advances and must be repaid in future seasons. In 1985 for instance, the wool guarantee price to farmers was 129 p. per kg and the average market prices was 103 p. per kg, leaving a 26 p. deficit to cover using the Special Account. In recent years, the total value of the wool clip has been £35–40 million.[41]

### 3.3.5 Commodity Commissions and Authorities

The government set up the *Meat and Livestock Commission* and the *Home Grown Cereals Authority* in 1967 to inform and administer rather than to regulate markets. Their influence on markets and marketing is therefore less easy to detect than that of the Boards. Also, whilst they are supported by levies on farmers and the trade, and contain government, trade and farming members on their governing bodies, producers cannot vote them out of existence as they can the Marketing Boards. They are therefore food chain-oriented rather than farming-oriented, and play an important role in policing and interpreting EC policies and regulations. The *Eggs Authority* was established in 1971 after an unsatisfactory attempt to regulate eggs through an Egg Marketing Board and the Authority functioned as a promotional body until being wound up in 1986. Individual egg producers have become sufficiently large-scale to prefer their own promotional efforts to those of an Authority.

The *Apple and Pear Development Council* was set up in 1968 to promote fruit marketing and does so on the basis of a levy raised from all commercial producers. As with the Boards, it can be voted out of existence and there have been occasional attempts by some growers to do this. The APDC established the Kingdom Scheme in 1980 to improve quality and raise the average price received by growers. This operates under the APDC umbrella but has its own headquarters and Board. The Scheme has gradually expanded and now dominates the supply of better quality English Cox apples, particularly those supplied to supermarkets.

There are many grower associations formed to encourage standardised grading, and to promote products such as mushrooms, tomatoes, flowers, etc. Often they relate to groups of growers and cooperatives with specific segments of a market (onions, potatoes, lamb, bacon, etc.). Few have much, if any, power to regulate marketing beyond a system of voluntary quality and packaging standards.

## 3.4 Stages in the Food Chain: (2) Food Manufacturing

Firms involved in processing and manufacturing food range from the small scale joint ventures with farmers for frozen foods to multinational and multi-million-pound corporations such as Unilever and Cadbury-Schweppes. Table 3.1 demonstrates that the manufacture of food is important in terms of employment and output. We consider how this sector is structured and the implications of activities in this part of chain for other sectors.

The first step is to set out a framework within which food manufacturing can be described and analysed. The theory of the firm and of competitive markets does not take us very far because of the oligopolistic structures that characterise manufactured food product markets. Oligopoly means 'the rule of the few' and a few firms dominate most markets for manufactured food products. Each firm bases its strategy and action upon how other firms will react to its action and an oligopolistic market therefore displays an anxious and uncertain stability, as each firm watches the others. From time to time there are sudden price wars, with price cuts by one firm matched by others and a proliferation of promotions and advertising campaigns until the market settles down uneasily once more. Many consumer goods, besides manufactured foods, are now produced under conditions of oligopoly in Britain because takeovers and mergers have concentrated production and reduced the number of firms. As we see later, oligopoly has also come to characterise food retailing over the last 20 years as well as food manufacturing.

An alternative approach to using the theory of competitive markets is to use the theory of workable competition and this is the framework that we adopt for both the manufacturing and the distribution sectors.[42]

### 3.4.1 The structure of food manufacturing

Food, drink and tobacco manufacturing in Britain employs over half a million people, more than 80 per cent of them in food businesses. Table 3.3 illustrates the types of business which predominate in Britain. The level of employment has declined in all sectors of the industry, a point we return to later under the heading of Performance. The size of business varies both within and between sectors. The slaughterhouse sector, for instance, is characterised by a lot of small firms. Over 80 per cent of these employ fewer than 100 people and account for over 50 per cent of the total labour force. In biscuit and crispbread manufacturing, two-thirds of the total labour force is employed by one-fifth of the firms, each with over 1000 persons per firm.

Concentration is more marked in terms of total sales, with two-thirds of sales in 1979 accounted for by the largest 30 firms. The average size of business is larger than for manufacturing as a whole and both size and concentration have been growing, though the extent varies according to the product. Concentration is high for bread, breakfast cereals and flour; in the late 1970s the largest 3 firms in each sector accounted for 76 per cent, 85 per cent and 79 per cent respectively of the UK market and, typically, the dominant firm alone held 25–40 per cent of the market. However, the proportion of the market held by the leading firm has in some sectors been eroded, making oligopolistic competition the norm.[43] Similar developments have taken place in the rest of the EC. The UK accounts for a quarter of the employment in the Food, Drink and Tobacco industries in the European Community (EUR-9) and in size is second only to West Germany.[44] We consider first why there has been such a high and growing

**Table 3.3**
**Food Manufacturing, 1984**

|  | employment | (%) | businesses | (%) |
|---|---|---|---|---|
| 1. Bread, flour, etc. | 119 950 | 25.8 | 642 | 22.0 |
| 2. Meat processing | 97 712 | 21.0 | 756 | 25.9 |
| 3. Sugar, soft drinks, etc. | 81 480 | 17.5 | 200 | 6.8 |
| 4. Dairy | 38 930 | 8.4 | 332 | 11.4 |
| 5. Fruit, vegetable and fish processing | 41 545 | 8.9 | 252 | 8.6 |
| 6. Other | 85 094 | 18.3 | 740 | 25.3 |
| 7. **Total food** | 464 711 | 100 | 2922 | 100 |
| 8. Drink and tobacco | 91 969 |  | 360 |  |
| 9. **Total FDT** | 556 680 |  | 3282 |  |

SOURCE   Compiled by the Business Statistics Office, Business Monitor PA1003 (London: HMSO) 1985, table 2.

concentration in food manufacturing and second why it has been so marked in the UK.

The high level of concentration cannot be accounted for simply by increasing investment and economies of size and scale. There was rapid technical change in the food industry during the 1960s and 1970s which made larger plant more efficient and encouraged increases in size, but this alone cannot account for a concentration rate of more than about 10 per cent. It seems to be the static nature of the market for food which forces firms to compete for market share and to merge into larger companies.[45] Oligopoly reduces the scope for price competition and instead there is competition through advertising and new products. Launching new products is expensive and risky; success rates for new products were as low as 2–10 per cent in the 1970s and the life cycle for a food product (i.e, from launch to demise) is becoming shorter, with only 40 per cent of new products lasting for more than 5 years.[46]

The second structural question posed above is why the UK in particular should have a large food processing industry. A number of reasons may be offered for this. First, there were historical accidents – the free trade era of cheap food and growing population and incomes, the availability of the City of London for financial services, and the relatively stringent hygiene regulations in Britian which raised a barrier to entry for new firms wishing to enter the market. Second, there has been a continuing strong demand for processed foods in Britain and a priority for consumers has been to be able to buy low priced foods. Third, there has been a 'soft' policy by governments towards oligopoly in the food and drinks manufacturing industries, accepting very high concentration ratios in different food markets.[47]

A final point to note concerning structure is the tendency for food companies to diversify into other businesses. For instance, the Census of Production reveals that only 17 per cent of the gross output of grain milling companies came from grain milling in 1977–9.[48] Mergers have created multinational conglomerates with interests in food, drink, tobacco, hotels, and sports and leisure. In 1982 it was estimated that only half of the sales of the ten largest food manufacturers in the UK consisted of food.[49] A number of advantages arise from conglomeration including spreading risks, cross-subsidising activities and avoiding tax. Conglomerates also avoid the stigma of monopolies which horizontal integration in the same sector is likely to attract.[50]

### 3.4.2   Conduct in the food manufacturing industry

As concentration has increased in many consumer goods markets in Britain, so the government has attempted to ensure that competition continues to survive. The Monopolies Commission was set up in 1948, and became the Monopoly and Mergers Commission (MMC) in 1965. This has since reported on the state of competition and conduct in food markets such as tea, cornflakes, pet food, frozen foods and bread, and we consider their report on discounts and retailers in more detail later on. The MMC became more actively involved in advising the government on monopolistic conduct after the Office of Fair Trading was set up in 1973. Practices that are not permitted include charging a lower price to retailers who stock only that firm's brand, and using advertising as a means of preventing other firms from entering the market. The power of food manufacturers to prevent retail price competition was eliminated following the Resale Prices Act of 1964, and there has been a steady erosion of the price-fixing and market-fixing powers of food manufacturers ever since.

The conduct of manufacturers regarding pricing policies has also been affected by systems of price control used in Britain. These were highly restrictive during and after the war until the mid-1950s and thereafter the government took steps to free the markets from both official controls and monopoly power. However, after 1965 and until 1979 price controls of one form or another were introduced, particularly between 1973 and 1979 under the Price Commission. During the period 1973–7, a Price Code covered a number of consumer goods, including several foods, and price increases for these had to be justified in terms of changes in costs and related to specified profit margins. These controls were opposed by food manufacturers who argued that profits were too low anyway.[51]

The major reason behind price controls surveillance during the 1970s was government policy towards inflation, which was 6 per cent in 1970, reached a peak of 25 per cent in 1975 and declined to 8 per cent in 1979. After 1979, government policy shifted towards using monetary policy to control inflation, leaving prices free to find their own levels and relying upon a high exchange rate (and cheap imports) to keep domestic market prices low. In 1985 the inflation rate was 5 per cent.

### 3.4.3 Performance in food manufacturing

The rate of profit in food manufacturing has been no worse than the rate for manufacturing in general and was sometimes better in the 1960–77 period. Profits recovered after the 1980–2 recession, though they were still only 3 per cent of sales in 1983 and 12–15 per cent of (historic) capital employed.[52] It is difficult to obtain one unambiguous measure of profit but it does seem that the combination of government surveillance of prices and the growing power of retailers has kept profits low in food manufacturing. Another indicator has been the slow rate at which food manufacturers in Britain have invested and replaced labour with capital. Whilst employment in food manufacturing declined by 18 per cent, 1972–82, employment in total manufacturing fell by 26 per cent.[53]

The degree of import penetration of the home market gives some indication of how successfully British food manufacturers have performed against foreign suppliers. Between 1975 and 1984, import penetration remained at 17 per cent for the food, drink and tobacco industries, whilst it increased from 18 per cent to 26 per cent for manufacturing industries as a whole.[54]

Innovation is another measure of performance and whilst on the one hand there have been many new products launched, they have often had a short life-cycle, as we have already noted. There has not been a particularly high level of expenditure on research and development, and advertising has been aggressively focused on trying to increase market share, rather than on providing consumers with more information.[55]

In summary, food manufacturing in Britain is a highly concentrated industry in which a few firms dominate. The government has attempted to take steps to ensure competition is relatively unhindered and to encourage price competitiveness in particular, although, as we explain later, the power of retailers has had probably more impact on prices than has government action. In terms of performance, the food manufacturing industry shares many of the problems of British manufacturing in general, and the continuing trend towards mergers and conglomeration does not seem to relate to either high rates of profit or high rates of return on capital.

This conclusion must be tempered by the knowledge that, under oligopoly, average rates of profit can be very misleading. A study by the Harvard Business School, for instance, found that the rate of net profit per sales for four brand leaders in food manufacturing varied widely according to market share. The profit rate was 17.9 per cent for the leading firm, 2.8 per cent for the second, 0.9 per cent for the third and *minus* 5.8 per cent for the fourth firm.[56]

## 3.5   Stages of the Food Chain: (3) Food Distribution

In this section we deal with both wholesaling and retailing. Wholesalers operate at several points in the food chain, buying from farmers and processors, and selling to manufacturers, other wholesalers and shops.

### 3.5.1   The structure of food distribution

The number of wholesaling firms has continued to decline with the growth of direct trading between retailers and farmers and retailers and manufacturers. The decline has been in general food wholesaling companies, but wholesalers remain important, as Figure 3.1 illustrates, accounting in 1979 for 58 per cent of purchases by caterers and for 54 per cent of purchases by retailers. Several wholesalers are also retailers and there are combines of retailers and wholesalers in networks such as Distributive Marketing Services Ltd (DMS).[57] In 1979 there were about 300 wholesale depots and another 50 with cash and carry sales as well. The business is dominated by 5 or 6 firms, with another dozen or so being important. Some wholesalers specialise in particular products (meat, fruit and vegetables, dry goods, bakery products, etc.), whilst others deal in many products but in a particular form (such as frozen foods). Fruit and vegetable wholesalers remain an important group, because wholesale markets still account for 40–50 per cent of this trade.

Turning to retailing, the proportions of food retailing accounted for by large food retailers has grown over the past 25 years (see Table 3.4). The largest five multiple food retailers (Tesco, Asda, Sainsbury, Argyll and Dee Corp.) accounted for 32 per cent of grocery sales in 1977; by 1982 this figure had increased to 50 per cent overall and to 55 per cent in the London television area.[58]

**Table 3.4**
**Food Retailers, 1982**

| | Business numbers | Number outlets | Persons engaged ('000s) | Gross Margin as % Turnover | Capital expenditure[1] (£ million) |
|---|---|---|---|---|---|
| Large grocery outlets | 113 | 10 892 | 379 | 17.7 | 473 |
| Other grocery outlets | 38 277 | 43 342 | 164 | 17.2 | 66 |
| Dairymen[2] | 6 985 | 8 350 | 53 | 29.9 | 26 |
| Butchers | 14 954 | 20 997 | 84 | 23.1 | 40 |
| Fishmongers, poulterers | 2 543 | 3 278 | 11 | 26.4 | 7 |
| Greengrocers, fruiterers | 13 262 | 16 360 | 59 | 25.0 | 24 |
| Bread and flour confectioners | 6 490 | 11 556 | 68 | 43.5 | 25 |
| **Total** | 82 625 | 114 774 | 818 | 20.0 | 661 |

NOTES  1. Exclusive of VAT.
       2. Including depots with roundsmen.
SOURCE  Compiled by the Business Statistics Office, Retail Business Enquiry, Business Monitor, 5DO25, Retailing 1982, HMSO, 1984.

In 1982 the Retail Business Enquiry found nearly a quarter of a million firms, of which 83 000 were classified as food retailers (accounting for 86–98 per cent of food sales, depending on the product). Table 3.4 shows that large grocery outlets accounted for 0.5 per cent of the number of businesses, 16 per cent of outlets and 67 per cent of sales. Table 3.4 also shows that large grocery outlets also accounted for two-thirds of investment and nearly half of employment. (However, total employment is greater than that shown in Table 3.1. This reflects the fact that grocery outlets retail other products besides food and drink.)

Although there has been a marked decline in the number of specialist food shops, there is no 'iron law' that such a trend must continue. Once the inefficient have been weeded out, specialist shops can benefit from trends towards health products and fresh foods. and retain their customers. Thereafter the supermarket chains will have to compete with each other for market share. An estimate made in 1980–1 suggested that the number of specialist food businesses would decline to 48 000 by 1985. Table 3.4 shows that the number had already fallen to 44 000 by 1982.[59] The number of specialist food outlets averaged about 1 per 1000 people.

### 3.5.2 Conduct in food distribution

An increasing proportion of sales by multiples is as own-brand goods, (about a quarter of all sales in 1981), which gives the retailers the power to pick and choose among suppliers. The Monopolies and Mergers Commission in 1981 reported on an investigation into the margins on food products added by different retailers (we discuss margins in more detail later).[60] They found that margins added by larger retailers were less than those by small retailers. For instance, on baked beans that mark-up on buying cost varied from 8 per cent for the four largest multiple stores, to 10 per cent for other multiples, and to 19 per cent for independent retailers and cooperatives. Larger retailers also paid lower buying prices. The four largest chains paid 95 per cent of the average buying price whilst the cooperative stores paid 120 per cent. The big stores were found to have a range of bargaining advantages, such as obtaining discounts based on their total business with a particular manufacturer or on the size of each delivery to regional depots. The multiple retailers also obtained longer trade credit, and had greater access to promotions and offers paid by the supplier.

The large retailer is also better placed to survive a price war. In 1977, when Tesco launched 'Operation Checkout', abandoning trading stamps and cutting prices, the other chains followed suit and stores that were unable to do so rapidly lost business. The independent retailers' share of the grocery trade fell from 43 per cent in 1971 to 31 per cent in 1979.[61]

An example of the power of retailers can be found in the flour and bread sector, in which manufacturing is highly concentrated. The practice in the 1960s was to sell bread to shops at a published retail list price, less 12–15 per cent discount, with scope for larger discounts to special customers. The declining market for bread, plus the ability of retailers to play the three bread suppliers off against each other, led to discounts increasing. By 1974 discounts of up to 35 per cent were being obtained by multiple retailers. The three bakery groups tried to agree not to offer such large discounts but they failed. The situation was partly resolved in 1974–5 with the bread subsidy scheme which limited discounts to 22.5 per cent. The bakery groups said in 1981 that any resumption of large discounts would bring about the closure of small plants and force them to charge even higher prices to small retailers to compensate for the larger discounts to the multiples.

The MMC concluded that the practice of giving larger discounts to the multiples was not against the public interest, but rather a sign of flourishing competition. It consequently did not feel any important action needed to be taken beyond monitoring the continued expansion of the multiples. They noted that wholesalers were also becoming large concerns (often with links to retailers), who could look after their own interests in a competitive market. Small retailers were less able to fend for themselves, and price discriminatory practices against them by food manufacturers left them even less able to compete with the multiples.

### 3.5.3 Performance in food distribution

The performance of the three big retail chains (Sainsbury, Asda and Tesco) appears satisfactory to judge from reported pre-tax profits, share prices and the readiness of investors to finance capital for expansion. Several smaller multiple chains and wholesalers have also performed well, though links with other businesses make the performance of their food operations alone difficult to judge. The increased power of retailers over all other sectors in the Food Chain raises potential problems of oligopolistic power, but this has been used to the benefit of the consumer.[62]

Small individual food shops are handicapped by low turnover (which leads to high margins) and expensive supplies (which means they must charge high prices). Small wholesalers often obtain insufficient profits to justify investing in the expensive cool-chain facilities (coders, stores and lorries) required by multiple retail buyers, and some wholesalers have moved vertically into pre-packing and preparing food products for supermarkets.

It seems unlikely that the revolution in food distribution will continue *ad infinitum*. The trend continues to be towards out-of-town satellite shopping centers. However, the growth of huge shopping malls such as have been built in France and the USA is limited in Britain by space and planning constraints. Perhaps a new trend in the 1990s might be the reoccupation of town centres by food shops. This will depend in part on consumers' demand for food, and in part on the willingness of local authorities to permit free market forces to operate on the use of property in towns.

We have looked at each stage in the food chain separately, and conclude this chapter by looking at the food chain in its entirety. We will do this by employing the concept of the marketing margin and consider how it is calculated and what it shows. In the following chapter we turn to international aspects of the food chain and use a second approach to consider the whole food chain, input-output analysis.

## 3.6　Gross Marketing Margins

The spread of prices between the farmgate and retail levels is often called the **gross marketing margin**, which may be defined as the difference between the retail price and the farmgate price expressed as a percentage of the retail price. Thus, if carcase meat sells from the farm at £1.50 per kilo and is sold retail for an average price of £2.20 per kilo, the gross marketing margin is 31.9 per cent. The 'mark-up' expresses the difference as a percentage of the farmgate price (thus in this example the mark-up is 46.7 per cent). Yet another way of stating the difference is as 'the share of the farmer in the retail value of food' (in this case 68.2 per cent). However it is stated, the margin is of interest because it indicates how much consumers are paying for marketing services and can therefore be compared with the number of functions performed during the food chain. The question becomes more interesting if the margin changes over time because we can then assess how far an increase in the margin relates to more services being provided, and how far it relates to less efficiency (and thus higher costs) in providing an unchanged number of functions.

The gross marketing margin can be viewed as firstly the overall margin for food, secondly as the margin for a particular type of commodity (or an individual product) and thirdly as the margin which exists in one part of the food chain. This last may involve, for instance, comparing large multiple retail chains and small individual shops. We briefly consider these in turn.[63]

We estimate gross marketing margins in Table 3.5, using the following procedure.

> *Gross marketing margin* = household expenditure on food (a)
> +0.33 of expenditure on catering (b)
> *less* Farm Output + Imports – Exports
> expressed as a percentage of (a + b)

This assumes that waste is negligible and that stocks are unchanged from year to year. 'Catering' includes food, accommodation and services and that is why we only allow a third of expenditure as final expenditure on food. An alternative for estimating expenditure on food is to use retail sales, adjusted for taxes and subsidies. Table 3.5 suggests that the gross marketing margin has changed little in the 1980s, but that the period of price controls reduced the margin during the 1974–7 period.

The United States Department of Agriculture publishes calculations of the farmers' share of a dollar's expenditure on food, divided according to whether expenditure is on household food or on eating out. These suggest the gross marketing margin in 1984 varied from 67 per cent of a dollar spent on household food to 85 per cent of a dollar spent on eating out. Overall, the marketing margin in the USA has averaged about 70 per cent in recent years,[64] higher than the 47 per cent estimated for Britain.

**Table 3.5**
**Estimated Gross Marketing Margin UK, 1974–84**

| | with $\frac{1}{3}$ catering | without | | with $\frac{1}{3}$ catering | without |
|------|------|------|------|------|------|
| 1974 | 34.4 | 29.1 | 1980 | 51.2 | 45.7 |
| 1975 | 38.7 | 32.8 | 1981 | 50.6 | 45.1 |
| 1976 | 37.4 | 31.3 | 1982 | 47.0 | 41.0 |
| 1977 | 38.8 | 32.6 | 1983 | 48.6 | 42.3 |
| 1978 | 45.4 | 39.8 | 1984 | 46.9 | 40.0 |
| 1979 | 46.3 | 40.5 | | | |

SOURCE   Calculated from CSO, *Annual Abstract of Statistics* (1986 edn, London: HMSO) Tables 9.1, 12.4, 14.10.

The gross marketing margin may be expected to increase for reasons such as the following:[65]

1. a greater volume of goods handled
2. a relative increase in the consumption of existing products which embody more services (and are therefore more 'useful')
3. a change of consumption to new products with greater levels of service attached
4. increased costs and inefficiencies of marketing firms
5. changes in export and import patterns
6. inflation which raises costs in the food industries more than the final prices of foodstuffs.

By looking at the demand and supply curves for marketing services we can assess how far such factors as these may have caused the margin to increase or not. During the 1960s, shifts to the right in the demand curve for marketing services, mainly because of reasons 2 and 3, were matched by shifts to the right in the supply of marketing services, mainly because technology improved and costs were lowered. In this way, we can account for the apparent stability of the marketing margin in Britain during the 1960s[66] and perhaps extend the argument to apply to the post-1977 period as well.

The gross marketing margin for particular commodities can be calculated by using the average price spread, rather than by using as above a ratio of expenditures, since unit values can be calculated for an individual foodstuff. Individual gross marketing margins vary from 15 per cent for products such as fresh potatoes, where few marketing services are performed, to 40 per cent for perishable horticultural products and 60 per cent for carcase meat and fresh milk. They would be found to be much greater for manufactured foods if it were possible to identify an individual food in the final product. However, the measure becomes meaningless for highly processed foods, where the agricultural raw material is a negligible part of the final product. The gross marketing margin on crisps is in the order of 1000 per cent if potatoes alone are considered, but the consumer of crisps is paying for a product far removed from the farm output from which it originated.

The third way in which the marketing margin can be used is to compare different ways of selling food (e.g., supermarkets which buy direct from producers *versus* shops using wholesalers and perhaps wholesale markets). Table 3.4 above suggests that large and small retailers apply similar average margins to their grocery business, though this similarity masks the very different nature of the services provided. As we have seen, large retailers obtain food at lower prices and can therefore afford to sell for lower prices. They have a higher level of capital expenditure than small retailers. It is interesting to analyse the margins at the processing, wholesaling and retailing levels of particular commodities. There are many examples of this being done both in the UK and abroad and we complete this section on marketing margins by considering three examples.[67] The first is bacon in Britain, the second poultry in Britain and the third fruit and vegetables in Europe.

The margins on bacon are used to illustrate the problem posed when businesses in the food chain deal with several commodities. In 1976, when the Price Commission considered bacon, on average the processors relied on bacon for a quarter of their sales, wholesalers for 39 per cent of their sales, commission agents in the wholesale trade for virtually all their business, and retailers for less than 5 per cent of their sales (see Table 3.6).

Table 3.6 illustrates the much lower margins that characterise wholesaling than is the case for processing or retailing where more value is added to the product. Also agents obtain a smaller margin than wholesalers because wholesalers take ownership and carry out whatever butchery operations on a side of bacon the retailer requires, whereas an agent (often importing) delivers sides in containers to processors and multiple retailers earning a much smaller percentage commission. It can be seen that Table 3.6 provides a useful summary of how the price of bacon in the shops is derived through value added during the food chain. However, it is more problematic to use it as diagnostic data. As the report on bacon notes, 'Because suppliers, wholesalers, and retailers do not deal exclusively in bacon, and because bacon is not a homogeneous product, we have not been able to determine exactly what margins are being earned in every case. From the

**Table 3.6**
**Sales and Margins on Bacon, 1976**

| Type of business (number in sample) | Average sales of bacon (£ million) | Bacon % total sales | % on bacon commission/ margin | % on non-bacon | Overall net profit % sales | % average overall margin |
|---|---|---|---|---|---|---|
| Curers (7) | 28.0 | 24.2 | 19.5 | 33.3 | 3.4 | 30 |
| Agents (10) | 198.6 | 98.8 | 1.6 | – | 0.6 | – |
| Wholesalers[a] (26) | 110.1 | 38.8 | 7.8 | 8.1 | 1.1 | – |
| Retailers | | | | | | |
| Multiples (5) | 52.7 | 3.5 | 17.1 | – | – | – |
| Independent (7) | 45.8 | 4.5 | 20.9 | – | – | – |
| Retailers | | | | | | |
| Butchers[b] (6) | 3.92 | 5.4 | 16.8 | – | – | – |
| Independent (9) | 0.06 | 4.8 | 21.4 | – | – | – |
| Bacon specialists (2) | 0.11 | 89.9 | 17.0 | – | – | – |
| Multiples (10) | 91.21 | 2.6 | 17.7 | – | – | – |
| Total (27) | 95.31 | 2.7 | 17.7 | – | – | – |

NOTES  a.  Dominated by one major wholesaler.

b.  Three of the four independent (column 3) butchers recorded margins of 25 per cent.

SOURCE  Price Commission (1978) *Prices, Costs and Margins in the Importation and Distribution of Bacon* (London: HMSO) Appendix 2.

**Table 3.7**
**Margins on Poultry, 1974**

| | Producers' surplus (deficit) as % sales | % Margins | | Average selling price (p/lb) | | |
|---|---|---|---|---|---|---|
| | | Wholesalers | Retailers | Producer | Wholesale | Retail |
| Frozen chickens | (8.39) ⎫ | 8.8 | 17.8 | 19.41 | 20.0 | 25.3 |
| Fresh chickens | - ⎬ (6.16) | 12.8 | 25.5 | – | 17.0 | 29.9 |
| Frozen turkeys | (1.74) ⎫ | 8.7 | 17.1 | 24.13 | 25.9 | 30.0 |
| Fresh turkeys | - ⎬ (8.84) | 7.5 | 23.2 | – | 23.4 | 44.0 |

SOURCE Price Commission (1975) *Prices and Margins in Poultry Distribution*, Report No. 11 (London: HMSO) Ch. 5.

**Table 3.8**
**Gross Distribution Margin, Excluding Taxes: Fruit and Vegetables,**
**Wholesale and Retail, 1978**

| | Average (for 1978) | 3 month range 1978 (%) | large supermarket (%) | traditional (%) | % share of market by traditional large | other |
|---|---|---|---|---|---|---|
| France | 49 | 46–50 | 45–48 | 49–51 | 30 | 56 |
| W. Germany | 50 | 43–46 | 36–41 | 46–49 | 25 | 55 |
| Holland | 45 | 42–51 | 36–46 | 41–51 | 18 | 44 |
| England | 47 | 48–50 | 43–49 | 48–50 | 47 | 20 |
| Belgium | 54 | 50–57 | 46–53 | 50–60 | 20 | 56 |

SOURCE   Derived from Table 4 in Montigaud J.-C., (1982) 'International comparison of fruit and vegetable distribution systems in Europe', 21st International Horticultural Congress, Hamburg, Vol. II *Proceedings*, vol. II, pp. 999, 1000.

information we have received however, we are satisfied that profits made in the distribution of bacon are not excessive.' (Price Commission (1978) p. 19)[68] Note also, that profit (or the net margin) is naturally less than the gross margin. In another survey, looking at shops in outlying areas, the Price Commission found that the gross retail margin on groceries was 14–15 per cent, whilst the net margin or profit (before deducting a charge for the proprietors' salary) was 5–7 per cent.[69]

The second example is poultry, again from the Price Commission. This permits a comparison of the margin with processors' profits and the effect of a year in which prices and costs rose but sales fell. Retailers and wholesalers maintained their margins and processors made losses (see Table 3.7). This example illustrates how the margin on fresh chickens was much more than the margin on frozen chickens, though it does not reveal whether this was the result of more added utility (a risky, more perishable product) or of less efficiency than in the marketing of frozen chickens.

The third example looks at fruit and vegetables in various countries and considers gross marketing margin for different types of store during three months in 1978. Table 3.8 is derived from Montigaud's Table 4 by expressing the three results he cites as a range. As can be seen, margins on fruit and vegetables appeared to be generally lower in the large supermarkets, but also they varied greatly over the three months (June, September and December) and from country to country. This final example illustrates the way looking at margins can lead to ideas for further research. For instance, the impact of different farmgate and processor supplied prices, and of different levels of demand for fruit and vegetables will affect the margins shown in Table 3.8. The table also raises the question of the size of the margin, considerably higher than the margins on bacon and poultry in Britain. Calculating margins can thus be used to compare different parts of the food chain.

# Notes

1. Mordue, D. in Burns, J. *et al.* (eds) (1983) *The Food Industry* (London: Commonwealth Agricultural Bweaux and Heinemann).
2. EC (1982) *The Agricultural Situation in the EC* (Brussels).
3. Food and Drink Federation (1985) *Bulletin* 1 (January) 10.
4. Bateman, D. I. (1976) 'Agricultural marketing: a review *J. Ag. Econ.* **27**:2 (May) 171–224. This review includes over 300 references.
5. Kohls, R. L. and Uhl, J. N. (1985) *Marketing of Agricultural Products* (6th edn, New York: Macmillan).
6. See, for instance, Ritson, C. (1985) 'Agricultural marketing: the scope of the subject', Part I of Jollans, J. L. (1985) (ed.) *The Teaching of Agricultural Marketing in the U.K.* CAS Report No. 8 (University of Reading).
7. British Institute of Marketing, quoted in Giles, G. (1972) *Marketing* (2nd edn, London: ELBS) p. 1.
8. Ritson (1985) op. cit., p. 18.
9. Barwell, quoted in Ritson (1985) op. cit.
10. Scitovsky, T. (1977) *The Joyless Economy, An Inquiry into Human Satisfaction and Consumer Dissatisfaction* (Oxford: Oxford University Press).
11. For a critical discussion of grading see Bowbrick, P. (1977) 'The case against compulsory grading', *J. Ag. Econ.* **28**:2 (May) 113–18.
12. The Agricultural Produce (Grading and Marketing) Act, 1928.
13. Report of the Committee appointed to review the working of the Agricultural Marketing Acts (London: HMSO, 1947) (*Lucas Report*).
14. For a critical view of FFB, see Pickard, D. (1982) 'The role of government in agricultural marketing', *J. Ag. Econ.* **33**:3 (September) 361–8.
15. Milk Marketing Boards (1985) *Dairy Facts and Figures*.
16. Whetham, E. H. (1972) *Agricultural Marketing in Africa* (Oxford University Press) Chs 1 and 2.
17. MacFadyen, J. Tevere (1984) *Gaining Ground: The Renewal of America's Small Farms* (New York: Holt, Rinehart, Winston).
18. PYO is discussed further in Barker, J. W. (1981) *Agricultural Marketing* (Oxford University Press).
19. *British Country Foods Directory*, January 1985. Joint publication of the NFU, FFB, MAFF and FDF.
20. Metcalf, D. (1969) *The Economics of Agriculture* (Harmondsworth: Penguin).
21. Haines, M. (1982) *An Introduction to Farming Systems* (Harlow: Longman) p. 200.

22. Plunkett Foundation (1985) *Directory of Agricultural, Horticultural and Fisheries Cooperatives in the UK* (Oxford). See also from the same source, *Statistics for Agricultural Co-operatives in the UK, 1984/5*.
23. Gasson, R. (1977) 'Farmers' approach to cooperative marketing', *J. Ag. Econ.* **29**:2 (May) 109–17.
24. Sargent, M. (1982) *Agricultural Cooperation* (Atdershot: Gower). The EC encouraged cooperative action under Regulation 17/64 (and subsequently under the revised Regulation 355/77, extended for ten years in 1984 as regulation 1932/84). Under this regulation, many cooperatives have been able to improve their facilities for storage, meat handling and vegetable processing. New cooperatives (called producer associations in the EC) can obtain start-up grants worth up to 5 per cent of turnover in the initial years of operating, conditional upon them attracting the majority of producers in their area to join. See Green Europe (1985), *The New Agricultural Structures Policy*, No. 211 (Brussels). See also Revell B. (1985) *EC Structures Policy and UK Agriculture* (As Report No. 2). (University of Reading). Also, Tracy, M. (1982) *People and Policies in Rural Development* (Langholm: Arkleton Trust).
25. Bonner, A. (1961, 1970) *British Cooperation* (Hannover Street, Manchester: Cooperative Union Ltd.).
26. Departmental committee on distribution and prices of agricultural produce. Final Report (London: HMSO, 1924) (*Linlithgow Report*). The committee's terms of reference were 'To enquire into the methods and costs of selling and distributing agricultural, horticultural and dairy produce in GB and to consider whether, and if so by what means, the disparity between the price received by the producer and that paid by the consumer can be diminished.'
27. There was a quickening of interest in cooperation after 1954. Two committees of enquiry into the efficiency of agricultural marketing were positive and supportive of cooperation (the Runciman and Verdon-Smith Committees) and the government increased support for cooperatives over the 1960–4 period. In 1962, the government set up the Agriculture Marketing and Development Committee (AMDEC) which gave support and advice to producer groups, then being set up by farmers to negotiate contracts with processors. The NFU was also active in this period and in 1962 established Agricultural Trading Ltd. (ACT) designed to handle bulk purchases for cooperatives. Several horticultural cooperatives were flourishing in the 1950s and again in 1962 a notable development in the fruit sector was the setting up of Home Grown Fruits Ltd., a second-tier cooperative providing marketing services for primary cooperatives and large individual growers, each with their own packing and grading facilities but marketing under one name and to one set of standards. In 1972, AMDEC was absorbed by the CCAHC and there was confusion between CCAHC and a new NFU-backed initiative, ACMS, during the 1972–9 period. See Barker (1981) op. cit.; Morley, J. (1975) *British Agricultural Cooperatives* (London: Hutchinson Benham); and Wormell, P. (1978) *Anatomy of Agriculture* (London: Harrap Kluwer).
28. Barker (1982) op. cit.
29. Sargent (1982) op. cit.
30. Dodds, P. R. (1965) 'Group Marketing', *J. Agr. Econ.* 16:3, 366–89.
31. Report of the Committee of Enquiry into Contract Farming (London: HMSO, 1972) (*Barker Report*). See also Allen, G. R. (1972) 'An appraisal of contract farming', *J. Ag. Econ.* 23:2, 89–98.
32 Newby, H. (1979) *Green and Pleasant Land?* (Harmondsworth: Penguin).
33 Malcolm, J. in Burns, J. A., McInerney, J. P. and Swinbank, A. (eds) (1983)

*The Food Industry: Economics and Policies* (London: Heinemann).
34. These and other milk statistics are taken from MMB, *Dairy Facts and Figures*, various years, but especially 1983 and 1985.
35. Wilkinson, G. A. in Swinbank, A. and Burns, J. (eds) (1984) *The EEC and the Food Industries* Food Economics Study No. 1 (University of Reading).
36. Rayner, A. J. (1977) 'The regional pricing policy of the MMB and the public interest', *J. Ag. Econ.* **28**:1, 11–26.
37. COMA (1984) *Diet and Coronary Heart Disease*, DHSS (London: HMSO) See Chapter 2 for a discussion of this report.
38. The EC milk market is documented annually in MMB, *European Dairy Facts and Figures*.
39. Wormell (1978) op. cit., p. 492.
40. Several aspects of the PMB and the potato market are covered in Ritson, C. and Warren, R. (eds) (1984) Agriculture's Marketing Environment (London: PMB). See especially Taylor, J., 'Background to Intervention in the British Potato Market'; Anderson, J., 'The Role of the Futures Markets'. Also see Marsh, J. (1985) 'Economics, Politics and Potatoes – the changing role of the PMB in Great Britain' *J. Ag. Econ.* **36**:3, 325–5.
41. Wormell (1978) op. cit., p. 501. MAFF (1986) *Annual Review of Agriculture*. See also, *The British Wool Marketing Board, its organization and operation* (Oak Mills, Bradford; BD14 65D).
42. This framework considers the **structure** of resource use in an industry (the way resources are used and allocated) **conduct**, and the effects of the structure and conduct on the **performance** of the industry. Performance is judged by setting up criteria. Then, if performance is found to be unsatisfactory, the reason is sought in the structure and conduct of the industry. By identifying how these should be altered, performance can then be improved. One advantage of this approach is that there is no initial condemnation of monopolistic structures without regard to the overall effects as there is in the neoclassical theory of the firm. The advantages of large scale for research and development and lower unit costs can be set against the potential disadvantages of monopolistic market power, price-fixing and profiteering. See Sosnick S. H. (1968) 'A critique of concepts of workable competition' *Quart. J. Econ.* **72**:3, 380–423.
43. Howe in Burns, J. *et al.* (1983) op. cit., pp. 106, 122.
44. Eurostat (1985) *Basic Statistics of the Community* (23rd edn, Luxembourg), Table 3.17. See also Swinbank A. and Burns J., (eds) (1984) *The EEC and the Food Industries*, Food Study No. 1, (University of Reading). This reference also contains a summary of the complex legal and policy framework within which the European Food Industry operates.
45. Howe (1983) op. cit., p. 108.
46. Cannon T. (1985) 'Marketing problems of the Food Chain, an overview', *Food Marketing* **1**:1, 3–19.
47. Burns in *Burns*, J. *et al.* (1983) op. cit., pp. 10–12.
48. Mordue in Burns, J. *et al.* (1983) op. cit., p. 32.
50. Greig, W. S. (1971) *The Economics of Food Processing* (Westport, Conn.: Avi Publishing) pp. 48–57.
51. Stocker T. (1983) in Burns, J. *et al.* (1983) op. cit.
52. Stocker T. (1985), Food and Drink Industries Council Bulletin, no. 1, p. 10.
53. CSO (1984) *Annual Abstract of Statistics*.
54. CSO (1986) *Annual Abstract of Statistics*, Table 12.3.
55. Howe (1983) op. cit.
56. *Financial Times*, 11 October 84.

57. Tanburn J. (1981), *Food Distribution in the 1980s* (London: CCAHC) p. 35.
58. Food and Drink Industries Council (1984) 'Retailer concentration', *FDIC Bulletin* No. 26 (March) pp. 16–20.
59. Tanburn (1981) op. cit.
60. Monopolies and Mergers Commission (1981) *Discounts to Retailers* (London: HMSO). See also Burns J. A. (1983) 'UK food chain: inter-relationships between manufacturers and distributors', *J. Ag. Econ.* **34**:3, 361–78.
61. MMC (1981) op. cit.
62. Sturgess I. (1984), 'Food retailing and agricultural adjustment', *J. Ag. Econ.* **35**:3, 365–78.
63. See also Hallett G. (1982) *The Economics of Agricultural Policy* (2nd edn, Oxford: Blackwell), Ch. 9.
64. USDA (1984) *National Food Review* 3:29, 38.
65. Baron P. J. (1977) 'Marketing analysis of the food processing and distributive margin in the UK', *J. Ag. Econ.* **28**:3, 221–32. See also O'Connell J. (1979) 'A critique of attempted analysis of aggregate marketing margins', *J. Ag. Econ.* **30**:2, 125–30.
66. Baron (1977) op. cit.
67. For other examples consult Bateman (1976) op. cit. and Maunder in Burns, *et al.* (1983) op. cit. For a consideration of the Price Commission from which these examples are drawn, see Mitchell's chapter in Burns, J. *et al.* (1983), and Maunder P. (1984) 'The food and drink industries and the second Price Commission (1977–9), *J. Ag. Econ.* **35**:3, 331–40.
68. Price Commission (1978). *Prices, Costs and Margins in the Importation and Distribution of Bacon* (London: HMSO).
69. Price Commission (1975) *Food Prices in Outlying Areas*, Report No. 10 (London: HMSO) Table 2.

# The Role of International Trade in the Food Chain 4

## 4.1 Introduction

Agricultural imports join the food chain at different stages, depending on the form in which they enter the country. Britain has traditionally relied upon imports of agricultural products and other raw materials and later in this chapter we use an input-output table to investigate at what stages of the food chain agricultural imports arrive. Agricultural exports have been increasing and the input-output table also helps to identify at what stages these leave the food chain. Before we examine the position in the UK, we consider some general aspects of international trade in agricultural products.

It is a characteristic of agricultural commodities that only a small proportion of world production is traded internationally, a consequence of the transport costs associated with trade involving bulky and relatively low-value commodities. Less than 20 per cent of world cereal production is traded internationally and less than 5 per cent of world beef production. A few commodities are produced mainly for export (notably tropical beverages, fruits and sugar cane, and tobacco) and individual countries specialising in agricultural exports will export a higher proportion of their production, but taking the world as a whole, international trade is a fairly marginal activity relative to world production.

There are important implications of this. First, fluctuations in the volume of trade are more pronounced than fluctuations in production. If 10 per cent of world production enters world trade and production declines by 5 per cent, exports are likely to fall by up to 50 per cent as exporting nations redirect supplies from the foreign to the domestic market. Second, these fluctuations in export volume will cause large movements in world prices because of the inelastic nature of demand for agricultutal products (see Chapter 2).

For these reasons, the revenue that is obtained from agricultural exports is often uncertain and exporting countries try to negotiate international agreements which can lessen the uncertainty and increase market stability. The unreliability of world markets has also encouraged importing countries to seek a measure of food security through greater self-sufficiency, despite the higher costs that this entails. We discuss the self-sufficiency issue later.

A number of attempts have been made to control the volume of world exports and to stabilise prices through international agreement for commodities such as sugar, tea, coffee, cocoa and wheat. Whilst each has succeeded for a time, changes in the structure of world supply and demand have eventually caused stresses which have led to the collapse of the agreement. We do not consider international agreements in detail.[1]

There has been a rapid increase in total world trade since 1945 (the period up to the 1970s has been called 'The Long Boom'[2] and agricultural trade grew in volume, though the proportion of agriculture in total trade has gradually declined, to a quarter in 1970 and 15 per cent in 1984. Barriers to trade in industrial goods have been reduced through negotiations within the framework of the General Agreement on Tariffs and Trade (GATT, established in 1947) and this has encouraged the gradual integration of countries into the world economy. However, agreement to reduce barriers to trade on agricultural products has proved much more difficult because of the high levels of protection afforded to agriculture in Europe, North America and Japan. This has led to a number of trade disputes between developed countries concerning access to markets for agricultural exporters, and the impact of subsidised trade on world markets. We return to these issues later in this chapter, but first we discuss Britain's agricultural trade and how it has changed in volume, direction and structure.

## 4.2 Britain's International Trade in Agricultural Products

Britain's traditional pattern of trade changed in the 1970s, and not only have exports of services and oil grown in importance relative to manufactures, but Britain has also increasingly imported manufactured goods. In addition, there has been a trend towards importing fewer, and exporting more, agricultural goods. The proportion of agricultural products in imports has declined to 14 per cent whilst the proportion of agricultural products in exports has grown to 8 per cent. Table 4.1 illustrates that in recent years, whilst the unit values of imports and exports have risen to a similar extent, the volume of imports has remained largely unchanged

**Table 4.1**
**Trade Indices for the UK 1971–3 to 1984**

|  | 1971–3 | 1983 | 1984 |
|---|---|---|---|
| *Imports of food, feed and alcoholic beverages* | | | |
| £ million | 2417 | 8237 | 9401 |
| Volume index 1900 = 100 | 105.2 | 107.1 | 109.3 |
| Price index 1980 = 100 | 36.8 | 119.9 | 134.4 |
| *Exports of food, feed and alcoholic beverages* | | | |
| £ million | 658 | 3938 | 4457 |
| Volume index 1980 = 100 | 59.0 | 109.9 | 119.2 |
| Price index 1980 = 100 | 38.0 | 122.3 | 128.3 |
| Self-supply[1] . . . all food consumed in UK (%) | 5.1 | 62.2 | 62.1 |
| Self-supply[1] . . . indigenous food consumed milk (%) | 78.2 | 81.8 | |

NOTE   1. Value of home-produced food as percentage of . . .
SOURCE   MAFF (1983, 1986) *Annual Review of Agriculture* (London: HMSO).

whereas the volume of exports has doubled. Self-sufficiency (domestic production as a proportion of consumption) has risen to over 60 per cent in the 1980s, and over 80 per cent for indigenous types of agricultural products.

Table 4.2 sets out changes in production, imports and exports for the main commodities and illustrates the trend to increased self-sufficiency in ·British supply. And increase in self-sufficiency can result from an increase in home production and/or a decline in consumption. Two major influences on Britain's self-sufficiency recently have been, first, entry to the EC in 1973 and, second, the growth of agricultural productivity which has been encouraged indirectly by protection under the Common Agricultural Policy (CAP). We discuss these in detail in Chapters 11 and 13 and here concentrate on their implications for trade and self-sufficiency in the UK.

*Cereals*   There has been a rapid increase in the production of cereals in the UK, leading to a decline in imports and a growth in exports. There are still gaps in indigenous production, and the country continues to import bread-making wheat (notably from Canada) and maize from the USA and EC. Exports of barley have increased a great deal: following the bumper harvest of 1984, 35 per cent of production was exported, and in 1985 28 per cent was exported. About half of cereal exports are to non-EC countries and such exports are only profitable with the benefit of a subsidy (called an export restitution and paid to exporters by the European Agricultural Guidance and Guarantee Fund, EAGGF, in Brussels).

*Potatoes* There is limited scope for trade in potatoes, except in the case of early and out-of-season potatoes when the retail price in Britain is sufficiently high to justify the cost of transport.

*Sugar* There is a managed market for sugar in the UK under European Community regulations, which restrict both home production of sugar beet by quotas, and imports of cane sugar by agreement under the Lomé Convention. Self-sufficiency has increased as beet yields have risen and acreage expanded, and the combination of a static or declining market with fixed imports has led to exports from Britain and other EC countries.

*Oilseeds* Britain imports most of the vegetable oil it requires. Oils such as soya, ground-nut, cottonseed and sunflower are imported, with production at home dominated by rape, which is grown with the aid of a production subsidy. Production of rapeseed oil now exceeds consumption, and 20 per cent of production has been exported in the 1980s.

*Apples and Tomatoes* Self-sufficiency has declined over the period shown in Table 4.2 as a result of the opening up of the UK market to French and Dutch competition after accession to the EC. Imports are generally in the same or overlapping seasons as British production and competition from imports is particularly intense in seasons when the pound exchange rate is high, such as between 1978 and 1981. During that period, acreage in Britain was reduced and subsequently, when the competitive position improved, growers were unable or unwilling to restore production to its former level. Another feature of the trade in apples has been the partial replacement of imports from the Southern Hemisphere during the British winter by stored apples from both domestic growers and growers in France and Italy.

*Beef and Veal* There are often quality differences between beef imports and exports which partly explain why imports and exports of beef balance out. Imports from South America have been replaced to some extent by imports from the EC.

*Mutton and Lamb* There is a quota arrangement with New Zealand which covers frozen lamb imports. Domestic production of Fresh lamb has increased, particularly since the Common Sheepmeat Regime of the EC was introduced in 1980, and domestic consumption has declined, resulting in an increase in self-sufficiency. In recent years there has been a growing export trade from the British Isles to the Continent. A border tax (called the 'clawback') prevents British farmers from retaining the whole amount of the higher prices obtained on the Continent whilst also benefiting from a price subsidy (the variable premium) and a ewe subsidy (in hill areas).

**Table 4.2**
**Trade and Self-sufficiency in the UK for Selected Products**

| '000 tonnes | average 1971-3 | | | | 1984 | | | |
|---|---|---|---|---|---|---|---|---|
| | Production | Imports | Exports | Self-sufficiency[1] (%) | Production | Imports | Exports | Self-sufficiency[1] (%) |
| Cereals | 15 258 | 8 297 (1 586) | 184 (151) | 65 | 26 590 | 2 706 (1 128) | 6 076 (3 002) | 114 |
| Potatoes (human consumption, incl. Channel Isles in UK) | 5 724 | 388 (–) | 83 | 95 | 6 273 | 742 (429) | 100 (–) | 90 |
| Sugar (refined basis) | 978 | 2 769 (53) | 297 (10) | 36 | 1 314 | 1 269 (126) | 258 (14) | 57 |
| Apples (excl. cider) | 401 | 280 (114) | 13 | 60 | 325 | 401 (254) | 20 | 46 |
| Tomatoes (incl. Channel Isles in UK) | 173 | 147 (45) | 1 | 35 | 152 | 236 (97) | 7 | 33 |

| | | | | | | | |
|---|---|---|---|---|---|---|---|
| Beef and veal (inclusive) | 930 | 327 (107) | 77 (71) | 79 | 1 131 | 198 (152) | 203 (159) | 100 |
| Mutton and lamb | 230 | 317 (5) | 25 (22) | 44 | 296 | 146 (0) | 57 (54) | 77 |
| Bacon and ham | 272 | 348 (300) | 2 – | 44 | 212 | 268 (264) | 7 – | 45 |
| All meat | 2 734 | 1 034 (442) | 117 – | 75 | 3 191 | 700 (495) | 336 – | 90 |
| Butter | 87 | 361 (156) | 8 (3) | 20 | 206 | 161 (71) | 25 (16) | 60 |
| Cheese | 176 | 151 (71) | 5 (2) | 55 | 245 | 145 (131) | 42 (12) | 68 |
| Wool (clip) | 34 | 174 (16) | 27 (15) | 25 | 39 | 131 (20) | 45 (28) | 39 |

NOTES   1. Calculated after stock changes (omitted in the table). Figures in brackets show trade with the EC (9), where data are available.

SOURCE   MAFF (1983, 1986) *Annual Review of Agriculture* (London: HMSO).

Therefore, the trade position for mutton and lamb is more or less the artificial creation of international agreements and EC support arrangements.

*Bacon and Ham*   There is a long-standing pattern of imports of pigmeat products from Denmark, Ireland and the Netherlands, based on the low relative costs of efficient family farms, organised into cooperatives to regulate quality and price to suit the British market. As a result, self-sufficiency has remained low (though it is higher for fresh pork, where transport and storage difficulties favour domestic producers).

*All meat*   Self-sufficiency is now very high for meat taken as a whole, with imports amounting to less than a quarter of domestic production. Imports have declined, particularly from non-EC countries.

*Butter*   Self-sufficiency has risen sharply and, without the impact of milk quotas in 1984, domestic production would have undoubtedly replaced all imports. Britain now imports less butter from the rest of the EC than it did before 1973.

*Cheese*   Imports have increased, especially from the EC, and consumption has risen as well.

*Wool*   Production has increased a little and the market has contracted, leading to an increase in self-sufficiency. However, the market is still largely supplied from abroad.

Summing up, self-sufficiency in Britain has been rising and the EC has grown in importance as both the source of our diminishing agricultural imports and the destination for our expanding agricultural exports.

So far we have considered trade by commodity, and we now use an input-output table for the UK (Table 4.3) to indicate the ways in which imports and exports of agricultural products are linked into the food chain. In Table 4.3, each of the food industries is shown in a matrix with purchases from, and sales to, other food industries, and the table includes import purchases (row 102), and export sales, (column 106). Using this matrix, it is possible to estimate, for example, the impact on trade of an increase in agricultural production in Britain. Two relationships are particularly relevant:

1. the strength of the linkages along the food chain, forwards to the consumer and exports, and backwards to farming and imports. These are vertical linkages, If they are strong, changes in agricultural production will have important effects on industry and trade.
2. the strength of the **Production Multiplier**, which is the effect on the production in the whole economy caused by a change in one of its parts. If agricultural production rises, so will the earnings of factors of production in this industry (wages, salaries, rents, interest receipts and profits). Some of this increase will be spent and will increase the demand for other goods, stimulating other industries to increase their production.

The trade effects will be to increase imports and redirect exports on to the domestic market.

The production multiplier sums up the strength of both the direct linkages and the indirect effects. (When the initial stimulus is an increase in export demand, it is called the Foreign Trade Multiplier since it shows the effect on production of an increase in exports.) Estimates of the value of the multiplier in industrial countries range from 1.5 to 1.9 for agriculture, and from 2.0 to 2.8 for food manufacturing. The greater multiplier effect for manufacturing reflects the relatively stronger linkages of the food industry with the rest of the economy.[3]

The strength of the multiplier allows an economic planner to consider the overall effects on employment and the balance of payments of an expansion in one industry. Of course, input-output analysis cannot give all the answers; in particular it does not show how far increased demand leads to higher prices and what effect that has on domestic and foreign trade.

Table 4.3 is divided into four quarters or quadrants, each giving a different set of information. In the top left or **second quadrant** the inter-industry matrix is shown. (Only selected industries are reproduced, and, whereas the term input-output should strictly be applied just to this quadrant, we use the term informally to refer to the whole of Table 4.3.) Each row shows how much an industry *sold* to other industries, (e.g., agriculture and horticulture sold £1910 million to itself, £36 million to the oils and fat sector, etc). The top right (called the **first quadrant**) shows final demand for each industry. Thus to take agriculture and horticulture again, £1027 million of output was sold to consumers, £609 million was exported, and total final output was £1727 million. If the sales to other industries are added, the sum is total agricultural output, (£8178 million).

The columns show the purchases and, as we note above, agriculture purchased £1910 million from itself, £19 million from the machinery industry, and so on down the industry rows: intermediate purchases total £4355 million for agriculture. The bottom left (**third quadrant**) shows imports, (£463 million for agriculture in 1979), and expenditure by the industry on labour and taxes, together with gross profits. In this way, the agricultural industry column accounts for all the £8178 million of income found by adding up sales along the agriculture row. Also in the third quadrant, we can see how much each industry purchased from abroad. Finally the bottom right (**fourth quadrant**) shows components of National Income, Expenditure and Product. In 1979, Gross Domestic Product was £247 984 − £54 616 = £193 368 million, and the contribution to GDP of both the food industries and agriculture can be derived in the same way, by deducting imports from final use. GDP is the sum of value-added in the domestic economy and the contribution to GDP and value-added of the milk and milk products industry in 1979 was £2499 − £212 = £2287 million.

## Table 4.3
## Industry × Industry Flow Matrix, UK, 1979 (£ million)

| Code | 1 | 21 | 34 | 57 | 58 | 59 | 60 | 61 | 62 | 63 | 64 | 65 | 66 | 67 |
|---|---|---|---|---|---|---|---|---|---|---|---|---|---|---|
| 1 | (1 910) | 11 | – | 36 | 2 537 | 1 872 | 83 | 462 | 9 | 152 | 5 | 573 | 296 | 240 |
| 21 | 444 | (98) | – | 1 | – | – | – | – | – | – | 1 | 1 | 1 | 1 |
| 34 | 19 | – | (7) | – | – | – | – | – | – | – | – | – | – | – |
| 57 | 5 | – | – | (162) | 15 | 22 | 16 | 1 | 67 | – | – | – | – | – |
| 58 | 4 | 1 | – | 45 | (421) | – | 15 | – | 18 | – | – | – | – | – |
| 59 | 2 | – | – | 4 | 1 | (55) | – | – | 6 | – | – | – | – | – |
| 60 | 3 | – | – | – | 3 | 8 | (111) | – | 21 | – | – | – | – | – |
| 61 | 12 | – | – | 2 | 30 | 2 | – | (26) | 359 | 4 | 38 | 173 | 47 | – |
| 62 | – | – | – | – | 1 | – | – | – | (6) | – | 1 | 1 | 1 | – |
| 63 | – | – | – | 1 | 2 | 8 | 40 | 2 | 56 | (13) | 84 | 24 | 47 | 25 |
| 64 | – | – | – | – | – | – | 1 | – | 32 | – | (146) | 1 | 5 | – |
| 65 | 1869 | – | – | 6 | 6 | 4 | 1 | 2 | 20 | – | 4 | (41) | 7 | 1 |
| 66 | 1 | – | – | – | 76 | 6 | 15 | 4 | 44 | – | 4 | 41 | (77) | 18 |
| 67 | 3 | 1 | – | 1 | 1 | – | 2 | – | 1 | 1 | 1 | 15 | 5 | (361) |
| 68 | – | – | – | – | – | – | 2 | – | – | – | – | – | – | 4 |
| Total (10–87) | 2 701 | 102 | 460 | 135 | 284 | 253 | 326 | 52 | 854 | 37 | 458 | 657 | 527 | 429 |
| 89 | 554 | 40 | 77 | 41 | 219 | 35 | 45 | 14 | 75 | 36 | 66 | 66 | 61 | 240 |
| 90 | 17 | 1 | 1 | 1 | 7 | 4 | 2 | 1 | 3 | 1 | 3 | 2 | 4 | 4 |
| 92 | – | 29 | 8 | 18 | 18 | 23 | 42 | 11 | 19 | 30 | 44 | 39 | 39 | 56 |
| 93 | 1 | 11 | – | 9 | 5 | 4 | 9 | 7 | 2 | 4 | 6 | 14 | 6 | 3 |
| 101 | 4 355 | 325 | 642 | 278 | 3 267 | 2 297 | 776 | 623 | 1 122 | 320 | 771 | 1 509 | 1 160 | 1 771 |
| 102 | 463 | 153 | 126 | 337 | 197 | 212 | 212 | 272 | 168 | 230 | 305 | 463 | 345 | 169 |
| 103 | 14 | 1 | 1 | 1 | 7 | 4 | 2 | 1 | 3 | 1 | 2 | 3 | 3 | 5 |
| 104 | –302 | 13 | 14 | 5 | 36 | 30 | 11 | 75 | 46 | 6 | 16 | 64 | 57 | 74 |
| 105 | 1 257 | 138 | 226 | 54 | 424 | 150 | 244 | 77 | 638 | 80 | 328 | 172 | 301 | 515 |
| 106 | 2 391 | 109 | 88 | 34 | 351 | 110 | 62 | 65 | 151 | 59 | 138 | 139 | 228 | 544 |
| 107 | 8 178 | 739 | 1 097 | 709 | 4 282 | 2 803 | 1 307 | 1 113 | 2 128 | 696 | 1 560 | 2 350 | 2 094 | 3 075 |

SOURCE   CSO (1983) Business Monitor, PA1004 (London: HMSO) input-output tables for the UK

KEY:

*Rows and columns*

| | | | |
|---|---|---|---|
| 1 | Agriculture and horticulture | 64 | Confectionery |
| 21 | Fertilisers | 65 | Animal feeding stuffs |
| 34 | Agricultural machinery and tractors | 66 | Miscellaneous foods |
| 57 | Oils and fats | 67 | Alcoholic drink |
| 58 | Slaughtering and meat processing | 68 | Soft drinks |
| 59 | Milk and milk products | 89 | Distribution, etc., and repairs |
| 60 | Fruit, vegetables and fish processing | 90 | Hotels, catering, and public houses, etc. |
| 61 | Grain milling and starch | 92 | Road and other inland transport |
| 62 | Bread, biscuits and flour confectionery | 93 | Sea transport |
| 63 | Sugar | 101 | Total intermediate |

| 68 | 89 | 90 | 92 | 93 | 101 | Consumers expend | Genl. govt. final consum | Gross fixed capitl formn. | Physcl increase in stocks | Export goods and services | Total final output | Total output | Code |
|---|---|---|---|---|---|---|---|---|---|---|---|---|---|
| 5 | 4 | 122 | – | 2 | 6 452 | 1 027 | 87 | 26 | –22 | 609 | 1 727 | 8 178 | 1 |
| 1 | 1 | – | – | – | 578 | 30 | 11 | 6 | –8 | 123 | 161 | 739 | 21 |
| – | 103 | – | – | – | 139 | 21 | 2 | 228 | 25 | 681 | 957 | 1 097 | 34 |
| – | 1 | 28 | – | – | 410 | 211 | 12 | 1 | 14 | 60 | 299 | 709 | 57 |
| – | 29 | 340 | – | 6 | 689 | 3 091 | 199 | 2 | 48 | 253 | 3 593 | 4 282 | 58 |
| 4 | 15 | 241 | – | 2 | 304 | 2 177 | 106 | 1 | –13 | 227 | 2 499 | 2 803 | 59 |
| 6 | 5 | 106 | – | 1 | 188 | 837 | 59 | 1 | 22 | 200 | 1 119 | 1 307 | 60 |
| 2 | 270 | 13 | – | – | 990 | 94 | 6 | 1 | –1 | 23 | 123 | 1 113 | 61 |
| – | 11 | 208 | – | 3 | 227 | 1 688 | 74 | 1 | 38 | 101 | 1 901 | 2 128 | 62 |
| 69 | 8 | 28 | – | – | 413 | 235 | 11 | 2 | 6 | 29 | 283 | 696 | 63 |
| – | 7 | 57 | – | 1 | 108 | 1 132 | 57 | 4 | 26 | 234 | 1 453 | 1 560 | 64 |
| – | 13 | 5 | – | – | 1 950 | 328 | 8 | – | 13 | 50 | 400 | 2 350 | 65 |
| 1 | 17 | 279 | – | 2 | 524 | 1 127 | 69 | 2 | 163 | 208 | 1 570 | 2 094 | 66 |
| 2 | 10 | 11 | 2 | 1 | 120 | 2 037 | – | 3 | – | 917 | 2 958 | 3 078 | 67 |
| 18) | 5 | 214 | – | 1 | 230 | 428 | 30 | – | 14 | 63 | 536 | 766 | 68 |
| 10 | 3 851 | 1 902 | 632 | 187 | 51 985 | 25 948 | 6 635 | 11 365 | 2 088 | 33 249 | 79 286 | 131 271 | Total |
| 40 | (894) | 594 | 265 | 23 | 10 140 | 19 561 | 729 | 1 524 | – | 2 264 | 24 078 | 34 218 | 89 |
| 2 | 70 | (5) | 11 | 7 | 466 | 6 178 | 558 | – | – | 998 | 7 724 | 8 190 | 90 |
| 1 | 1 569 | 157 | (24) | 16 | 3 918 | 1 596 | 117 | 222 | – | 250 | 2 185 | 6 103 | 92 |
| 2 | 30 | 32 | 3 | (9) | 657 | 266 | 35 | 72 | – | 3 216 | 3 589 | 4 246 | 93 |
| 34 | 10 931 | 4 190 | 1 809 | 899 | 114 707 | 86 494 | 36 846 | 30 031 | 2 491 | 52 611 | 208 473 | 323 180 | 101 |
| 53 | 641 | 466 | 61 | 2 269 | 33 072 | 12 950 | 1 555 | 5 507 | 359 | 1 173 | 21 544 | 54 616 | 102 |
| 1 | 44 | 9 | 7 | 5 | 1 126 | 2 748 | –2415 | –2 355 | – | 896 | –1 126 | – | 103 |
| 13 | 1 731 | 260 | 331 | 45 | 6 338 | 14 879 | 2 338 | 1 286 | 145 | 445 | 19 093 | 25 431 | 104 |
| 13 | 14 273 | 2 115 | 2 545 | 646 | 115 131 | – | – | – | – | – | – | 115 131 | 105 |
| 12 | 6 598 | 1 150 | 1 350 | 382 | 52 806 | – | – | – | – | – | – | 52 806 | 106 |
| 6 | 34 218 | 8 190 | 6 103 | 4 246 | 323 180 | 117 071 | 28 324 | 34 469 | 2 995 | 55 125 | 247 984 | 571 164 | 107 |

*Rows only*

| | |
|---|---|
| 102 | Imports of goods and services |
| 103 | Sales by final demand |
| 104 | Taxes on expenditure less subsidies |
| 105 | Income from employment |
| 106 | Gross profits and other trading income |
| 107 | Total input |

To use input-output analysis in economic planning, it is necessary to recalculate the inter-sectoral flows as percentages, showing input require-ments per £1000 of total output. Such tables are also published by the government. This analysis does have its limitations, because of the need to assume constant marginal coefficients (in other words, if £10 of extra sugar output requires £3 of extra agricultural output, this relationship has to be regarded as constant whatever the level of expansion contemplated). For small changes, input-output analysis is a useful way of highlighting the effects of changes in sectoral output on the rest of the economy.

The input-output Table can also be used to calculate the difference between purchases and sales, (though not gross marketing margins because the products are changed during manufacture). For instance, the soft drinks industry paid £69 million to the sugar industry, £33 million to other food industries (including itself), purchased £63 million of imports and made final sales worth £536 million. Assuming all the imports were agricultural products, we may conclude that the soft drinks industry converted £165 million of sugar and raw materials into £536 million of final products, a three-fold increase in value.

Table 4.3 does not reveal very much about the distributive trades in the Food Chain but this deficiency is made good in Table 4.4 where two sets of inter-industry transactions are shown, one relating to 1979 and the other to 1963, recalculated at 1979 prices. Whilst the definitions have altered, Table 4.4 does appear to confirm two trends which we noted in this and the previous chapter. First, the decline in the importance, both absolutely and relatively, of wholesaling and, second, the increasing importance of domestic agriculture rather than imports in the supply of food in Britain. (Data for 1984 for Table 4.4 can be found on page 43 Figure 3.1.)

## 4.3   Agricultural Trade and the European Community

It is impossible to get a clear view of Britain's trade in agricultural products without an account of the position of the EC as a whole, and the role of the CAP as a major determinant of EC export policy. The European Com-munity is the world's largest importer of agricultural products and the largest agricultural exporter after the USA. In this section we consider the pattern of trade and how it has been changing over recent years. Our focus is on Community trade with the rest of the world rather than on trade among EC member states, although intra-EC trade has grown rapidly, and countries such as France and the Netherlands are only net earners of foreign exchange from agriculture as a result of intra-EC exports.

Britain has relatively less agricultural trade with other member states than is average for the EC, largely because of a continuing commitment to Commonwealth suppliers, yet it still imported 44 per cent of food and

**Table 4.4**
**The Food Chain, £ Billion, 1979 prices**

| | Purchases by Manufacturers | Wholesalers | Caterers | Retailers | Consumers | Total Domestic Sales | Exporters |
|---|---|---|---|---|---|---|---|
| UK agriculture | 3.1 (1.4) | 2.4 (4.7) | – | 0.2 (–) | – | 5.7 (6.2) | 1.8 (0.1) |
| Importers | 2.9 (2.4) | 1.9 (4.2) | – | 0.3 (–) | – | 5.2 (6.6) | – |
| Manufacturers | – | 3.9 (8.0) | 1.0 (3.3) | 7.2 (17.0) | – | 12.1 (28.9) | – (0.6) |
| Wholesalers | – | – | 1.4 (–) | 8.5 (–) | – | 9.9 | – |
| Caterers | – | – | – | – | 3.2 (3.3) | 3.2 (3.3) | – |
| Retailers | – | – | – | – | 20.4 (22.2) | 20.4 (22.2) | – |
| **Total purchases** | 6.0 (3.8) | 8.2 (16.9) | 2.4 (3.3) | 16.2 (17.0) | 23.6 (25.5) | 56.5 (67.2) | 1.8 (0.7) |

NOTE   Figures in brackets are for 1963, inflated by the retail price index for food to 1979 prices.

SOURCES   Derived from Wollen, G. H. and Turner, G. (1970) 'The cost of food marketing', *J. Ag. Econ.* **21**:1, 63–77, and Tanburn, J. (1981) *Food Distribution in the 80s* (London: CCAHC).

agricultural products from EC countries and exported 49 per cent of such goods to the rest of the EC in 1983. The position of the UK with regard to Community trade with the rest of the world in 1983 was that Britain accounted for 17 per cent of EC imports from the rest of the world (food and agricultural products) and 10 per cent of EC exports; farming in Britain provided 10 per cent of agriculture's net value-added in the EC in 1983 as well.[4]

Table 4.5 sets out the main commodity imports and exports of the EC in 1983. Products which are listed as 'regulated' are those covered by the Common Agricultural and Common Fisheries Policies. The Community imports mainly raw materials and feedingstuffs, which, together with fruit and vegetables and tropical products, account for 70 per cent of agricultural imports. As Table 4.5 shows, half of the imports are products which are regulated under EC common policies (agriculture and fisheries). The cereal and cereal substitute products imported were calculated to be equivalent to 10 million hectares in 1982 (10 per cent of the EUR 9 utilised agricultural area). Sources of imports in 1983 were other industrial countries (46 per cent), Lomé LDCs (13 per cent), other LDCs (33 per cent) and Planned Economies (7 per cent) Lomé LDCs are those Third World countries which have signed the Lomé Convention with the EEC. They are also called ACP States).[5] The share of agricultural goods in Community imports has gradually declined with the expansion of other imports, notably oil. In 1958, 30 per cent of EUR-6 imports were agricultural but this had declined to 15 per cent of EUR-10 imports by 1980; from 1980 to 1983, the proportion remained stable.[6]

Imports into the EC are governed by two arrangements. First, the Common Agricultural Policy (CAP) governs the 'regulated products' (except fish) shown in Table 4.5. Second, products face tariffs which are the result of the Common Customs Duty (CCD). This is common to all member countries and provides EC producers with an advantage over non-EC producers within the EC market (the concept of Community Preference is basic to the EC). The CCD is reduced by agreement with a number of countries, notably Associate States in Europe, Mediterranean countries and the African, Caribbean and Pacific (ACP) countries that are linked to the EC through the Lomé Convention. Certain products also have zero tariffs that are bound under GATT and therefore must be allowed free entry from GATT signatories (such as the USA). The most important examples are cereal substitutes (manioc and corn gluten feed) and oilseeds (notably soyabeans and soyabean meal).

In 1983, 55 per cent of agricultural and food imports entered the EC free of duty, 34 per cent with a mixture of positive duties and levies, and 11 per cent with only an agricultural import levy. Agricultural imports from LDCs are almost all free of duty as a result of the Lomé Convention (and bilateral agreements with non-Lomé countries such as India), although the volume

**Table 4.5**
**EC Trade (External Only) in Food and Other Agricultural Products, million ECU,[1] 1983**

| 'CAP regulated' products | Imports (million ECU) | (%) | Exports (million ECU) | (%) | Balance (million ECU) |
|---|---|---|---|---|---|
| Oilseeds, cakes, etc. | 8 774 | 34.1 | 1 354 | 7.6 | −7 420 |
| Fruit and vegetables | 4 354 | 17.0 | 1 417 | 7.9 | −2 937 |
| Cereals, incl. manioc | 3 220 | 12.5 | 4 320 | 24.4 | 1 100 |
| Fish, etc. | 2 726 | 10.6 | 859 | 4.9 | −1 867 |
| Raw tobacco | 1 686 | 6.6 | 240 | 1.4 | −1 446 |
| Sugar and honey | 937 | 3.6 | 1 428 | 8.1 | 491 |
| Pigs and pig meat | 811 | 3.1 | 882 | 5.0 | 71 |
| Cattle, beef and veal | 802 | 3.1 | 937 | 5.3 | 135 |
| Milk products | 650 | 2.5 | 3 336 | 18.9 | 2 786 |
| Wines | 609 | 2.4 | 1 394 | 7.8 | 785 |
| Sheep, goats and meat | 501 | 1.9 | 16 | 0.1 | −485 |
| Flowers, etc. | 284 | 1.1 | 507 | 2.9 | 223 |
| Poultry and poultry meat | 144 | 0.5 | 548 | 3.1 | 404 |
| Eggs | 21 | 0.1 | 165 | 1.0 | 144 |
| Other[2] | | | | | |
| Total Regulated | 25 753 | 100 | 17 701 | 100 | −8 052 |
| Unregulated[3] | 24 601 | − | 8 992 | − | −15 609 |
| Total | 50 354 | − | 26 693 | − | 23 661 |

NOTES  1. In 1983, £0.58 = 1 ECU.
         2. Seeds, flax, hemp, hops, dehydrated fodders.
         3. Includes coffee, cocoa, tea, spices, wood and cork, hides and skins, alcoholic beverages, rubber, various crude agricultural raw materials and processed food.
SOURCE  Green Europe (1985) *Community Imports of Food and Other Agricultural Products*, No. 213 (Brussels).

of trade is often limited by quota and seldom involves agricultural products which would directly compete with EC farmers.

EC exports of agricultural and food products have grown rapidly in recent years, twice as fast as imports (15 per cent per annum for exports 1973–82, 8 per cent per annum for imports). Little of this growth can be ascribed to the 'comparative advantage' of producers with relatively low costs of production because of the high prices and market protection of the CAP. The main engine of growth in exports has been the high and stable prices offered producers in the EC, which have encouraged investment and expansion. Member states have gained from exports at a national level because of the Community-financed subsidies available, and during the 1970s, when oil prices were high, many EC countries sought to expand agricultural exports in order to obtain foreign exchange and partially offset the effect of more expensive oil imports. (The French President in 1978 spoke of France's 'Green Oil'.) The problem, that an individual country reaps the whole benefit but only pays one-twelfth of the cost, is called the 'Free Rider' problem and has encouraged the upward spiral of EC agricultural production and exports.

EC exports of agricultural products are primarily processed foods and agricultural commodities in surplus under the CAP; in 1983, cereal and milk products accounted for 29 per cent of total agricultural exports and together, CAP-regulated products accounted for 66 per cent of exports (see Table 4.5). Because CAP prices are held above world market prices, exporters must be subsidised if they are to compete successfully. There has been a long-running controversy concerning the fairness of these subsidies, and the effects on world markets of selling large quantities of subsidised cereals, dairy products, meat and sugar.

The architects of the CAP envisaged occasional subsidised exports of food products, in years when harvests were good, but did not design the system to accommodate continuing subsidised exports. Our discussion of the CAP would be incomplete without a consideration of three important consequences of the growth in agricultural exports from the EC: first, the cost to the EC itself; second, the use of Food Aid to dispose of surpluses; and third, the friction with other agricultural exporting countries, notably the USA. All three issues have become of increased relevance to British farmers and taxpayers in recent years.

## 1 The costs of EC agricultural exports

A proportion of EC exports of agricultural goods would take place at world market prices on the basis of the efficiency of production and level of costs alone. Examples include some to the output of French wheat growers, Dutch dairy farmers and Spanish fruit growers. However, high CAP prices have tempted more resources to be used than cost competitiveness alone could justify. The surpluses which are exported are largely the result of the CAP, not of a cost advantage. The true 'cost' of EC exports is therefore estimated in the same way as the cost of the CAP as a whole.[7] There is the trade cost, which is composed of the cost to consumers (of paying higher prices), and the cost of resource misallocation by producers. Then there is the budget cost to the EC of paying for CAP support and surplus disposal.

In this section we shall concentrate on the budget cost and return to the trade cost in the section on self-sufficiency later in this chapter.

Surplus production in the EC, which has been removed from the market under one of the CAP regulations, poses a problem of disposal. Perishable products which cannot readily be stored are destroyed and others are marked with coloured dyes to prevent them from being used for human consumption. (Fruit and vegetables are examples of the former, potatoes and cereals of the latter.) Some products are subsidised for use in industries (cereals for starch, cereals and wine for alcohol distillation). Some can also be resold within the EC once market prices recover. However, for a substantial proportion of surplus production, the only possibility of obtaining a financial return on intervention stocks is to export with a subsidy. For example, if the EC intervention agency buys up beef at a price of 75, and an exporter can find a world market at 50, then the exporter will be willing to buy the beef from the EC (and thus reduce stocks), with a subsidy of 75–50 = 25 (plus a margin which we ignore). The cost to the EC of destroying the beef would be 75, whereas by exporting with a subsidy the cost is only 25. Clearly, the EC will prefer to export, rather than destroy, the beef, offering to pay an export subsidy up to 74.

In the 1980s, 30 to 40 per cent of expenditure under CAP market regimes has been on export subsidies (called refunds or restitutions).[8] Of this, 80 to 90 per cent has been accounted for by export refunds on cereals (primarily wheat and flour), milk products, beef, veal and sugar. During the 1980–3 period, exporters were obtaining about a third of their total returns as subsidies in these products. In some cases, the proportion rose to more than half.

The growth of production, and the reduction in imports, has also caused a financial problem for the CAP, since receipts from agricultural import levies are much less than the subsidies paid. During the 1980–5 period, levies and receipts from import levies and the CCD varied between 24 and 38 per cent of the cost of agricultural export subsidies.

It is apparent that the EC will remain an important world exporter of food and agricultural products so long as it produces, and can afford to dispose of, surpluses. The Commission has adopted various measures in an attempt to halt the growth of surpluses and eventually reduce them, and foresees a continuing problem in the 1990s unless policies are radically changed. Under present policies, the Commission foresees over 20 per cent of production being surplus to requirements in the EC by 1990 for cereals, wine, sugar and fresh fruit, and over 10 per cent being surplus for milk products and vegetables.[9] These projections omit Spain and Portugal, which will contribute to surpluses of wine, fruit and vegetables, and olive oil, though not necessarily to exports. (Many of these products do not have a ready export market, and the Commission has preferred to dispose of surpluses by distillation and destruction rather than by offering export subsidies).

## 2   Surpluses and food aid

One method of disposing of surpluses is to give them to countries as food aid. In 1984, a total of 10 million tonnes of grain (about 5 per cent of world exports) was shipped by 25 donor countries to over 100 recipients. Some of this was emergency relief, in response to natural disasters, and we do not discuss this aspect of food aid here. Rather, we consider the question of food aid as an on-going method of disposing of surpluses and an alternative to subsidised exports.[10]

The EC food aid system gives food to Third World countries on the basis of applications made, and an annual allocation for food aid is made in the EC Development budget. For instance, the food aid allocation in 1984 by the EC and member countries was for 2 million tones of cereals, 155 000 tonnes of dairy products, 20 000 tonnes of vegetable oil and 13 500 tonnes of sugar.[11] These figures can be put in perspective by comparison with EC exports in 1983–4 of 23.8 million tonnes of cereals and in 1984 of 2.2 million tonnes of dairy products and 4.5 million tonnes of sugar.[12]

How effective is EC food aid? If valued as a source of calories, dairy food aid is not cost-effective (i.e. both the EC and the recipient could gain if the EC disposed of the food another way and paid the recipient money to buy cereals on the world market). Cereals are more cost-effective but even so may not be to the overall benefit of the recipient. Positive effects in a recipient country include saving foreign exchange otherwise required for food imports, providing 'incentive goods' for skilled workers (butter-oil, for instance), and releasing agricultural resources from domestic food production that can be redirected to production for export. Food aid can also be used to establish emergency and stabilisation stocks geared to the needs of the poor, and be linked to 'Food For Work' schemes, which provide employment and improve the infrastructure. The negative effect on the recipient country is that the local food supply is discouraged, and returns from agriculture are reduced, with the likely result that in the future the country will be even less able to feed its own population. Thus, the various benefits must be balanced against a major long-term cost, unless the recipient country has little scope for domestic food production and uses food aid as a means to facilitate industrialisation.[13] There are attempts underway in the EC to improve the food aid system in ways that reduce the negative effect on local food supply.

## 3   The USA

Agricultural exports have played an important role in the US economy ever since the Pilgrim Fathers, but during the late 1950s and 1960s, when world markets were over-supplied with food and a large proportion of US exports were as concessionary aid, it seemed as though the role had finally disappeared. It was only in the 1970s that world markets again became under-supplied and US agricultural exports once more expanded rapidly to a peak of $43 billion in 1981. Since then exports have declined; in 1985, US exports of agricultural products totalled $29 billion and imports had risen to $20 billion. The US has dominated the grain market (over 50 per cent of world exports in the early 1980s) and, with Brazil, the international oil seeds market. Since 1970, agricultural exports have generally accounted for about 20 per cent of total US exports, and up to recently, the surplus on agricultural trade helped pay for the deficit on manufactured trade.[14]

The agricultural lobby is important in the USA for reasons not dissimilar to those in Europe (see Chapter 14). There are over 2 million farmers in the USA, with particular groups able to exert political pressure and seek government support on questions of trade. Thus, dairy, sugar and rice producers are protected against imports, and the government is quick to defend the interests of Florida exporters of citrus products, Mid-Western grain producers and the 45 000 farmers growing soya beans. (Two-thirds of US agricultural exports in recent years have been grain and oil seeds.) During the 1980s agricultural recession in the mid-west, farm indebtedness became a political problem and the importance of the farm vote was demonstrated by government measures designed to increase grain exports.

The three biggest markets for US agricultural exports during 1980–3 were the EC ($10 billion per annum), Japan ($5 billion) and the USSR ($2 billion). There are a number of ways in which the USA has viewed the CAP as working against its interest as an agricultural exporter. First, the use of levies on imports reduces market access for the USA. Second (and this is a complaint against the EC rather than just the CAP), Community Preference ensures that US exporters cannot compete on an equal footing in EC countries with EC competitors. Third, subsidised EC exports are unfair competition of world markets. The enlargement of the EC to include the Iberian peninsula in 1986 makes it likely that all three effects will increase in the 1990s unless the CAP is radically reformed: agricultural exports to Spain and Portugal in the 1980s have amounted to $1–2 billion per annum from the USA and $1 billion from EUR-10. There are also a number of countries wishing to join the EC in which the USA has an agricultural interest (notably Turkey), and any expansion of EC pre-ferential agreements will have further effects on US agricultural exports.

A final and long-standing bone of contention has been the wish of many in the EC to tax soyabean imports as a way of increasing EC consumption of chronically surplus olive oil. Soyabeans and soyabean meal enter free of duty as a result of GATT negotiations in the 1960s, when the USA agreed to accept the CAP mechanism of import levies on certain products in return for free entry on some agricultural products and reduced tariffs on others (citrus products and tobacco are important examples).

As a result of the contest for export markets there have been a number of areas of trading conflict in the 1980s, involving flour, pasta, citrus products, poultry and sugar. A measure of agreement on several of these products was reached in 1986. Fluctuations in the dollar are important in their effect on world prices (particularly for grain), and, consequently, they affect the size of the subsidies that the EC is required to offer to exporters. A factor which is likely to be of increasing importance in agricultural relations between the USA and the EC in the future is the transnational character of companies in the food industry. Food companies in the USA have invested in processing and distributing food in the EC and vice versa. Several US companies are also well established in Latin America, producing mainly for export to the USA (e.g., coffee, beef and fruit), whilst European companies have retained their historical interests in agricultural exporting from the Caribbean and Africa. The operations of some of these multinational food companies have become increasingly integrated, and this blurs the divide between US and EC interests at the industry level. An example is fruit processing, where US companies operating within the EC have much to gain from the continuation of the CAP.

Whilst the USA is the largest competitor in world markets, other countries, such as Australia, have also been adversely affected by the expansion of EC agricultural production.[15] Imports by Britain and other EC countries of Australian cereals, meat, sugar and fruit and vegetables have been rapidly reduced since the 1970s, and weakened world markets for such commodities make it difficult for Australia to find alternative outlets for many of these products. (In 1983, 36 per cent of Australian exports were agricultural and food products.) Argentina and Canada are other countries which have been affected. Together with Australia they are also affected by US measures designed to match EC subsidies. The 1987 GATT negotiations are viewed as an important opportunity to stop the competitive subsidisation of agricultural exports by the EC and the USA.

## 4.4  Trade and Self-sufficiency

To what extent is it desirable to keep as much as possible of the food chain totally under domestic control? There are four main arguments in favour of greater self-sufficiency.

### 4.4.1  The case for increased self-sufficiency

#### (i)  Food security

The essence of this argument is that a country must be able to provide the minimum nutritional requirements for its population from local production in times of emergency. Achieving a level of self-sufficiency that can satisfy this requirement is a prudent insurance policy for any government, and the extra cost involved can be viewed as the insurance premium paid to ensure that the population will not go hungry even if world markets for food break down. Note that if most industrial countries aim to achieve such a degree of self-sufficiency, the effects of world catastrophes will be muted and world market prices will not rise as much.[16]

Therefore, food security can be used to justify supporting agriculture and discriminating against imports in order to achieve some basic minimum level of self-sufficiency. The precise level required would depend upon how rapidly domestic supply could be expanded during the emergency, and how effectively demand could be rationed to ensure minimum nutritional requirements were met. The absence of a free market in times of emergency permits less palatable and duller foods (e.g., powdered eggs) to replace preferred foodstuffs. In the 1970s, there were shortages on world markets of important foodstuffs, though not comparable with those during the Second World War. It was apparent that the levels of self-sufficiency achieved in the 1971–3 period were adequate to ensure that Britain could feed itself in an emergency. Therefore, whilst the food security argument could perhaps be used to justify 50 per cent self-sufficiency in Britain in 1970, it cannot justify the rise to 60 per cent self-sufficiency which occurred over the 1970–85 period.

## (ii) Agricultural support

Policies designed to maintain farm incomes and provide employment in rural areas are discussed later in the book and invariably require the protection of domestic producers against lower-priced imports. Thus, increasing self-sufficiency is one result of agricultural support policies.

The problem raised by this argument is, what happens once 100 per cent self-sufficiency is attained? As we can see in Table 4.2, Britain is 90 per cent self-sufficient in potatoes and meat (taken as a whole), and more than self-sufficient in cereals. For the European Community, self-sufficiency in the 1980s has exceeded 100 per cent for all but a handful of farm products and there have been substantial surpluses for more than a decade. It is simpler to attain self-sufficiency than it is to reduce subsidised production.

Those who wish to support agriculture beyond 100 per cent self-sufficiency sometimes argue that there is a growing gap in the world between the demand for food and its supply. However, the world's hungry have no *effective* demand for food and the surplus would have to be given to them free, with the disincentive effect on local food production noted above in the discussion of food aid. The argument is static and ignores the potential for food production in poor countries which can be stimulated by a buoyant international demand for food.

### (iii)   Multiplier effects

A third argument for increasing self-sufficiency relates to the multiplier effects of extra agricultural production on the domestic economy: these were discussed earlier in relation to the input-output Table. A £1 increase in exports of food will raise output and employment. In contrast, a £1 increase in food imports at the expense of local producers will reduce output and employment.

A related argument to justify a high degree of self-sufficiency is that it enables firms making agricultural machinery, etc., to expand production and, through concentration, obtain economies of scale and hence increase exports. In 1984, for instance, exports from Britain of agriculturally-related products (chemicals, machinery, consultancy services, etc.) amounted to £1031 million, or nearly a quarter of exports of food, feed and alcoholic drink. The accession of Spain and Portugal to the EC in 1985 could be seen as an export opportunity for British agriculturally-related goods and services, not just an import opportunity for the British consumer.[17]

### (iv)   The balance of payments

An increase in self-sufficiency reduces the net import of food and therefore the foreign exchange requirement to pay for food imports. This argument relies for its validity upon the reduction in food imports being the least inefficient way of achieving the objective of saving foreign exchange. Other options might include protecting forestry or some other industry from imports, giving subsidies to exporters of manufactures, or encouraging the inflow of capital by offering tax concessions to foreign investors. However, the balance of payments gain from increased agricultural self-sufficiency is unlikely to be assessed in this way, because some degree of farm support is viewed as inevitable for reasons other than the balance of payments. More likely, the argument will be about the marginal net gains from expanding agriculture, and how far extra foreign exchange can be obtained from a selective expansion without incurring any extra cost to the exchequer.

The balance of payments argument has another aspect for member countries of the EC, which follows from the principle of Common Financing. Member countries pay to the European Fund in Brussels a fixed contribution related to GDP (in 1986, each country paid 1.4 per cent of a theoretical value added tax), together with 90 per cent of receipts from the agricultural levies imposed on imports from outside the EC, and receipts from the Common Customs Duty. In recent years, about 70 per cent of this has been spent by the agricultural part of the Fund (FEOGA, or the European Agricultural Guidance and Guarantee Fund).

Member countries which succeed in producing an exportable surplus of agricultural goods covered by the CAP, and which also reduce their imports from non-EC countries to a minimum, are likely to be net beneficiaries from this arrangement, since they pay little into the Fund but attract FEOGA expenditure in order to intervene in agricultural markets and to subsidise exports. Typical examples in recent years have been the Netherlands, Denmark and Ireland. In contrast, Britain has continued to import agricultural goods from non-EC countries to a significant extent, and has only recently begun to achieve self-sufficiency in major CAP commodities. Consequently, Britain, along with West Germany and Italy, has been a net contributor to the budget of the EC. The contribution has been reduced in the 1980s as a result of the Luxemburg Agreement, under which countries which are net contributors and have a lower than average GDP per head (such as Britain) obtain a partial refund.

The UK can reduce its *net* foreign exchange cost in three ways, all of which are consistent with increased self-sufficiency. First, measures which expand home production and replace non-EC imports reduce the payment of import levies and duties to FEOGA. Second, agricultural policies can be adopted in line with EC Directives, that call for generous levels of expenditure in Britain which can be reclaimed from FEOGA. (This involves increased British government expenditure as well, since most Directives only allow for 25 per cent of expenditure to be recouped from FEOGA.) Third, the government can encourage production of those agricultural products which can obtain EC export subsidies.

The problem posed by the budgetary system of the EC is sometimes called the 'Free Rider' problem. Each country individually gains from following policies designed to limit its budgetary contribution, but in doing so the member states make their collective budgetary problem worse.

### 4.4.2 The case for more trade and less self-sufficiency

#### (i) Comparative advantage

The basis of the argument against increasing self-sufficiency is the case for free trade. There is a general presumption among economists that some trade is usually better than no trade, and more trade better than less trade. The intellectual basis for this view can be traced back to the theory of comparative advantage as expressed by Ricardo, and developed in the nineteenth century by neoclassical economists.[18]

Resources which are used wherever their opportunity costs are low, relative to those in other countries, can produce an exportable surplus. This, in turn, can be used to purchase a commodity, for which domestic opportunity costs are high, in greater quantities than could be produced at home under a policy of self-sufficiency. New Zealand, for instance, has very low costs of production in dairying and sheepmeat. The UK could obtain more butter and lamb by specialising in manufactured goods and exporting them to New Zealand in return for butter and lamb, than it could by attempting to be self-sufficient in butter and lamb itself. Both countries can gain from specialising in producing and exporting goods in which they are relatively more efficient, and importing goods of which they are relatively inefficient producers.

This argument requires several assumptions to be made and three are particularly important. The first two are that prices should reflect marginal opportunity costs and that resources should be mobile between uses, for instance moving from butter and lamb to manufacturing in Britain, and vice versa in New Zealand. The third assumption required is for international exchange rates to reflect the relative competitiveness of the countries engaged in trade. Short-term changes in exchange rates are mainly determined by speculative pressures and rates of interest, but over the longer term, exchange rates do seem to reflect underlying competitiveness.

The free trade argument is modified in the case of Britain by membership of the EC. There are two considerations; the effect of forming a customs union and the effect of the CAP.

A. *Comparative advantage and customs unions* Customs unions which impose duties against non-member countries obviously do not fully exploit comparative advantage, but there are circumstances in which the establishment of a customs union can bring about a more efficient use of resources.

Imagine there are three countries, (A), (B) and (C), which can produce a commodity at a cost of 10, 18 and 25 respectively. Under free trade, (A) will supply all three countries with the commodity. Moving from free trade to a customs union of (B) and (C), which is protected by a common customs duty of more than 8, *diverts* trade away from the most efficient producer (A), and is against the free trade philosophy.

In contrast, imagine that to begin with (B) protects itself with a duty of 9, and (C) protects itself with a duty of 16. A customs union of (B) and (C) with a common external duty of 9 will *create* trade from (B) to (C). Consumers in (C) now have access to the lower cost producers in (B). In this way, a customs union with a common external duty can be defended as a move towards freer trade. It is a second-best solution to completely free trade but still an improvement on complete protectionism.

B. *The EC as a customs union* The EC is a customs union that seeks to encourage a movement towards freer world trade. This is expressed in Article 110 of the Treaty of Rome: 'By establishing a custom union between themselves, Member States aim to contribute in the common interest to the harmonious development of world trade, the progressive abolition of restrictions on international trade and the lowering of customs barriers.

The common commercial policy shall take into account the favourable effect which the abolition of customs duties between Member States may have on the increase in the competitive strength of undertakings in those States'.

Treaty establishing the European Economic Community, Rome, 25.3.57. Text in force 1.1.73. Published by HMSO, London, 1973, pp. 44–5.

It was mainly this commitment to free trade that permitted the EC to establish its import levy system as part of the CAP during the 1960s whilst remaining within GATT.[19] The EC imposes a Common Customs Duty against non-member states to provide Community Preference. Reductions in duties have been negotiated through GATT covering many industrial goods.

Under the CAP, however, the internal price level is sustained by a combination of import duties and levies and there has been no gradual reduction in the level of protection against imports from efficient producing countries. We have already discussed the costs incurred in disposing of agricultural surpluses resulting from the CAP. We now see that the CAP imposes a significant trade diversion cost on consumers in the EC.

Unfortunately, there has been less trade creation in agricultural products within the EC than was initially expected. The Treaty of Rome envisaged that agriculture in the EC would become increasingly specialised by region, according to the principle of comparative advantage. By establishing a common market with unified prices, the most efficient farmers would gradually become dominant and regions with disadvantages would seek alternative means of employing their resources. In this way trade creation would occur to partly offset the trade diversion cost.

In fact, precisely the opposite has occurred, as a result of setting prices at a level designed to help the marginal, high cost producer to remain in business. Far from encouraging regional specialisation within the EC, the CAP has prolonged the inefficient production of agricultural commodities (such as fruit and vegetables in the north, cereals and milk in the south, and wine and olive oil in remote, mountain regions poorly served by roads).[20] Commission attempts to improve specialisation have invariably stopped short of steps to reduce the number of farms in disadvantaged areas, confirming the primacy of social goals in the CAP rather than purely agricultural ones. The forces which make self-sufficiency so attractive to member countries also makes any move towards Community-wide specialisation unattractive. Every member country has groups of high cost, inefficient farmers which it can protect at Community expense, thanks to the 'Free Rider' problem.

Summing up, comparative advantage offers a strong argument in favour of more trade and less self-sufficiency, even in the context of a customs union. Members of a customs union that adopts a common external duty will sustain a loss from trade diversion. They can also gain from trade creation if the common external duty reflects the costs of production of the most efficient producers within the customs union. However, the CAP has set agricultural prices at levels that reflect the costs of production of the least efficient producers. As a result, there has been little gain from trade creation to offset the cost of trade diversion.

## (ii) Retaliation

Another argument against policies designed to increase self-sufficiency is the pragmatic one that other countries will retaliate. The USA has in the past threatened the EC with quotas on steel imports if the EC in its turn interfered with the importation of US oilseeds and feedingstuffs into the EC. A country such as the UK, with 20 per cent of national income exported, and a similar proportion imported, is likely to lose far more in reduced exports, if other countries do retaliate, than British farmers can hope to gain from extra output and profits.

## (iii) International agreements

The member states of the EC are founder members of GATT, and, as we have seen, subscribe to the objective of free trade in the world. Within GATT, there is an agreed philosophy that protection should sooner or later give way to freer trade, and there is little scope for increasing levels of protection for agriculture. The mechanisms of the CAP were only agreed upon after long, and at times acrimonious, exchanges, and at the price of tariff concessions by the EC in other product markets. These arrangements are now fixed ('bound under GATT'), and can only be changed by the EC offering trade concessions in other directions.

To sum up, there are no very good economic arguments in favour of increasing the role of domestic farmers in the food chain beyond the level necessary for minimal food security. On the contrary, there are strong arguments for imports to continue to play a large part in the domestic food chain. There is the cost to consumers of paying higher prices, or to taxpayers for paying farm subsidies. There may be a benefit to the excheq-uer so long as imports continue. However, the experience of the EC has been that domestic production rises to squeeze out imports and create surpluses that themselves become a cost to the exchequer. Finally, there is the resource cost which is hidden but which may be as great as the costs of protection to taxpayers and consumers. Resources are diverted from other uses where output and employment are consequently reduced. Other countries with a comparative advantage in agriculture also experience a resource cost and diversion towards the less efficient that in turn reduces their ability to buy goods from other countries. Any political, social and strategic gains from inhibiting international trade are won at an economic cost. We return to the matter in Chapter 13, with the aid of a graphical analysis. In Chapters 14 and 15 we explore further the policy implications of choosing to protect and support domestic agriculture at the expense of international trade.

## Notes

1. Brown C. P. (1975) *Primary Commodity Control* (Kuala Lumpur: Oxford University Press).
2. Williamson J. (1983) *The Open Economy and the World Economy* (New York: Basic Books).
3. OECD (1982) *Problems of Agricultural Trade* (Paris). Employment multipliers can also be calculated. For an interesting agricultural example of this use of Input-Output analysis, see Craig G. M., Jollans J. L. and Korbey A. (1986) *The case for agriculture: an independent assessment.* Chapter 3 (University of Reading: CAS Report 10).
4. EC (1985) *The Agricultural Situation in the Community* (Brussels).
5. Green Europe (1985) *Community Imports of Food and Other Agricultural Products*, No. 213 (Brussels).
6. European File (1983) *The External Trade of the EC*, No. 19/85 (December).
7. Buckwell A. *et al.* (1982) *The Costs of the CAP* (London: Croom Helm).
8. EC (1982, 1984, 1985) *The Agricultural Situation in the Community* (Brussels).
9. Green Europe (1985) *Supply and Demand of Agricultural Products: Outlook to 1990*, No. 212 (Brussels).
10. Hopkins R. F. (1984) 'The evolution of food aid', *Food Policy* (November).
11. EC (1984) op. cit., p. 78.
12. EC (1985) op. cit., p. 254.
13. Clay, E. and Pryer, J. (1982) *Food Aid: Issues and Policies* (Sussex: IDS, July). The issues are also dealt with by Matthews, A. (1985) *The CAP and Less Developed Countries*, (Dublin: Trochaire).
14. EC (1984, 1985) op. cit. Other sources of data for the USA include USDA (annual) *Agricultural Statistics*, Washington and World Food Institute (1984) *World Food and US Agriculture, 1960–83* (Iowa, 4th edn.) See also the debate in *Food Policy*: Sanderson, F. (February 1983) 'US farm policy in perspective'; Paarlberg, D. (February 1984) 'Tarnished gold: US farm commodity programs after 50 years', Cathie, J. (February 1985) 'US AND EEC agricultural trade policies'.
15. Bureau of Agricultural Economics (1985) *Agricultural Policies in the EC. Their Origins, Nature and Effects on Production and Trade*, Policy Monograph No. 2 (Canberra, Australia).
16. Ritson, C. (1980) *Self-sufficiency and Food Security* (University of Reading: CAS Paper No. 8).
17. House of Commons Agriculture Committee (1985) *The Accession of Spain and Portugal, with special reference to British Agricultural and Food Imports and Exports*, 5th Report, 1984–85 Session (London: HMSO).
18. For example, Lindert, P. and Kindleberger, C (1982) *International Economics* (7th edn, Illinois: Richard Irwin).
19. Hine, R. C. (1985) *The Political Economy of European Trade* (Brighton: Wheatsheaf).
20. Bowler, I. (1985) *Agriculture under the CAP: A Geography* (Manchester University Press).

# Agriculture in the National Framework 5

This chapter explains, in the broadest of terms, what the UK agricultural industry is and how it fits into the national economy. The intention is to give an overall picture of the industry so that the more detailed chapters which follow can be set in context. Because UK agriculture is part of the broader agriculture of the European Community, to which a common policy is applied, it is valuable to compare the position of farming in Britain with that found in other member states. Many of the changes happening in British and European agriculture have parallels elsewhere in the world. Indeed, to appreciate the position of agriculture in the UK's national framework it helps to have an awareness of the process of general economic development, the role which agriculture plays in this process and the changes which occur in the agricultural industry as a country grows and becomes richer. We start, then, with an overview of British agriculture but progress to the European Community level and then the world scale to give the national scene some perspective.

## 5.1 A Broad Introduction to UK Agriculture

Some basic statistics are useful for setting the scene in our discussion of agriculture's place in the national framework. Agriculture currently occupies four-fifths of the UK surface area, engages 3 per cent of the working population, accounts for 3 per cent of total investment, is responsible for 4 per cent of all borrowings, generates 2 per cent of Gross National Product and supplies 62 per cent of the nation's food supplies, or 82 per cent if only those food products which are grown commercially in significant quantities in the UK are considered. Agricultural products in the form of food and feedingstuffs constitute 4 per cent of the country's total value of exports and 10 per cent of its imports.[1]

Exaggerated claims are often made for the national importance of agriculture. It is sometimes talked of as the country's biggest industry, and in some ways it is. In terms of land area, agriculture clearly uses more than other types of production, although it could be argued that tourism and recreation in the countryside are spread even more extensively. On the other hand, agriculture engages only one worker in 40 and creates an even smaller fraction of the nation's income. Obviously no single figure can adequately represent agriculture's relative importance; this will depend on which particular aspect of national life is being considered and will vary according to whether output, use of resources such as capital or labour, environmental impact or exchequer support is taken as the yardstick. In rural areas agriculture's importance is obviously greater than at the national level, although even here there are popular misconceptions which overestimate its role as a provider of employment or income. The importance of agriculture can also be seen differently when viewed dynamically: for example, in periods when the nation's balance of payments has needed improvement, farming has often been given a prominent expansionary role as an import-saving industry out of proportion with the contribution it normally makes the national economy.

The term 'agriculture' requires some clarification. It is used here to include horticultural activities but excludes forestry and fishing, although sometimes in official statistics these latter two activities are bracketed with agriculture as a primary-industry group, and as forestry is in some areas a competitor with farming for land it is necessary on occasion to consider agriculture and forestry together. 'Agriculture' is conventionally thought of as terminating at the farm gate, hence the alternative name 'farming industry', so that the ancillary industries such as animal medicine, farm machinery and chemical manufacture and distribution are excluded, as are also the food distribution and processing sectors.

In an increasingly integrated economy, sectoral divisions are often arbitrary and sometimes meaningless. However, the degree to which linked activities can be considered as dependent on or an integral part of agriculture must vary with the type of activity. For example, while food retailing undoubtedly uses some food produced in the UK, it is not dependent to the extent that no alternative sources of supply could be drawn upon. Thus when the National Farmers Union tries to demonstrate the importance of British agriculture as a provider of jobs, it has some justification in going beyond the 633 000 people who are farmers or farm workers.[2] It points out that there are 400 000 jobs in industries such as machinery and fertilisers which supply British agriculture with inputs – although one must add the caveat that they also export some of their output. Going further, some 375 000 jobs exist in the food and drinks industries, whose inputs come from UK farmers, and a further 820 000 work in the food retail and distribution sectors. However it would be a travesty simply to add all these together and conclude that the sum indicates the numbers of people (2.2 million, or 9 per cent of the employed labour force) who depend on agriculture for employment. One might as well include electricity workers and publicans in the extravagant claim – they too provide goods and services to farmers. The essential point is that activities in one sector of the economy are, to some extent, dependent on all others, directly or indirectly. While there is no denying that agriculture's state of prosperity will have implications for the rest of the economy, and particularly so in rural areas, there is nothing unique in this. Indeed, a depressed manufacturing or mining industry may have greater income and employment implications for country areas than farming, even if these industries are not close by.

In this section of the book, the conventional boundary of the agricultural sector as a farm sector is adopted, although this is relaxed a little in Chapter 15. There we will also see that simple figures are inadequate to show the importance of farming to the social or environmental aspects of the national framework or to the political process.

### 5.1.1  Income and employment

Here we will concentrate initially on two major aggregate indicators – the contribution to national income (more strictly, Gross Domestic Product) of agriculture, and the proportion of the working population it engages (both as hired workers and as self-employed farmers). Agriculture's relative national importance to land use, to investment, to borrowing, to government spending, etc., are covered in the chapters devoted respectively to those topics, as is also a more detailed consideration of labour.

In terms of its relative contribution to GDP and employment, UK agriculture has been declining throughout this century (see Table 5.1). The period immediately following the Second World War saw a temporary reversal of the long-term downward trend, but this was soon restored. Agriculture's share was relatively stable during the early and mid-1970s, with the exception of 1973 when farm prices rose and output increased but since 1977 the downward trend has continued.

In contrast, the volume of production in absolute terms has risen for about one hundred years, particularly during and shortly after the Second World War. It is estimated that the **gross output** of UK agriculture has almost tripled since 1946, and part of this extra output has replaced imports. The growth in output has happened despite a decrease in agricultural area of about 4 per cent and in number of employees of about 50 per cent. Greater volumes of inputs have been bought from other sectors of the economy to generate this extra output. Figure 5.1 shows that within the last three decades the upward trend in the volume of production has continued

**Table 5.1**
**Agriculture's Share of Employment and Gross Domestic Product –**
**Selected Years**

| Year | Percentage of GDP | Year | Percentage of total working population* |
|---|---|---|---|
| | – | 1871 | 17 |
| | – | 1881 | 14 |
| | – | 1891 | 12 |
| 1900 | 7 | 1901 | 9 |
| 1910 | 6 | 1911 | 9 |
| 1920 | 6 | 1921 | 8 |
| 1930 | 3 | 1931 | 8 |
| 1940 | 4 | 1941 | n.a. |
| 1950 | 6 | 1951 | 6 |
| 1960 | 4 | 1961 | 4 |
| 1970 | 2.6 | 1970 | 3.1 |
| 1971 | 2.5 | 1971 | 3.0 |
| 1972 | 2.6 | 1972 | 3.0 |
| 1973 | 2.7 | 1973 | 2.9 |
| 1974 | 2.7 | 1974 | 2.7 |
| 1975 | 2.5 | 1975 | 2.7 |
| 1976 | 2.6 | 1976 | 2.7 |
| 1977 | 2.5 | 1977 | 2.7 |
| 1978 | 2.3 | 1978 | 2.6 |
| 1979 | 2.2 | 1979 | 2.6 |
| 1980 | 2.1 | 1980 | 2.6 |
| 1981 | 2.1 | 1981 | 2.6 |
| 1982 | 2.2 | 1982 | 2.7 |
| 1983 | 2.0 | 1983 | 2.7 |
| 1984 | 2.1 | 1984 | 2.6 |
| 1985 (provisional) | 1.8 | 1985 (provisional) | 2.6 |

NOTES  * From 1970, percentage of total civilian manpower engaged in all occupations.
– Indicates changes in ways of measurement.
SOURCE  Historical Abstract, *British Labour Statistics* 1970; HMSO *Annual Review of Agriculture*.

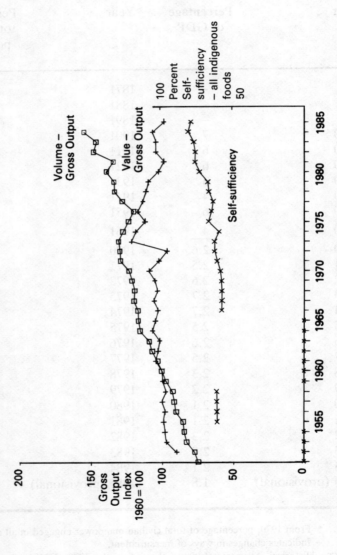

**FIGURE 5.1**
Agricultural Gross Output and Self-sufficiency (UK)

except for the dry conditions of 1975 and the drought of 1976. However, because the prices of farm products have been falling in real terms, that is after inflation has been taken into account, the *value* of UK agriculture's gross output has grown by much less – an annual average of less than 1 per cent in the postwar period. Following a jump above the trend in the early 1970s, the value of gross output has moved downwards sharply in the last decade. Taking the extra inputs into account suggests that the trend in the value of agriculture's gross product (gross output minus gross input) has been downwards, and was some 10 per cent lower in 1984 than in 1946 (see Chapter 11).

Clear increases in physical productivity have been exhibited by land and labour, shown by rising levels of output per unit area and per worker when looked at separately. There has been less improvement in the *overall* productivity of resources employed by British agriculture because of the much greater quantities of capital now used by farmers (machinery, buildings, etc.), although there are methodological difficulties in trying to measure its extent. From 1946 to about 1970 the volume of total output and total input increased more or less in step, implying no improvement in physical efficiency despite the undoubted technological advances that occurred. Since 1970 the volume of inputs has increased less rapidly than output, suggesting some real improvement in productivity. The generally accepted figure for the average annual increase in productivity of the industry over the decade up to 1983 was about 2 per cent, with the partial productivity measure for labour being of the order of 4 per cent, a lower figure than the 6 per cent experienced in the preceding decade.[3] These labour productivity figures compare favourably with the rest of the economy.

### 5.1.2 Other major characteristics: self-supply, output mix and types of farming

Other major characteristics we should note at this initial state include the proportion of the nation's food supply which UK agriculture provides and the types of crop and livestock products which form its output. Currently the UK industry produces rather more than half of the total UK supply of agricultural products, estimated as 62 per cent of the total value of UK food supplies in 1984. Considering only those foods which are grown commercially in Britain, the degree of self-supply has risen particularly with wheat, sugar, most meats and milk products (see Chapter 4). Of the major commodities, the UK is self-sufficient (or more) in cereals, potatoes, poultry-meat, beef, pork, liquid milk and eggs. However, this view of self-sufficiency does not show the imports of chemicals, machinery and other inputs used to produce the level of output, which a wider view of 'self-sufficiency' would embrace.

In terms of the commodity mix of UK agriculture, typically about two-thirds of the total value of output has come from livestock products and only one-third from crops (see Table 5.2) – a ratio which was temporarily reversed in the near-siege conditions of the Second World War but which was rapidly restored with the coming of peace. Since the UK joined the European Community and came under the influence of its Common Agricultural Policy the crop proportion has increased to nearer 40 per cent of total output. Currently the largest contributors, with a fifth of total output each, are milk and milk products (together) and cereals. Milk has for long been of importance, a situation which reflects not only the pattern of demand by consumers and the comparative advantage which areas of Britain possess in grass production but also the price stability afforded since the 1930s by the state-regulated marketing system for milk and dairy products. In contrast, cereals have doubled in relative importance since the early 1960s, especially after the UK joined the European Community. Cereals overtook milk and milk products for the first time in 1984, helped by a particularly good harvest. The cause of the cereals expansion is primarily the combination of a technical revolution resulting in much heavier yields, especially of wheat, with price support under the Common Agricultural Policy. Eggs, fruit and pigmeat have declined, but these longer-term trends are subject to considerable short-term variation and may not indicate the current direction of adjustment. In particular the

**Table 5.2**
**Composition of Total Output of UK Agriculture**

| | (%) 1959/60– 1961/2 | (%) 1973/4– 1975/6 | (%) 1982 | (%) 1983 | (%) 1984 | (£m.) 1984 |
|---|---|---|---|---|---|---|
| Farm crops | | | | | | |
| Cereals | 10.2 | 14.1 | 19.1 | 17.6 | 19.9 | 2 424 |
| Potatoes | 4.7 | 5.2 | 4.1 | 4.4 | 4.7 | 578 |
| Sugar beet | 2.4 | 1.6 | 2.3 | 1.9 | 2.0 | 244 |
| Other | 0.8 | 1.2 | 2.3 | 2.4 | 3.1 | 378 |
| Horticultural crops | | | | | | |
| Vegetables | 5.5 | 6.7 | 5.5 | 6.2 | 6.4 | 778 |
| Fruit | 2.9 | 2.1 | 1.9 | 2.0 | 2.0 | 240 |
| Other | 1.7 | 2.0 | 1.9 | 1.9 | 1.9 | 234 |
| *All crops* | 28.2 | 32.9 | 37.1 | 36.5 | 40.0 | 4 876 |
| Livestock | | | | | | |
| Cattle | 13.9 | 16.7 | 15.2 | 16.1 | 15.9 | 1 938 |
| Sheep | 5.3 | 4.0 | 4.7 | 5.1 | 4.6 | 557 |
| Pigs | 10.3 | 10.9 | 8.4 | 8.1 | 8.2 | 994 |
| Poultry | 4.5 | 5.9 | 5.5 | 5.5 | 5.5 | 674 |
| Other | 0.5 | 0.4 | 0.8 | 0.8 | 0.8 | 94 |
| *All livestock* | 34.5 | 37.9 | 34.7 | 35.6 | 34.9 | 4 257 |
| Livestock products | | | | | | |
| Milk and milk products | 23.1 | 21.1 | 21.7 | 2.1 | 19.2 | 2 338 |
| Eggs | 10.9 | 6.9 | 4.8 | 4.4 | 4.5 | 554 |
| Clip wool and other | 1.1 | 0.5 | 0.5 | 0.7 | 0.6 | 73 |
| *All livestock products* | 35.1 | 28.5 | 27.0 | 26.9 | 24.3 | 2 964 |
| Other not included above | 2.1 | 0.7 | 1.2 | 0.9 | 0.8 | 93 |
| **Total output** | 100.0 | 100.0 | 100.0 | 100.0 | 100.0 | 12 190 |

SOURCE  MAFF *Annual Review of Agriculture* (various years) (London: HMSO).

introduction of state support for sheep-meat in 1980 has appeared to arrest the slow decline in relative importance of this meat to UK agricultural output.

Most British farms have more than one enterprise and so might be termed 'mixed', although specialisation has been rising over the last quarter century as production has become concentrated into fewer but increasingly larger units. The overall picture is of a crop-orientated, arable East and a livestock-dominated, pastoral West in which the lowlands are characterised by dairy farms (augmented increasingly with cereals) and the uplands by extensive sheep and cattle farming (Figure 5.2). The geographical pattern of farming is important because, among other things, it has direct links with the appearance of the countryside, the size and location of the rurally employed population and the nature of the farming and land-owning political lobbies in different counties.

The type of farming which goes on in particular areas is influenced not only by 'natural' factors such as soils and climates but also by economic conditions, in particular on markets for agricultural produce. It was noticeable that, when the state marketing scheme for milk was introduced in the 1930s which gave farmers very similar prices per gallon irrespective of where they lived, milk production developed in those Western and remote areas in which it was previously unprofitable because of high costs of transporting the milk to major conurbations. Similarly, in the 1980s the support given to cereal prices by the Common Agricultural Policy, combined with new heavier-yielding varieties, has been given as the cause of the recent spread of cereal growing to land which has traditionally grown grass, the change carrying major environmental implications.

### 5.1.3  Constituent countries of the United Kingdom

In generating its output, UK agriculture uses some 18.8 million hectares of land, or 78 per cent of the total area. A detailed description of its quality and the way farming uses this scarce resource will be made in Chapter 6; here it is sufficient to point out that differences in climate and weather, soil type and socioeconomic factors result in England alone accounting for the overwhelming majority of the UK's agricultural activity (over three-quarters when measured by the proxy of standard gross margin, which will be explained later, or gross product) (Table 5.3).

**FIGURE 5.2**
**Dominant Systems of Farming in the United Kingdom**

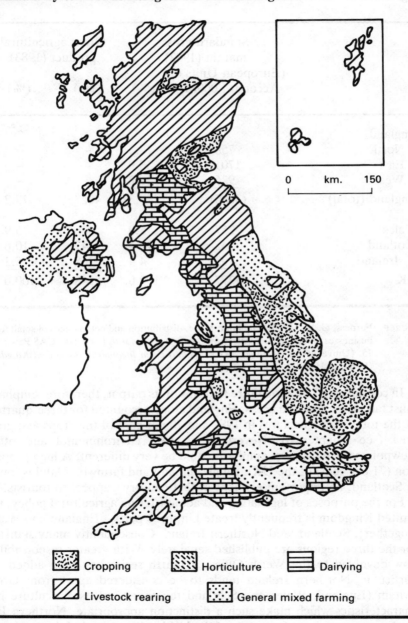

0   km.   150

Cropping   Horticulture   Dairying

Livestock rearing   General mixed farming

SOURCE Edwards, A. and Rogers A. (1974) *Agricultural Resources* (London: Faber).

Table 5.3
Agricultural Activity in the Constituent Countries in the UK

| | Standard gross margin (1982) (European Units of Account ['000M]) | (%) | Gross agricultural product (1983) (£m) | (%) |
|---|---|---|---|---|
| England | | | | |
| North | 758 | 17.3 | – | – |
| East | 1701 | 38.9 | – | – |
| West | 926 | 21.2 | – | – |
| England (total) | (3885) | (77.3) | 4027 | 78.2 |
| Wales | 257 | 5.9 | 306 | 5.9 |
| Scotland | 542 | 12.4 | 547 | 10.6 |
| N. Ireland | 193 | 4.4 | 268 | 5.2 |
| UK | 4377 | 100.0 | 5148 | 100.0 |

SOURCE   Furness, G. W. (1983) 'The importance, distribution and net incomes of small farm businesses in the UK' in *Strategy for family-worked farms in the UK*. CAS Paper No 15. (University of Reading). MAFF (1986) *Farm Incomes in the United Kingdom* (London: HMSO).

In considering UK agriculture in terms of its output, therefore, emphasis must fall on the lowland areas. England alone accounted for three-quarters of the total arable hectares in 1982 and 70 per cent of the crops and grass area (see Chapter 6). Nevertheless, from environmental and other viewpoints the balance of emphasis could be very different. A high proportion (71 per cent) of the UK's rough hill land and farmwood land is found in Scotland, an important element in that country's appeal to tourists.

For the purposes of legislating and administering agricultural policy, the United Kingdom is frequently treated in three parts – England and Wales (together), Scotland, and Northern Ireland. Consequently many statistics for the three regions are published separately. With greater responsibility now devolved to the Welsh Office, a fourth series has been added. In particular, Northern Ireland tends to be considered apart from Great Britain (England, Wales and Scotland together) and its agriculture has characteristics which make such a distinction appropriate. Northern Ireland represents about 6 per cent of the UK land area, and of the total area of 1.4 million hectares some 1.1m. ha. are used for agriculture, contributing 6–7 per cent of the UK output. However, agriculture in Northern Ireland is relatively more important than in Great Britain, generating approximately 6 per cent of the Gross Domestic Product of the region (as opposed to 2 per cent for the whole UK). It also engages a higher proportion of the working population (8.8 per cent of the employed workforce as opposed to 1.8 per cent in England, 2.6 per cent in Scotland, and 4.5 per cent in Wales), has a unique tenure pattern and a different farm size structure.

Physical factors of soil, drainage and the climate of Northern Ireland tend to favour grassland rather than arable farming. Cereal and root crops account for less than 10 per cent of the crops and grass area, whereas for the UK as a whole the figure is over a third. Dairying, beef production and livestock rearing are the most important activities. Intensive livestock enterprises based on pigs or poultry have traditionally been important subsidiary enterprises in the region, mainly as a means of increasing business size on family farms with limited land area, but in recent years these have come under economic pressure, in particular from increases in feed prices.

## 5.2  Agriculture in the European Community

The declining role of agriculture as an employer and generator of income in the British economy with the passage of time and with growing living standards is mirrored in all the member countries of the European Community. Indeed, as we will show later, this pattern of sectoral change is a worldwide phenomenon characteristic of high-income industrialised countries. Furthermore, a decline in the importance of agriculture is part of the general pattern of development so that, by and large, the prosperity of a country, in terms of average income per head, is inversely related to the relative importance of its agricultural sector.

Both absolute numbers of persons and the proportion of the total working population engaged in agriculture has been falling in each Community country, as Table 5.4 shows. Population figures give snapshots at certain dates of the size of the workforce and cannot reveal the dynamics of the situation: in any one year some people will die, retire or transfer to other industries, while others will leave school and join the workforce or transfer in from other industries. What is clear from the figures is that the flow out over the period shown has been greater than the flow in. During the 1960s and 1970s, when most economies were growing rapidly, a substantial proportion of the labour leaving agriculture transferred to other industries, aiding economic growth by moving from an industry with low labour (marginal) productivity to ones where the extra labour generated

**Table 5.4**
**The Size of the Agricultural Sector in the EC**

Numbers of persons engaged in agriculture ('000s)

| | W. Germany | France | Italy | Nether-lands | Belgium | Luxem-bourg | UK | Ireland | Denmark | EEC 6 | EEC 9 |
|---|---|---|---|---|---|---|---|---|---|---|---|
| 1955 | 4285 | 5041 | 7740 | 533 | 310 | 26 | 1171 | 442 | 505 | 17 935 | |
| 1965 | 2966 | 3538 | 4956 | 388 | 230 | 19 | 850 | 340 | 385 | 12 097 | |
| 1977 | 1656 | 2022 | 3149 | 289 | 123 | 9 | 661 | 236 | 218 | | 8363 |
| 1979 | 1544 | 1867 | 3012 | 279 | 118 | 10 | 632 | 220 | 208 | | 7890 |
| 1982 | 1382 | 1758 | 2545 | 248 | 106 | 8 | 632 | 196 | 207 | | 7082 |

(Greece 1083 (1982))

Proportion of population engaged in agriculture, forestry and fishing
(as % of total civilian employment)

| | W. Germany | France | Italy | Nether-lands | Belgium | Luxem-bourg | UK | Ireland | Denmark | EEC 6 | EEC 9 |
|---|---|---|---|---|---|---|---|---|---|---|---|
| 1955 | 18.5 | 26.9 | 40.0 | 13.2 | 9.3 | 19.4 | 4.6 | 38.9 | 24.9 | 26.1 | |
| 1965 | 11.1 | 18.2 | 26.1 | 8.8 | 6.4 | 13.7 | 3.4 | 32.0 | 17.0 | 16.5 | |
| 1976 | 7.0 | 10.9 | 15.5 | n.a. | 3.4 | 6.0 | 2.7 | 23.8 | 9.3 | | 8.4 |

| | W. Germany | France | Italy | Netherlands | Belgium | Luxembourg | UK | Ireland | Denmark | EEC 6 | EEC 9 |
|---|---|---|---|---|---|---|---|---|---|---|---|
| 1979 | 6.2 | 8.9 | 14.9 | 6.0 | 3.1 | 6.4 | 2.6 | 21.0 | 8.3 | | 7.7 |
| 1982 | 5.4 | 8.2 | 12.1 | 4.9 | 2.9 | 4.7 | 2.7 | 17.1 | 8.9 | | 6.7 |

(Greece 31% (1981))

Gross national product derived from agriculture (%)*

| | W. Germany | France | Italy | Netherlands | Belgium | Luxembourg | UK | Ireland | Denmark | EEC 6 | EEC 9 |
|---|---|---|---|---|---|---|---|---|---|---|---|
| 1956 | 7.5 | 10.2 | 19.7 | 10.7 | 7.3 | 9.0 | 4.5 | 27.2 | 18.5 | 10.7 | |
| 1966 | 4.2 | 7.4 | 12.5 | 7.4 | 5.7 | n.a. | 3.3 | 19.6 | 10.2 | 7.0 | |
| 1976 | 2.8 | 5.2 | 8.6 | 4.9 | 3.2 | 3.3 | 2.6 | 16.5 | 5.3 | | 4.4 |
| 1979 | 2.1 | 4.8 | 7.5 | 4.1 | 2.6 | 2.8 | 2.2 | 13.7 | 4.8 | | 5.0 |
| 1982 | 2.2 | 4.3 | 6.3 | 4.4 | 2.6 | 3.4 | 2.3 | 11.1 | 5.5 | | 3.6 |

(Greece 17% (1982))

NOTES  * 1956–66 data refer to Gross Domestic Product at factor cost in national currencies and at current prices (include forestry, hunting, and fishing).
1976–7 data refer to Gross National Product at factor cost and at current prices (exclude forestry and fisheries).

SOURCES 1955–66: OECD, Agricultural Statistics 1955–68 (Paris).
1976: EUROSTAT, National Accounts, 1978 (Luxembourg).
1977–82 figures: EC, The Agricultural Situation in the Community (annually) (Brussels).

much more additional output in value terms. This was especially the case in those countries which started the period with large farming populations such as France and Italy, but was of little importance in the UK.

Higher levels of general unemployment in the 1980s mean that labour displaced from farming may not have another job to go to. Consequently attention has switched from encouraging labour to leave agriculture to finding ways of retaining it in farming as a means of keeping down the rate of unemployment and the social problems which that brings. Alternatively, one might say that the opportunity cost of retaining labour in farming, which was formerly high, is now lower. This view is taken less in the UK than in other countries, one explanation being the smallness of the UK's agricultural labour force in relation to its total working population.

A second feature of the table is the wide disparity seen between countries, not only in the proportions of the Gross Domestic Product generated by agriculture but, particularly, in the percentage of the labour force engaged. In 1955, when only 4.6 per cent of the UK's working population was in farming, comparable figures for Italy were 40 per cent, Ireland 39 per cent, France 27 per cent and Germany 18 per cent, with the figure for the six original EC members averaging 26 per cent. Large differences still exist. The unusually low figure for the UK stems from historic and legal factors such as the earlier occurrence of the industrial revolution, the enclosing of land, the development of Britain as a free trading nation (especially with its Empire) so that home agriculture was subject to strong competition from abroad, and a tradition of passing property between generations without splitting it equally among offspring. UK agriculture was exposed to strong international competition in the third quarter of the nineteenth century, but even before then the proportion of the population in agriculture was comparatively small; in 1871 the figure was only 17 per cent, much lower than France or Italy in 1955 (27 and 40 per cent respectively) and even in 1800 was probably no more than 35 per cent.

A third feature of the table is the relationship between the proportion of the population engaged in agriculture in each country and the proportion of GNP derived from agriculture. For the four EC member countries with the largest farm populations (West Germany, France, Italy and now Greece) the percentage of the population in agriculture is much greater than the proportion of GNP derived from agriculture, about double. This suggests strongly that the incomes of people in farming in those countries may be lower than those working in other sectors, and that a 'farm income problem' may exist, necessitating some form of national or Community policy to raise incomes. These simple percentages should not be taken to imply too much about the size of any income problem; for example, farmers may have off-farm jobs, earnings from which contribute to their living standards but do not show up in the agricultural sector's income *from farming*. However, it is broadly true that disparities of incomes between farmers and non-farmers lie at the root of why many governments have become involved with their agricultures and why the Common Agricultural Policy was formed and continues to exist. The whole farm income question will be referred to in a separate chapter.

### 5.2.1 Implications of differences between countries in the sizes of their agricultural sectors

The scale and significance of the change experienced by the farming sector, often described as agricultural adjustment, has been much greater in countries other than the UK because far higher proportions of the working population (and their families) have been required to leave farming. Between 1955 and 1982 in Italy over a quarter (28 per cent) of the entire working population shifted from agriculture to other activities, so that the farming population fell to a third of its 1955 level. France saw 9 per cent of its population shift and Germany 13 per cent; again, the farm population ended up at only about one-third of its earlier level. In contrast, in the UK the fall was proportionately much less; the numbers in agriculture fell by less than half in the period 1955–82 but, more important, this involved only 2 per cent of the working population.

Taking a job outside agriculture frequently involves moving one's place of living, so that there is not only a disruption to work patterns but to domestic life as well. Young single people are more mobile in this respect than people with families. The result of large scale decline in agricultural employment is often rural depopulation and an ageing of those left in the countryside; with this go the problems of contracting rural services (public transport, shops, education) and debilitated, if not dying, villages. As we shall see in Chapters 14 and 15, while the underlying causes of agriculture's decline as an employer are largely outside the control of society, steps can be taken to alleviate the effects, sometimes by providing alternative sources of employment and income in the countryside and sometimes by simply cushioning existing farmers, for example by supporting the prices of their produce.

As a broad generalisation, it would seem that other EC member countries, especially France, are more actively concerned about the social implications of agricultural adjustment than is the UK, probably a reflection of their proportionally greater levels of agricultural employment. In the UK the number of people living in the countryside has been rising over the last two decades despite the fall in agricultural jobs. Depopulation has turned into repopulation in most areas, but the social composition is changing and a new set of problems is evident, such as friction arising from conspicuous inequalities between the living standards of well-off newcomers and the remaining rurally-employed wage-earners.

Differences between member countries in the magnitude of their farming sectors are an important source of conflict within the Common Agricultural Policy, which is supposed to operate commonly across the EC. The balance of political power between the farm sector and other groups, notably consumers and taxpayers, will vary between countries, although not in any simple proportional relationship with population numbers. Countries like the UK with a small farming population are likely to object to policies resulting in high food prices which are favoured by the governments of countries where the agricultural vote is of higher political significance. And support for farmer incomes financed by Community funds will benefit some more than others, so that some countries will be net beneficiaries while others are net payers. The position of the UK as a net contributor to Community funds (most of which are spent on the Common Agricultural Policy, between 60 and 70 per cent in the years 1980–4) has been a notable source of friction within the Community. The EC and the problems of the CAP will be returned to in Chapter 12.

## 5.3   The Changing Role of Agriculture in the Process of Economic Development

This book is primarily concerned with agriculture in Britain and other industrial economies, especially those of the European Community. But to understand these countries it is necessary to give them the context of agriculture as a worldwide activity. British and European farming can then be seen as fitting into a pattern of general economic development, although with some special features.

Table 5.5 illustrates the proportion of the working population engaged in agriculture for a range of low income, middle income and high income countries. Broadly, the percentage of the population involved in agriculture is inversely related to the economy's level of development, as indicated by income per head. Many developing countries still have over 70 per cent of their population in agriculture – largely as subsistence or peasant farmers – and these are countries with very low incomes per head. The correlation is not exact (Switzerland with 5 per cent has a higher income per head than the UK with 2 per cent) but the relationship is valid as a generality.

The reduction in the relative importance of agriculture is not something which just 'happens'. Rather, agriculture plays a vital role in the development process itself. While the impetus for growth may in some cases come primarily from other sectors (for example, the sale of oil or other minerals) even here agriculture generally has an important secondary part to play. Consequently it is possible to draw up a general model of development

**Table 5.5**
**Agricultural Employment and GNP per Head – Selected Countries**

| | Percentage of labour force in agriculture | | GNP per capita |
| --- | --- | --- | --- |
| | 1960 | 1980 | ($ 1980) |
| Low income economies | | | |
| Chad | 95 | 85 | 120 |
| Bangladesh | 87 | 74 | 130 |
| Malawi | 92 | 86 | 230 |
| India | 74 | 69 | 240 |
| Pakistan | 61 | 57 | 300 |
| Middle income economies | | | |
| Ghana | 64 | 53 | 420 |
| Indonesia | 75 | 58 | 430 |
| Philippines | 61 | 46 | 690 |
| Morocco | 62 | 52 | 900 |
| Peru | 52 | 40 | 930 |
| Turkey | 78 | 54 | 1 470 |
| Algeria | 67 | 25 | 1 870 |
| Greece | 56 | 37 | 4 380 |
| Industrial market economies | | | |
| Ireland | 36 | 19 | 4 880 |
| Spain | 42 | 15 | 5 400 |
| UK | 2 | 2 | 7 920 |
| Japan | 33 | 12 | 9 890 |
| Austria | 24 | 9 | 10 230 |
| USA | 7 | 2 | 11 360 |
| Norway | 20 | 7 | 12 650 |
| Switzerland | 11 | 5 | 16 440 |

SOURCE    World Bank (1982) *World Development Report 1982* (New York: Oxford University Press).

which embraces countries at all stages of development and which can be used to predict patterns for growing economies. One such model, in graphic form, is given in Figure 5.3.

## FIGURE 5.3
## Relationships between Growth Rates in the Whole Economy and in the Agricultural Sector

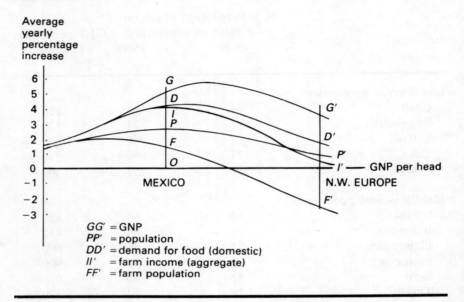

SOURCE   FAO (1962) *Agricultural Commodities – Projections for 1970* (Rome) p. A-49.

NOTES   1. Both the *GG'* (change in GNP) and *PP'* (change in population) lines reach a peak of annual change and subsequently decline. GNP is a measure of income *before* the annual consumption of capital through wearing out and obsolescence is deduced.

2. The growth in GNP per head is indicated by the distance between *GG'* and *PP'* lines (i.e. *PG* or *P'G'*). It will be less than GNP growth if population growth is more than zero, and in the diagram is higher among middle-income countries than high or low income ones.

3. The growth in demand for food (*DD'*) is the result of combining population growth (*PP'*) with the rise in average incomes (*PG*) modified by the fraction of that income spent on food (i.e. the income elasticity of demand for food). This elasticity falls with rising incomes, so that at higher incomes the *DD'* line closely approaches the *PP'* line.

4. Rising incomes per head among the farm population is indicated by the vertical distance between *FF'* (change in farm population) and *II'* (change in aggregate farm income). If the relative position of average agricultural incomes is to remain constant with respect to average incomes in the economy as a whole, *FI* must be the same as *PG* (or equivalent at higher levels of GNP per head). A smaller *FI* will indicate a declining relative income position for farmers.

This model of the changing significance of agriculture, drawn up by the Food and Agriculture Organisation of the United Nations, is based on empirical evidence although individual countries will show variations around the generalities which the lines represent. The stage of development in terms of income per head of the population corresponds with the horizontal axis, poorer countries being on the left and richer ones on the right. The lines indicate percentage annual changes in important development parameters: in Gross National Product ($GG'$), population ($PP'$), demand for food ($DD'$), income of the farming industry ($II'$) and the numbers of people engaged in agriculture – the farm population ($FF'$).

The diagram has the properties of an economic model, in that it tries to encapture the essential elements of a real-world situation. It attempts to isolate and keep the most important features and discards the rest; in this instance no account is taken of the legal or social institutions of countries, or even the nature of their economic system, planned or market-based. Although this may mean that the model is 'unrealistic' in that it does not completely describe the real-world of individual countries, it may still give far more insight into a problem and enable better predictions to be made than would a less abstract approach.

Detailed comments on the diagram are presented in the notes which accompany it. How a country can shift itself to the right towards high average incomes is beyond the scope of this book, although it should be borne in mind that once incomes per head have been increased to some critical level the process of economic growth tends to be self-sustaining. Here we can simply note the main points to be drawn from the diagram, concentrating on the right-hand, high income section.

Firstly, while among low income countries any annual increase in GNP is largely matched by population increases, implying that income per head remains little changed, among richer countries growth in GNP outstrips population growth. Among the highest income countries this difference is perhaps smaller than in the middle income group, implying that the rate of economic growth (rise in income per head) slows down a little. Middle-income countries can often grow rapidly by bringing unused resources into production or by borrowing technological ideas from more developed countries, and do not run into severe problems of diminishing returns to their investments. Richer countries, however, are much more dependent on inventing new processes and products.

Secondly, for middle income and rich countries the demand for food expands at a slower rate than does the country's overall economic activity, implied by GNP. In other words consumers will use an increasing amount of spending power on non-food items.[4] This will cause the non-farm sectors of the economy to expand more rapidly than agriculture. While, when income levels are low, much of any additional spending power goes on food, this diminishes with increasing incomes so that, among high income countries, the annual demand for food is hardly more than the increase in population, as is indicated in the right-hand extremity of the diagram by the close proximity of the $DD'$ and $PP'$ lines. The low annual growth in the demand for food in the UK has already been discussed in Chapter 2.

Thirdly, the annual increase in the income of the farming industry as a whole declines as countries become richer (line $II'$). In the diagram at the extreme right end the rise in farming industry income is close to zero, indicating an almost static total income to farmers. It is certainly growing at a much slower rate than the economy in general, as indicated by GNP. Principally this comes about because: a) the demand for food expands less quickly than general demand; b) farmers in rich, industrialised countries persistently expand output by taking up new technology, and this depresses the real prices they receive for their produce; c) with more affluence consumers devote an increasing proportion of their food spending to packaging and processing, with a declining fraction going to the farmer; and d) in industrialised economies an increasing fraction of the revenue coming to farmers from selling their produce is passed over to other industrial sectors by the purchase of fertilisers, pesticides, fuel and machinery. The substitution of bought-in inputs for 'home grown' ones is a feature of farming in the industrialised countries (chemical fertilisers in place of manure, fuel and engines in place of horses and so on). A fourth feature of the diagram, therefore, is a divergence of the demand for food change line $(DD')$ and the income change line $(II')$.

The fifth point relates to the line representing annual changes in the size of the farm population $(FF')$. In the initial stages of growth agriculture has to absorb almost all the rising national population. It is the principle employer and can make use of extra labour to help with its new, greater productivity. This is shown by the closeness of the PP' and FF' lines at the left-hand end of the diagram, and the implication is that the number of people in farming rises. However, with economic growth the annual change in farm population soon becomes negative and, among high income countries, there is a substantial annual outflow of people. Hence we witness a fall in the numbers engaged in farming. Those fewer left in will have to share the almost-static farming income, and whether the average farmer income will manage to keep up with incomes in the rest of the economy will depend on how fast numbers in farming can be reduced.

Britain corresponds with the extreme right-hand end of the diagram; if anything, the lines should be projected a little further. Short term factors (such as entry to the EC and its effect of farmers' incomes and food prices, and the general economic slump of the late 1970s and early 1980s) and its own peculiar history causes the UK presently not to fit the various lines exactly, although the directions of change and magnitudes are of the expected order. Currently the UK's annual average growth trend in GNP is only some 1–2 per cent, the total population is virtually static, the demand for food expands at less than 1 per cent annually and the farm population has consistently seen a negative change of over one per cent. The income of agriculture as a whole has declined in real terms, a situation not incompatible with the model if extrapolated to the right or if due allowance is made for historical accident. The main value of the model is to point out that Britain's agriculture has changed and is changing in ways which are broadly experienced elsewhere and for which explanations can be offered which depend on established economic principles.

While economic forces dictate the inevitability of the long-run decline in the relative importance of agriculture and, as we shall see later, much of its related structural detail such as the sizes of farms, governments in industrial countries have the power to slow the decline. Agriculture policies taken overall usually aim to defend farmers from the squeeze between falling (real) product prices and costs, and the structural changes to which this gives rise. They use national resources to prolong a level of farm employment and a mix of farm sizes and types which are incompatible with the underlying market situation. The incompatibility grows as farmers adopt technical advances and supply greater quantities to a static market; the budgetary cost of maintaining the retarded structure also increases.

The EC's Common Agricultural Policy is clearly of this defensive type. So too was the UK's national policy from the late 1940s to the early 1970s, although in a less obvious form. There may be sound arguments for moderating the rate of agriculture's decline – strategic, social, political and, particularly in the short-term, economic – but the inevitability of the long-term trend is inescapable. The more and longer it is resisted the greater the cost, both in terms of public spending and in the opportunity cost of excess resources retained in agriculture. When budget limits are reached and reforms of policy are demanded, as in the EC of the mid-1980s, the speed of reform is constrained by the implications for farmers, their families and rural communities of a too-rapid return to a position nearer the long-term trend; many farms could not survive in a less protected market.

Farmers and their pressure groups acting in their own interests can be expected to encourage governments to support agriculture against the long-term trend and, once secured, to campaign against the reform of that policy. Their arguments tend to be narrowly confined to the farming industry and the short-term. However, if a case for public support of agriculture is to be made it should be broadly based because the implications of giving assistance stretch beyond the farm gate – to other industries, especially those engaged in exporting, to the markets for capital, labour and land, to the appearance of the countryside and the composition of rural society, among others. The rationale for agriculture policy will be examined in Chapter 14.

From this broad description of agriculture's role in the economy we move to a detailed examination of the UK farming industry, starting with the resources it employs.

## Notes

1. Figures for 1984. Broadening the list of items classed as agricultural products, as in the HMSO *Annual Review of Agriculture*, raises the proportion of food, feed and alcoholic beverages in UK exports to 8 per cent, and to 16 per cent of exports in 1984.
2. National Farmers Union (1983) *British Agriculture – The Facts*, Press Notice 149.
3. MAFF (1985) *Annual Review of Agriculture 1985* (London: HMSO) and Economic Development Committee for Agriculture (1977) *Agriculture into the 1980s: resources and strategy* (London: National Economic Development Council).
4. A link should be made with the concept of income elasticity of demand for food and Engel's Law; see Chapter 2.

# Resources in Agriculture: Land 6

## 6.1 The Quantity and Quality of Agricultural Land

Land bears an obvious significance to any study of agriculture and the rural economy. Its physical characteristics largely govern what farming activities can be carried on and, in turn, its appearance is affected by farming. The ownership of land is highly influential in determining who may use the land for farming and other purposes, which again has environmental and social consequences for the countryside. Land is an important asset in the business of farming and to the wealth position of farmers. Land ownership is associated with social and political power, even in an industrial society like the UK.

The most natural point at which to start is an inventory of the quantity and quality of land available. In broad quantity terms the UK uses about 19 million hectares of agricultural land, 78 per cent of the total land surface. England accounts for just over half the area (51 per cent) followed by Scotland (33.8 per cent), Wales (8.8 per cent) and Northern Ireland (5.7 per cent).

The MAFF Land Service and its Scotland and Northern Ireland equivalents distinguish between different land qualities using a set of physical criteria such as height, slope, climate, soil and drainage, and the extent to which these factors constrain agricultural use. The two highest grades, which can grow a wide variety of farm or horticultural crops, account for only 12 per cent of the UK's land (see Table 6.1). There is a wide middle band (36 per cent of the area) but just over half the total is in the two lowest grades with little potential beyond extensive sheep and cattle rearing. Poor land is a characteristic of those parts of the UK designated as Less Favoured Areas (LFAs, more recently termed Disadvantaged Areas); in these Areas farmers receive special treatment under the European

131

**Table 6.1**
**An Estimate of Agricultural Land by Grades, UK, 1976**

| A '000 hectares | England | Wales | Scotland | N. Ireland | UK | (England as % U.K.) |
|---|---|---|---|---|---|---|
| Grade 1 | 327 | 3 | 20 | – | 350 | (93) |
| 2 | 1 652 | 39 | 155 | 36 | 1 882 | (88) |
| 3 | 5 343 | 295 | 880 | 457 | 6 975 | (77) |
| 4 | 1 553 | 745 | 660 | 531 | 3 489 | (45) |
| 5 | 1 019 | 604 | 4 765 | 62 | 6 450 | (16) |
| **Total** | 9 894 | 1 686 | 6 480 | 1 086 | 19 146 | (52) |

| B Percentages | England | Wales | Scotland | N. Ireland | UK |
|---|---|---|---|---|---|
| Grade 1 | 3.3 | 0.2 | 0.3 | – | 1.8 |
| 2 | 16.7 | 2.3 | 2.4 | 3.3 | 9.8 |
| 3 | 54.0 | 17.5 | 13.6 | 42.0 | 36.4 |
| 4 | 15.7 | 44.2 | 10.2 | 49.0 | 18.2 |
| 5 | 10.3 | 35.8 | 73.5 | 5.7 | 33.7 |
| **Total** | 100.0 | 100.0 | 100.0 | 100.0 | 100.0 |

SOURCE   Agriculture EDC (1977) *Agriculture into the 1980s: land use* (London: National Economic Development Office).

Community's Common Agricultural Policy, being given higher rates of grant towards investment than are awarded to other farmers, and subsidies on production to compensate for the natural handicaps they face. In 1984, 9.8m.ha were officially classed as being in LFAs, about 53 per cent of the UK's utilisable agricultural area. There are marked regional differences of land quality. In Wales and Scotland over 80 per cent of agricultural land is in the poorest two grades, while in England the figure is only 26 per cent. At the other extreme, nearly half of the Eastern Region of England is Grade 1 or 2. This is reflected in significant differences between

England and the other countries in the farming systems, types and sizes of farms to be found there. As four-fifths of the UK Grade 1 and 2 land is found in England, one would rightly expect England to dominate arable crop production, especially horticultural types, with the other countries containing most of the extensive livestock production.

### 6.1.1 Land lost from agriculture

In the last twenty years much concern has been expressed about the amount of land which has been lost from farming. Precise information on the quantities involved is not readily available[1] and attempts to work with 'second-hand' data create problems; the UK's annual June Census of agriculture has a slightly varying response rate and coverage so changes in the total acreage it records do not necessarily imply transfers out of (or into) farming. Bearing in mind the problems of estimation, the concensus is that in the 1960s and early 1970s about 30 000 ha per year left agriculture, in crude terms less than 0.2 per cent per annum of the total agricultural land or 0.25 per cent of the crops and grass area. It would be wrong to conclude that all this land has become covered with concrete. About half (15 000 ha) went into urban use and within this fraction about one-fifth was used for reservours and mineral extraction. Another 4000 ha per year was used by forestry; in the 1960s more land went into forestry than to urban use.[2] A further 10 000 ha per year was difficult to trace confidently.

Contrary to popular belief, the annual amount transferred to urban use has been declining since the 1930s, and averaged only 9500 ha per year in the period 1975–80 (see Table 6.2). This containment reflects the effectiveness of planning control after 1947 and, in the most recent years, the depressed state of the UK economy. In terms of the quality of land lost, there seems to have been no great bias towards the better land having been built over. Agriculture occupies about four-fifths of the total UK land area, and some 90 per cent of Grade 1 land and 85 per cent of Grade 2 is still in agricultural use. Since the Second World War, the most highly productive agricultural areas, like East Anglia, have been relatively untouched by urban encroachment. However, in highland zones in Scotland development has shown a marked preference for better land, presumably because existing settlements are situated in favourable farming localities.

Current rates of land loss from agriculture do not pose any significant threat to the overall level of food output. Increases in productivity mean that output per hectare has been rising by about 3 per cent per annum, a figure at least ten times greater than the annual land loss; total farm output has thus been increasing despite using less land.

Much of the land lost has been of low agricultural quality; afforestation has been largely on land of Grades 4 and 5. Even better quality land which

**Table 6.2**
**Annual Average Transfers of Agricultural Land to Urban Use in England and Wales for Stated Periods**

| Years | '000 hectares |
|-------|---------------|
| 1922–26 | 9.1 |
| 1926–31 | 21.1 |
| 1931–36 | 25.1 |
| 1936–39 | 25.1 |
| 1939–45 | 5.3 |
| 1945–50 | 17.5 |
| 1950–55 | 15.5 |
| 1955–60 | 14.0 |
| 1960–65 | 15.3 |
| 1965–70 | 16.8 |
| 1970–75 | 14.9 |
| 1975–80 | 9.3 |

SOURCE   Best, R. (1983) 'Urban growth and agriculture', *British Association for the Advancement of Science*, Annual Meeting.

is clearly marked for a change in use, e.g., farmland adjacent to towns for which plans are in hand or likely, is frequently under-utilised or idle, so its transfer from agriculture does not imply much loss in production. On the positive side, land on which houses are built can frequently still produce much food from gardens[3] and land 'lost' to purposes such as open-cast mining can be reclaimed and eventually re-enter agriculture.

If land loss is rarely simple or absolute, neither should land which is retained in agriculture be regarded as available with no constraints. Farmers in water catchment areas, Sites of Special Scientific Interest, or on land owned by the Ministry of Defence may be restricted in their cropping patterns. In these sorts of areas the multiple-use characteristic of land becomes of practical importance, and interests of non-farmers have to be recognised.

## 6.1.2 *Alternative measurements and evaluations*

The official land classification described above is based essentially on land's potential flexibility for farming uses rather than what it actually produces; the latter would seem to be the more relevant in assessing the implication of land-using projects such as the siting of a new airport. Productivity is also affected by less permanent characteristics ignored in the classification, such as the amount of fixed equipment on the land (farm buildings, fences etc.), the standards of farm management and the size of farms. Changes in farming technology will allow some land to become more productive – powerful machinery permits heavy clay lands to be used for arable crops, and these may yield more heavily than lighter land – changing its value to farmers. Better transport raises the value of previously inaccessible land and possibly changes the types of crops that farmers will grow and their degree of specialisation. Land good for one type of high value crop in not necessarily good for other high value ones. In short, it is difficult to rank land from good to bad from a farming viewpoint in any absolute linear way.

But criteria for valuing land other than by its qualities for farming exist and are of increasing interest. To a tourist, the rugged hill land of Exmoor or Scotland may be excellent to look at and to tramp through, but agriculturally it is of very low productive potential. What is good for one purpose may or may not imply 'goodness' for another. Ecologists and their pressure groups have made the rest of society acutely conscious that areas such as the Somerset levels are uniquely valuable in the biological sense, although concern about their survival has only become popular within the last decade.

Land can have value in terms of potential urban uses. Worries about the loss of farm land have a longer history than has concern over wildlife habitat destruction, and since the Second World War a system of development planning has ensured that houses, roads and factories are not constructed in locations that society feels to be inappropriate.[4] Planning control has helped create a large price differential between 'agricultural value' and 'development value' which dwarfs the margin between the market values of land of high and low farming quality and makes such factors irrelevant in determining whether the owner is willing to sell his land for development (although such qualities may be taken into account by planning committees). In urban use the land, because of its location, could well be of much greater value to society than if it remained in agriculture. This, coupled with the fact that even in agriculture land productivity is influenced by factors other than physical characteristics, should cause us to question the whole basis by which we measure land quality and the assumption that our stock of land is fixed or finite.

The assumption that our economy is working with a finite stock of land resource is a mistake; it grows out of too simplistic a definition. The scientific community which produces advances in farming technology is devoted to the process of creating new resources, but these resources are not physical quantities. In a real sense land is being "created" in the laboratories, and the economic and social systems make this new resource available to producers. Indeed, some argue that science functions to create the resources of the next generation. Those who concern themselves that the nation has only a fixed supply of agricultural land, which in area terms is constantly being eroded by urban development, are trapped in a belief in a fixed technology, which can be interpreted as an arrogant assumption that this generation has made the last contribution to the definition of land resources – that technological advance stops here.

The conclusion is, therefore, that measures of area and land classifications are only partial parameters of resources, but carry the danger that they appear to be absolute measures. In essence, land resources cannot be expressed satisfactorily in physical dimension and physical characteristics alone because resources can only be defined in economic terms – scarcity with respect to needs under conditions which are not unchanging. The recently awakened awareness in much of society of the aesthetics of countryside appearance and ecological diversity leads to just one more manifestation of this phenomenon; without the appreciation of these characteristics it would not be possible to label Areas of Outstanding Natural Beauty or Sites of Special Scientific Interest and to classify them as worthy of protection. The concept of resources, such as land, is ultimately a social invention and a cultural creation. 'Their existence rests on need, desire, perception, and concepts of usefulness.'[5] Resources are fundamentally economic entities, and they cannot be interpreted or defined or the stock of them measured in the absence of a human intellect. Historical studies find that the stock of land and other productive assets, such as labour skills and capital assets, has been continuously redefined as succeeding generations, with different technologies, attempt to evaluate their own stock of resources.

### 6.1.2 Rights over land

Land, even that used for farming and forestry, is not a resource which is exclusively used by farmers and foresters. It is looked at by tourists, and so is an 'input' for the tourist industry; any number of people may view a field from a distance without harming its farming productivity or each other's enjoyment. The farmer would find it impossible to prevent a field being used in this way – viewers are 'non-excludable' – and, indeed, he would have no incentive to try to erect walls or fences because simply looking does not take away anything of substance from the farmer.

But multiple use goes further than merely looking on passively. Walkers and riders frequently have rights of access across land which is privately owned. In law this is quite feasible since ownership of land is really the possession of a set of rights, and the owner of a piece of land may have the right to use the land for farming or to let it to others, or to pass it to his descendants, but he does not possess *all* the rights associated with that land. For example, he may not own the right to extract coal or oil which may be under the land, and certainly he will not have the right to change the land from agricultural use to housing (development rights); these are now nationalised and he has to seek public planning permission if he wishes to build on land. Similarly, there may be public rights of access. If walkers wish to exercise their rights, the farmer may suffer some inconvenience. In this instance the uses compete in some way with each other and each party – the walkers and the farmer – will try to discourage the other from exercising their rights. Some uses may benefit each other; a case could be made that fox hunting benefits the farmer by controlling vermin and at the same time is enjoyed by huntsmen.

Since the Second World War attention has been focused increasingly on the relationship between farming and the 'natural environment', with modern practices decried on the grounds that they encroach on the rights of others, costing society dear in terms of polluted water courses, denuded hedgerows and reduced wildlife. As a result society has taken steps to protect rights to enjoy what it considers public assets (flora, fauna and natural features of the landscape) against further depletion, and if possible to reverse the process. Society has formalised rights, which it thought it always had enjoyed but which were under present threat, in a variety of ways; for example, pollution regulations involve legal sanctions, and the establishment of Sites of Special Scientific Interest, National Parks and Areas of Outstanding Natural Beauty, all attempt to promote the rights of society to determine what land is used for. Because some observers feel that the nature of the countryside is being too radically and irreversibly altered by farming practices, proposals have been made for removing the rights of farmers to change the use of land, especially old pasture, by extending the formal planning procedures which currently apply to housing and other development to include major agricultural changes.[6]

## 6.2 The Ownership of Agricultural Land

Who owns land is a matter of great importance because owners are in a position to control or influence many of the important changes occuring in the farming industry, earning land ownership the label of the hidden force within agriculture. In broad terms ownership determines which persons can become farmers and thus who can derive an income from the use of land. In an industry in which land plays such a significant role as a factor of production, and where, as will be shown later, land has become relatively more expensive, entry to the industry is severely restricted to those who are children of farming families. Even where land is not owned but rented, the landlord (usually) is in a position to select his tenant, who will frequently be already a farmer and come from a farming family. Farmers are not necessarily the people who would be the most capable at farming but rather the people who are in the most fortunate position with respect to access to land.

Land ownership affects the rate of structural change in agriculture, especially farm enlargement. Most farmers want to increase their areas – expansion seems almost to be a fundamental behavioural pattern and is seen among both small farms, where more area might be a pressing business necessity, and large farms, where it is not. But the ability of either to expand is severely constrained by the lack of available land to purchase or rent. Many farmers who wish to reduce their area will retain hectares under low levels of utilisation because they may be reluctant to lose control of it (as would happen if it were let under existing tenure legislation) and because for most of the postwar period it has been an attractive investment.

Agricultural land is a store of wealth and, in addition, has bestowed upon its owners real capital gains, i.e. it has been a 'growth stock' and its value has generally risen faster than the rate of inflation – at least since the Second World War. Rising land values also mean that landowners (including owner-occupier farmers) have become among the wealthiest sectors of society, frequently in an unplanned way. They thus find themselves the subject of capital taxes, and engage a sizeable group of lawyers and accountants to plan ways of minimising tax liability. (Tax legislation is described in the Appendix to this chapter.) Because of concessions for owners of agricultural land, given in large part to offset the potential effect of capital taxes on family farms, its purchase has been an attractive way of minimising capital taxation both to outsiders and to high income farmers. The lower rates of tax applied to capital gains than to current income have reinforced its attractiveness; businessmen search for ways of switching income from current to capital gain forms and investment in agricultural land has filled their requirements admirably.

Finally, land ownership has a strong link with the rural environment, both natural in the sense that woodland hedges and ponds can be destroyed or created only by and with the agreement of landowners, and also with the social environment. In rural areas, land means power and, frequently, prestige and social influence. In the past, landed estates exerted influence over rural housing, the non-agricultural economy, and the appointment of clergy, magistrates and Members of Parliament, and more than a vestige of those proprietorial rights remain.[7] Ownership generally implies control over how the land is used, and is particularly responsible for influencing, within the constraints of planning control, the rate at which land moves from agricultural to urban uses.

### 6.2.1 Who owns Britain's agricultural land?

Although land ownership is so important to not only agriculture viewed narrowly but also to the whole of the economic, social, political and environmental fabric, information on who owns the land of Britain is regrettably scarce. There is no comprehensive register of who owns agricultural land[8] and, among private owners, knowledge is extremely sketchy. Consequently we lack much of the detail, such as the size distribution of estates owned by individuals or family trusts, which is vital to the use of the data for policy purposes, for example when assessing the likely impact of capital taxation on landowners or of changes in ownership on land improvements (drainage or new buildings) which carry implications for the environment.

Knowledge has not always been so scant. In the Victorian period of prosperous farming, landowners were stung by criticisms from Liberal reformists that much of Britain was owned by a handful of families. Lord Derby was one landowner who advocated a comprehensive survey in order to disprove the accusation, and the results were given in Parliamentary Papers of 1874–6 as the Return of Owners of Land in England and Wales (1874), Ireland (1876) and Scotland (1879). Popularly known as the New Domesday Book, these Returns were the basis of statistical collation and analysis[9] which revealed the opposite from the aristocratic landowners' assertions, much to their embarrassment. Ownership *was* heavily concentrated in a few hands: excluding land holdings of less than one acre, a quarter of the whole territory was held by only 1200 persons with an average of 16 200 acres each; another quarter was held by 6200 persons with an average of 3150; another quarter by 50 770 persons averaging 380 acres and the remaining quarter by 261 830 persons averaging 70 acres each. Landowners were categorised according to the sizes of their estates: aristocrats held 10 000 acres or more, gentry 1000–10 000 acres and yeomen 100–1000 acres. These terms have more lately been used to describe social class and origins rather than in their size-of-estate meaning. Concentration of ownership varied widely; in some areas (Rutland, Northumberland, Nottinghamshire, Dorset, Wiltshire and Chesire) 'aristocrats' controlled over a third of the total area, whereas they were least important (10 per cent or less of the area) im Midlesex, Essex and Surrey. Despite the increased awareness of inequalities of wealth distribution in the UK, and the introduction of capital taxes designed, at least in part, to achieve a

**Table 6.3**
**Ownership of Agricultural Land, GB, 1978**

| Owners | '000 ha | % of agricultural land, Great Britain |
|---|---|---|
| Private individuals, companies and trusts: | | |
| domestic | 16 000 | 90.3 |
| (of which held by foreign nationals) | (200–300) | (1.0) |
| Financial institutions | 215 | 1.2 |
| Traditional institutions: | (1 515) | (8.5) |
| Central government | 462 | 2.6 |
| Local authorities | 365 | 2.1 |
| Nationalised industries, etc. | 225 | 1.3 |
| Other | 464 | 2.6 |
| **Total** | 17 730 | 100.0 |

SOURCE  Northfield (1979). *Report of the Committee of Inquiry with the Acquisition and Occupancy of Agricultural Land*, Cmnd 7599 (London: HMSO). A more detailed breakdown of institutional ownership is given in Burrell, A. Hill, B. and Medland, J. (1984) *Statistical Handbook of UK Agriculture* (London: Macmillan).

redistribution, no systematic attempt has been made during the twentieth century to achieve a coverage of information on the scale of the New Domesday. In the words of the American economist, Raup, 'We are blind where we need to see most'.

In the UK, only the broadest indications of the ownership pattern is available (see Table 6.3). In the late 1970s in Great Britain, private individuals, private trusts and companies together owned 90.3 per cent of the agricultural area, with overseas nationals holding just over 1 per cent. Of the remainder (9.7 per cent), most was held by a wide variety of institutions broadly describable as public, such as central or local government, or semi-public, such as religious institutions or educational charities. Some of these, notable the Crown, the Church and colleges of Oxford and Cambridge universities, have been owners of agricultural land for many hundreds of years; the overwhelming majority of their land is not farmed directly by them but is let to tenant farmers, and the net rents (i.e., after the costs of estate administration) form part of the institution's investment income – in the case of the Church of England to support the incomes of clergy. Other institutions, such as central government departments (e.g., the Ministry of Defence), the National Coal Board and regional water authorities, hold agricultural land because they need to do so in order to carry out their main operations, and again nearly all of it is let. Local government in England and Wales owns land as the result of past government policy for the creation of small-holdings; in Scotland small-holdings are the responsibility of the Secretary of State for Scotland.

A further type of owner is represented by the financial institutions, a group of private corporate bodies consisting primarily of insurance companies, pension funds and closely related organisations. This section of landowner received much attention during the later 1970s, engendering the setting up by the government of the Committee of Inquiry into the Acquisition and Ownership of Agricultural Land, because of these rapidity with which their landholding had been acquired. In total they owned only 1.2 per cent of the agricultural area of Great Britain in 1979 (1.9 per cent of the area of crops and grass), thought to have increased to about 2.0 per cent by 1982;[10] since then no significant increase has occurred. With the growth in recent years of the number and size of superannuation schemes there has been a steady flow of money seeking investment and fund managers have seen land as an asset with long-term growth prospects which matched their long-term liabilities. Land is seen as one form of a range of possible investments and of relatively minor importance in their portfolios. Most of their land is let to tenant farmers and the rewards to their landownership come in the form of rental income and in the rise of land values; as will be seen later, both rents and values have a history of rising at a rate faster than inflation. However, the decline in the real price of land during the 1980s curtailed the interest of financial institutions in further land ownership.

## 6.3 The Use of Land for Farming – Land Tenure

It is of course not necessary for a person to own land to be a farmer, nor do people who own agricultural land need to farm it themselves, although in the European Community most of the land is farmed by its owners, that is, they are owner-occupiers.[11]

Throughout the developed world a variety of legal arrangements has been devised to enable land ownership and farming to be carried on by separate people or legal entities; principally these are *share cropping*, involving a division of the crop according to some agreed fraction or of the sharing of the proceeds of its sale or a share of the profits through partnerships, or *cash tenancy*, in which the tenant pays the landlord an agreed sum which is known in advance and does not vary with any short-term fluctuations in the prosperity of the farming activity. The landowner, in return for giving up his right to occupy and use the land, at least for a time, receives compensation from the tenant farmer, who thereby has the opportunity to use a productive asset without the need to purchase it, with the heavy financial outlay that might be involved.

### 6.3.1 Land tenure and farm tenure in the United Kingdom

In the UK two forms of land tenure predominate, owner-occupation and cash tenancy, also called renting. (Land occupied by a tenant is also termed 'let land'.) In addition, there are arrangements in Great Britain for the short-term letting of grazing (0.25 million acres or 2.5 per cent of the England and Wales area) and other minor tenure forms exist, such as conacre in Northern Ireland (a system of annual letting) and crofting in the Highlands of Scotland, which has local importance. Here discussion will be restricted to the broad tenancy picture, although even in such general terms considerable confusion can be found.[12] Northern Ireland is now virtually all owner-occupied (although conacre exists there) following a series of Land Purchase Acts (the last in 1935).

The United Kingdom is generally regarded as the country in which the division of functional responsibility between the owners of land and those engaged in the activity of farming saw its apogee. In quantitative terms, a long process of accretion of land into rented estates in the eighteenth and nineteenth centuries culminated in 89 per cent of the holdings in Great Britain being rented or mainly rented at the outbreak of the First World War. Later advocates of the landlord and tenant system look back at the Victorian period and cite how well the arrangement coped both with the period of high farming in the 19th century's third quarter and with the agricultural depression which lasted from the late 1870s, with periods of short relief, up to the First World War. In the prosperous period high farm profits led to higher rents; new buildings were provided and, stimulated by cheap loans from the government, landlords undertook extensive drainage schemes. During the subsequent agricultural depression landlords accepted part of the burden in the form of reduced rents or remissions. Some, with financial resources outside agriculture, continued to spend heavily on fixed equipment to facilitate a switch by farmers to livestock production as a reaction to the collapse in cereal prices and as a way of constraining the fall in rents.

Landlords who wished to sell were thwarted by falling land prices and the adverse economic conditions, although their resolve to sell can only have been strengthened by Lloyd George's 1908 Budget which contained taxes aimed at breaking up the concentration of land ownership. A short-lived agricultural boom occurred after the First World War which encouraged land buyers, especially sitting tenants. Landowners were willing to sell, and in the period 1918–22 one quarter of the area of England changed hands. The result was that the proportion of rented land shrank from about 90 per cent at the outbreak of the First World War to 64 per cent in 1927.

The second great period of agricultural depression occurred from 1927 onwards and this time it was accompanied by a general economic depression. These circumstances, and the state control of farming during the Second World War, effectively prevented much further reduction in the proportion of let land. However, the decline continued after the war, encouraged by legislation which provided a high degree of security of tenure for tenants. This was thought necessary for an urgently required increase in national food output, but it also prevented agricultural rents from rising in response to higher farm profits (resulting from product price support) and created a price premium for land with vacant possession. In 1976 legislation extended to the family of tenants in England and Wales the right to succeed to the tenancy, effectively removing the landlord's ability to gain possession of his land for up to three generations; this increased the likelihood of landlords selling any let land which might come to hand with vacant possession, or of farming it themselves either directly or in partnership with an experienced farmer. While this right has now been modified (see Appendix to this chapter which describes landlord tenant legislation and capital taxation) and the disincentive to letting in the form of higher income tax rates previously applied to rents as 'unearned income' removed, the tendency is still for landlords not to relet, continuing the slow decline in the area of tenanted land (see Figure 6.1).

### 6.3.2 Statistics on tenure

According to official statistics, in 1985, 39 per cent of the agricultural total area of England and Wales was rented.[13] This reflects a continuing decline from the 62 per cent of 1950, 51 per cent of 1960, and 47 per cent of 1970. However, an important distinction must be drawn between *land* tenure and *farm* tenure, since farming businesses frequently hold both owner-occupied and rented land. Three farm tenure categories must therefore be distinguished – wholly-owned, wholly-rented and mixed-tenure farms. The decline in the relative importance of the rented sector has been accompanied not only by a rise in owner-occupation, but also, since the 1950s, by a rise in importance of mixed tenure farms, especially among holdings of 200 ha and over (Figure 6.2). Since 1975 there seems to have been some retreat of mixed tenure and advance by wholly owned holdings of 120 ha and over. Circumstantial evidence points to mixing of tenures (owners taking on rented land or tenants buying, the latter being more common in recent years) as having been an important but ill-documented way in which farms have grown in size in the postwar period.

There is an association of tenure pattern with farm size. In Great Britain

144

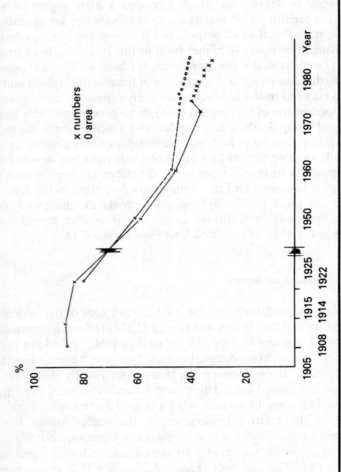

FIGURE 6.1
Proportion of Holdings and Total Area in Rented and Mainly Rented Holdings, Great Britain

SOURCE   MAFF (1968) A Century of Agricultural Statistics (London: HMSO) and Annual Review of Agriculture, various years (London: HMSO).

owner-occupation is predominant among small farms – of those of under 2 ha, four holdings out of five were owned or mainly owned in 1981 – but rented holdings constitute almost half the total number of holdings above 200 ha. However, there is cause to treat these figures – indeed any tenure figures – with caution. A fundamental problem with statistics on tenure is that the categories in the classification, at least those used in England and Wales, are far too coarse to differentiate between the many varieties of landholding arrangements found. These arrangements mainly take place within farming families and are used especially to achieve the satisfactory transfer of land (and other assets) between generations by minimising liability to capital taxes. Frequently they involve the creation of a tenancy within a family which disguises a situation which is close to owner-occupation, for example when a father owns land but forms a farming partnership with his son and the partnership then rents the land from the father.[14] As a consequence of these arrangements, frequently of baroque complexity, it is difficult to be precise about the extent of renting in British agriculture, but it seems that about a third of the land is let commercially, the rest being farmed by the owner or his family directly, or through managers, partnerships or farming companies.[15]

### 6.3.3   Who are the landlords?

The popular image of the agricultural landlord is that of a member of the aristocracy whose family have been Lords of the Manor for countless generations and who wield influence, usually paternalistically and benevolently, over all types of rural matters. More recently the (erroneous) impression could have been gained from the farming press that these private landlords had been largely displaced by pension funds and other 'city money'. Fragmentary information, however, leads to the inescapable conclusion that, despite the recent attention given to financial institutions, the overwhelming majority of let land (at least two-thirds and probably nearer four-fifths) is still in the hands of private owners.[16] While a detailed inventory has been built up of the holdings of public and semi-public

**FIGURE 6.2**
**Distribution of Numbers of Holdings and Area by Tenure, England and Wales**

(a) Numbers

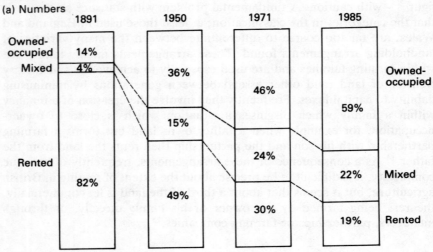

(b) Area

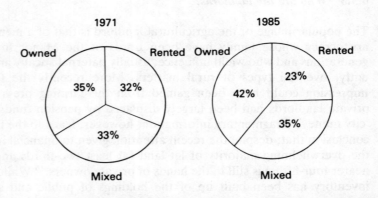

SOURCE   MAFF (1968) *A Century of Agricultural Statistics* (London: HMSO); MAFF (various years) *Agricultural Statistics for England and Wales* AH series (Guildford; MAFF).

bodies (central and local government, the Church, charities, etc.) very little is known of that much larger portion of tenanted land which is held privately. Two points, however, are evident. First, there is a concentration in the hands of a small number of individuals or family interests. Many landowning families have proved remarkably resilient to capital taxation, under professional advice using the various reliefs, exclusions and abatements permitted in the legislation to great effect. Second, a substantial portion of privately-owned tenanted land, at least a third and possible much more, is held in the form of trusts rather than in individual ownership. Trust are legal arrangements which bypass capital taxation and facilitate the handing down of estates complete (or with minimum diminution) between generations at some cost in terms of loss of direct control.

Trusts reflect the complex nature of landownership in Britain and the difficulty of defining categories and describing patterns which are interpretable in terms of policy. The establishment of some improved form of land registry (as recommended by the Northfield Committee Report in 1979) would at least be a start in throwing light on this major lacuna in agricultural information.

### 6.3.4 Economic implications of tenure

Owner occupation has grown to a dominant position in UK agriculture despite a persistent defence of the landlord and tenant system by well-connected elements of the agricultural establishment. Each tenure system has advantages and disadvantages in economic terms, but the debate also carries strong political and social overtones. Arguments for the protection of the landlord and tenant system have largely been conducted in terms of farming advantage at the individual holding or industry level, augmented by predictions of the likely social and environmental consequences which might ensue from the disappearance of (especially private) landowners – the disruption of established social patterns, a shrinkage in the availability of let accommodation for the low-income rural worker with the threat of extra housing demand falling on local authorities, fewer rural jobs and its social consequences, and a reduced concern with countryside conservation. Such arguments have succeeded in winning for landowners a number of important tax concessions.

The economic case for the landlord and tenant system rests on the supposed greater efficiency resulting from separating the business of farming from that of estate management, the easier provision of capital when the burden is shared, the greater flexibility with which farm sizes can be adjusted to suit the abilities of farmers and to accommodate new technology, and the ability of landlords to recruit as tenants able talent from a wide range of backgrounds so that entry to farming is not confined to the children of existing land-owning farmers.[17]

Evidence points to a far more complex situation occurring in reality. Tenants are not clearly more efficient than owner-occupiers except in certain types of farming in certain size groups (e.g., small specialist dairy farms).[18] Rather, it is mixed-tenure farms, especially those on which the farmer rents most but not all his land, which appear generally to be the most successful, notably among cropping and capital intensive types and medium sized farms;[19] however, large scale arable and mixed farming seems more efficient under owner-occupation. The association between efficiency, type of farming and size is complex, and no strong support is available on grounds of efficiency for the superiority of one form of land tenure or the other in anything like the simple way the respective advocates might wish.

Neither is there clear evidence that the landlord and tenant system is superior to owner-occupation in its provision of capital; rather than having more capital, tenants as a group have less. Owner-occupiers in England and Scotland have been found to be heavier investors in buildings and works than landlords and tenants combined (in the order of 2:1). Paradoxically, although having less fixed capital, as a system the landlord and tenant could be better at assessing the desirability of additional fixed capital since some owner-occupiers seem to over-invest in buildings which do little to enhance their farms' productivity, but the case is not proven.

Neither is there conclusive evidence on the greater flexibility of tenancy in responding to structural adjustment. In the late 1960s in England more rented land than owned land was changing hands, a surprising finding at a time when the proportion of rented land in the country was declining.[20] But much of this tenanted land would appear to have gone to expanding owner-occupiers (who then became mixed) rather than being retained in the wholly-rented sector. More recently, farm growth seems to have been mainly through purchase.

Perhaps the most socially attractive feature of the landlord and tenant system is the potentially easier entry it affords to those with farming ability but limited capital, especially from outside farming families. This is not of great practical importance under present conditions. The Northfield Committee (1979) conjectured that in recent years about 500 full-time holdings have been let each year to new entrants in England and Wales (somewhat more if Scotland is included). For an industry engaging some 224 000 full-time farmers, partners, directors and managers, the proportion represented by new entrant tenants is thus very small. And to start farming as an independent entrepreneur, finance is required. Even at the bottom end of the farm size spectrum – the county council smallholdings of which about 150 become available each year – Northfield reported that an entrant would require a minimum of £6000 to act as a borrowing base to stock a 32 ha holding. For a 40 ha dairy farm it would be £32 000 (1977 prices). Not only, then, are opportunities rare but the entry costs are high, effectively raising a barrier to tenant farming to all but those with successor rights or access to considerable capital. Under such circumstances any influence which landlords might have on the overall quality of tenant farmers by selection of the most able new tenants would seem to be very small.

## 6.4  The Use of Land for Farming – Cropping Pattern

The use which British agriculture makes of our countryside is not, for most areas, the product of deliberate and coordinated planning. Rather, it is the result of a myriad set of business decisions taken by individual farmers who, in turn, are responding within constraints – imposed by soil types, climate, location, other technical limitations, proximity to markets, personal preferences and objectives – to signals coming predominantly in the form of prices of farm products and costs of inputs. With few exceptions, farmers are free to use their land as they wish, and they could try to grow tulips or pineapples if this was their whim. Even those areas which have been designated as being of particular environmental interest (National Parks, Areas of Outstanding Natural Beauty, National Nature Reserves, Sites of Special Scientific Interest and similar categories, which together constitute less than one-fifth of the total area) there is little effective control over farming.[21] Owners of land are also normally free to increase its agricultural productivity, and the installation of new buildings, improved drainage, better fences and removal of impediments such as hedges and ponds can all be carried out with few restrictions. The freedom to erect concrete, steel and asbestos structures which can have major visual impact on the countryside finds a marked contrast in the planning control which an urban householder will meet if he tries to add a new porch to his house.[22]

The present pattern of farming land use (shown in Table 6.4) is thus a reflection of farmers' response to past economic conditions, and changes in the appearance of the countryside can largely be explained in terms of rational business behaviour as the economic and technical environment of farming alters. For example, the expansion in wheat acreage since UK entry to the EC is closely linked to the higher profitability of this crop, and of cereal farming in general compared with other types. If changes in the pattern of land use are occurring in the 1980s which are causing concern, we need generally look little further than the signals being given to farmers in order to explain them, although it is important also to know about the decision-making process at the farm level if farmer response is to be rationalised satisfactorily.

Agricultural land use is dominated by grass – 70 per cent of the total crop

**Table 6.4**
**Land Use by UK Agriculture**

| | Average 1971–3 ('000 ha) | Average 1982–4 ('000 ha) | Average 1982–4 (% of total area) | % change 1971/3–1982/4 |
|---|---|---|---|---|
| Wheat | 1 124 | 1 774 | 9.5 | +58 |
| Barley | 2 285 | 2 111 | 11.3 | −8 |
| Oats | 320 | 1 15 | 0.6 | −64 |
| All cereals | 3 791 | 4 015 | 21.4 | +6 |
| Potatoes | 231 | 195 | 1.0 | −16 |
| Sugar Beet | 192 | 201 | 1.1 | +5 |
| Oilseed Rape | 9 | 222 | 1.2 | +2 467 |
| Horticulture | 280 | 232 | 1.2 | −17 |
| Total tillage | 4 875 | 5 152 | 27.5 | +6 |
| Temporary grass (<5 years old) | 2 342 | 1 837 | 9.8 | −22 |
| Total arable | 7 217 | 6 989 | 37.2 | −3 |
| Permanent grass (>5 years old) | 4 977 | 5 109 | 27.2 | +3 |
| Total grass (excluding rough grazing) | 7 319 | 6 946 | 37.0 | −5 |
| Rough grazing (including common grazing) | 6 655 | 6 151 | 32.8 | −8 |
| Total crop area | 19 149 | 18 763 | 100 | +2 |

NOTE    The figures for 1984 are provisional.
SOURCE    MAFF, *Annual Review of Agriculture* (London: HMSO) (various years); there
have been some small definitional and coverage changes (see source).

area in 1982–4 – with just under half of that grass classified as 'rough grazing'. Rough grazing dominates in Scotland (67 per cent of its total agricultural area in 1982) and it is important in the North Region of England (30 per cent) and Wales (23 per cent). For England as a whole the figure is 8 per cent, with East Anglia and the West Midlands showing the lowest regional figures of 2 per cent. Of the remaining grass, a conventional division is made between 'permanent' and 'temporary' on the basis of whether the grassland has been ploughed within the last five years. 'Temporary' grass is grown in rotation with other crops, and so might more properly be classified under the 'arable' label. To avoid this problem the term *tillage* is used to mean arable land *other than* grass. The tillage area has varied with time, being particularly high during the Second World War when there was an urgent need to increase crop output. In England and Wales 27 per cent of the crops and grass area was tillage in 1938; by 1944 this had risen to 47 per cent. Although the proportion fell back after the war (in 1958 it was 37 per cent), by 1982 tillage was up again to 46 per cent of the crop and grass area, reflecting both changes in ways of feeding livestock and the encouragement of cropping rather than grass-fed livestock in the 1970s and 1980s under the Common Agricultural Policy.

Wheat and barley and other minor cereals together account for almost four-fifths (79 per cent) of the total tillage area. Even in East Anglia, a region well-suited to potatoes, sugar beet and horticultural crops, 69 per cent of the tillage area was taken by cereals in 1982. The wheat area has grown steadily in the postwar period, and rose by 58 per cent over the period 1971–3 to 1982–4, mainly in the latter years, reflecting in part the CAP's pricing policy. The biggest proportional expansions tend to have occurred in regions not thought of as traditional cereal areas, making the change particularly noticeable (the North, South West, Wales and Scotland). Barley expansion was notable over the 1950s and 1960s, but remained static over the 1970s, although in Scotland the barley area increased by 28 per cent between 1974 and 1982. Overall the cereal acreage rose by 6 per cent in the period 1971–3 to 1982–4 and total tillage also by 6 per cent.

Of the other tillage crops the most spectacular expansion has been by Oilseed Rape (more than a twenty-five-fold increase). Though it represents less than 1 per cent of the total crop area, the visual impact of its yellow flowers has gained it a reputation as a child fostered unwittingly by government support policy. Sugar beet showed a modest rise over the period 1971/3–1982/4, but this crop has a high degree of regional concentration: about half is in East Anglia, with a further 22 per cent in the East Midlands, a reflection of its climatic and soil requirements and the distribution of sugar processing factories. The potato area has followed a long-term decline and now occupies less than half the area it did in 1950.

## 6.5 Margins of Cultivation

It is a matter of UK history that in those periods when the prices of farm products have been high, such as in wartime and in the mid-nineteenth century, farmers tend to use land more intensively and to bring into cultivation land previously either not used, or used only for extensive livestock grazing. Conversely, when prices have fallen to low levels, as happened after the Napoleonic Wars, during the late Victorian depression and in the 1920s and 1930s, poorer land is left uncultivated and the better land is not cropped so heavily. In its most dramatic manifestation, fences tend to be put up on steep hillsides when farming is at its most profitable and such land is subsequently abandoned when farm product prices fall. This gives rise to the notion of the intensive and extensive margins of cultivation, both of which are seen to change with movements in farming prosperity.

The **intensive** margin of cultivation reflects, as the title implies, the intensity with which land is used. Increasing usage of variable inputs such as fertiliser will result in higher levels of crop yield but for technical reasons each additional unit of fertiliser will generate less additional crop response, that is, diminishing marginal returns will be experienced. A farmer who is aiming at maximum profit will keep increasing his application of fertiliser so long as the cost of the additional input (Marginal Factor Cost or MFC) is greater than the value of the extra crop yield (Marginal Value Product of MVP). At a given level of input costs and product prices he will be maximising his *total* profit if he uses that level of fertiliser where MFC = MVP. This level of fertiliser use represents the intensive margin of cultivation. If the price of the crop rises, say due to government price support policy, then the farmer will find it profitable to use more fertiliser, that is, the intensity of fertiliser usage which brings MFC and MVP into balance is greater. The intensive margin is said to have increased. A similar effect would have arisen if the cost of fertiliser was reduced by, say, a government subsidy. Sometimes the term 'intensive margin' is reserved for that intensity of land use found to be the most profitable *on the best land available*, but the notion of intensity changes resulting from cost or price movements applies to all qualities of land.

The **extensive** margin reflects the fact that not all land is of equal quality in terms of its capacity to grow crops; typically the yield which can be expected with a given quantity of seed, fertiliser and other resources will decline as the land is higher in altitude. At a given level of input costs and product prices, applying the rule of MFC = MVP to maximise profits to land of different quality will mean that the best land is used the most intensively (that is, it has the greatest quantities of fertiliser etc. applied to it per hectare), and generates the greatest profit margin for the farmer (margin being thought of as the difference between the total value per hectare of the crop minus the total costs of the variable inputs per hectare). Poorer land will be used less intensively and will generate less margin for the farmer.

If a farmer has a range of land qualities on his farm, as a hill farmer might, he would find that as he brought poorer and poorer land into cultivation a stage would come at which the poorest land generated such a small margin over its fertiliser and other variable costs that it was not worth the effort of managing. Land which was *just* worth farming would indicate the extensive margin of cultivation. If the prices of farm products rose, then land not quite worth cultivating under the previous conditions would become worthwhile to farm, and the extensive margin would be said to have extended. A similar result would have been brought about by a fall in the price of fertiliser or other input costs. Conversely, of course, a fall in farm product prices or a rise in input costs would cause the extensive margin to contract and land previously cultivated would be abandoned.

The way that government policy affects both the intensive and extensive margins of cultivation is of major concern to many outside the farming industry, such as fertiliser manufacturers and environmental interest groups. The latter point out that high product prices supported by Common Agricultural Policy measures, and to a less general extent but often more specifically by input subsidies, have extended both margins with environmental consequences. At the intensive margin, higher levels of fertilisers, herbicides and pesticides have been applied, much higher than an agriculture unmanipulated by government policy would have used, and this has been accompanied by higher levels of river pollution and the diminution of wild life. At the extensive margin, land which would not have been farmed or which would only have been used for very extensive systems, such as upland moors or lowland sandy heaths with cattle or sheep ranching, under the incentives offered by the CAP becomes worth considering for more intensive systems; the fences and fertilisers which accompany intensification change wild life and impede public access.

The notion that margins extend because of support policies can be applied to various agricultural conflicts. For instance, the combination of high product prices and subsidised drainage schemes make intensive arable farming of lowland marsh areas, such as the Somerset levels, attractive. They also help to explain the development of cereal growing on land of Grade 3 quality hitherto used mainly for permanent pasture, a less intensive farming system. When it comes to the practice of cultivating headlands and the removal of hedges there is a danger of confusing the implications of technical change, such as the development of large machinery which requires sizeable uncluttered areas to be effective, with the margin-extending effects of price support policy. But even here there is an indirect link, in that the pace of invention and development of machinery and chemicals is influenced by the prosperity of farmers who form the customers of machinery and chemical manufacturers. (See Chapter 11 for a discussion of technical change.)

It is tempting to assume that lowering farm prices or removing subsidies would restore many of the environmental features currently under pressure from the increased levels of output that support policies induce. This could be the long-term result, but in getting to that position there may be transition costs which could be harmful and irreversible. For example, one way in which farmers could respond in the short term to a reduction in their incomes resulting from lower product prices would be rapidly to remove hedges, to abandon conservation measures and to eliminate livestock enterprises which required hired labour and grassland in favour of all-arable systems workable by the farmer alone. To prevent such contra-intentional reactions, price reductions or subsidy removal aimed at the contraction of margins must be carefully planned and gradual.

### 6.5.1 Economic rent and contractual rent

One consequence of rising farming prosperity, when intensive and extensive margins increase, is a rise in the value of land. If let on a free market, the rents which farmers are willing to pay increase. Even during periods of steady incomes the owners of better quality land will command higher prices and rents. Elementary production economics shows that the margin left to a farmer from producing a crop is greater per hectare on good land than on poor land. Part of this will be the reward to the farmer for taking the risks and putting in management effort. If there is a ready market in land available for rent, farmers will compete against each other and bid up the rents until the residue which the farmer has on each quality of land, after paying the costs of production and the rent, is just enough to compensate him for the trouble involved. The amount paid to the landlord is termed **economic rent**; it is greater on good land and declines with the poorer land until, at the extensive margin of cultivation, economic rent is zero. The same explanation lies behind the fact that highly productive owner-occupied land commands a higher market price than poorer land.

In the real world rents paid by farmers consist of more than just economic rent. Land is rarely devoid of capital equipment (buildings, roads, fences, etc.) and part of the rent paid (termed **contractual rent**) is to cover the provision and maintenance of this equipment. Also the very process of administering rented land involves some costs which are passed on in the rent. Rents are also the subject of considerable legislation which has an impact on their levels. Of particular importance are the arbitration procedures used to settle disputes about rent levels between landlords and tenants and the criteria used by arbitrators. Very few rents go to arbitration (less than 1 in 200 in the late 1970s), but the levels set by arbitrators are yardsticks for the much more numerous rent revisions agreed between landlords and tenants. (See the Appendix to this chapter for a description of the legislation.) Prices of owner-occupied land similarly reflect the capital equipment found on it and other factors unrelated to its productive potential (as will be seen later). Nevertheless, the quality of the land (including its location) is reflected in rent levels and land prices, and rises in agricultural prosperity lead to higher rents and prices. But it is important to note the direction of causality; it is the increased farming prosperity which gives rise to the higher rents and land prices and not that higher rents cause farm product prices to rise.

## 6.6 The Market in Agricultural Land

### 6.6.1 Land as a non-wasting asset

Land, unlike machinery and buildings, is not an asset which wears out when used for agricultural production; it is a non-wasting asset. Indeed any system of production which does *not* preserve the ability of the soil to produce (by allowing its structure to deteriorate or by polluting it) is, according to one eminent agricultural economist, not really *farming* at all. Edgar Thomas believed that soil fertility 'is conserved under any system of farming provided such a system is technically capable of being continued indefinitely. Indeed a system of production from land which does not conserve soil fertility should rightly be regarded as belonging not to agriculture but to mining.'[23]

If the farming patterns as found in the UK (and Europe) can be regarded as the sort which could be continued indefinitely and are non-depletive of soils, and generally this would appear to be the case, then the prices which land fetches when sold on the open market, or the rents obtained, do not have a 'wearing out' factor built into them. Land will be valued by the contributions which it is felt that it can make to production this year and for succeeding years, augmented by other attributes which land may possess, such as the rights to enjoy country sports, domestic privacy or tax concessions which go with landownership.

### 6.6.2 The size of the land market

The land market exists to facilitate changes in land ownership. About 1.5 per cent of the agricultural land of the UK is sold annually. Sale is not the only method of ownership transfer, and inheritance plays a major part; in the late 1960s approaching half the land under owner-occupation in England had been inherited, possibly higher proportions in some regions, such as 60 per cent in East Anglia.[24] There are also sales between members of families at prices below prevailing open market levels and these transactions are hard to identify. Although the land market is very much a marginal one, with only a small fraction of the total land being sold each year in contrast with, say, potatoes where virtually all the commodity changes hands, nevertheless open-market land prices are of intimate concern to farmers, affecting the wealth and borrowing position of those who own land. Rising land prices may enable farmer-owners to borrow further to finance investment in working assets such as machinery and livestock. Prices influence farmers' ability to expand farm size, affecting in a complex way the ease of borrowing, the size of the sums which have to be raised to buy more land and the willingness of other farmers to release land to the market. Most of the land changing hands in the 1970s (various estimates suggest between 60 and 85 per cent) was bought by farmers enlarging their businesses. Private individuals dominate the market as buyers and sellers of both owner-occupied and let land, although in the smaller let market (one-sixth the size of the vacant possession market) financial institutions have been minor but significant purchasers, accounting for between a third and half the area purchased since 1977.[25] Land prices are seen by outsiders as a barometer of the prosperity of agriculture, a view for which there is some justification.

### 6.6.3 Land prices

A glance at Figure 6.3 will show that land prices have a history of fluctuation. Periods of farming prosperity have seen rising land values, such as during the Napoleonic Wars, when demand for home-produced food was high because of reduced foreign supply, and in the Victorian period of 'high farming' from about 1850 to the late 1870s when rising urban populations kept food demand high. On the other hand, farming depressions have seen falling land prices; the £130 per ha average recorded for England and Wales in the 1870s had fallen by more than 50 per cent by the turn of the century, and the market did not regain the £130 level until the 1940s. The late 1950s and 60s saw steadily rising prices, while the 1970s were characterised by quite violent price fluctuations, clouded by high rates of inflation, and in the early 1980s prices have declined. Figure 6.4 shows the movements over the 1970s and early 1980s in both nominal and real terms, that is after allowing for the falling value of money; choice of base year is somewhat arbitrary, but 1971 can be justified on the grounds that it preceeded UK entry to the European Community and the international boom in agricultural product prices of the early 1970s. Land prices, both vacant possession and tenanted, have clearly risen at rates faster than that of inflation (as indicated by the Retail Price Index); that is, owners have made real capital gains if they bought land in 1971 (or earlier) and

**FIGURE 6.3**
**The Cost of Farmland, 1790–1976**

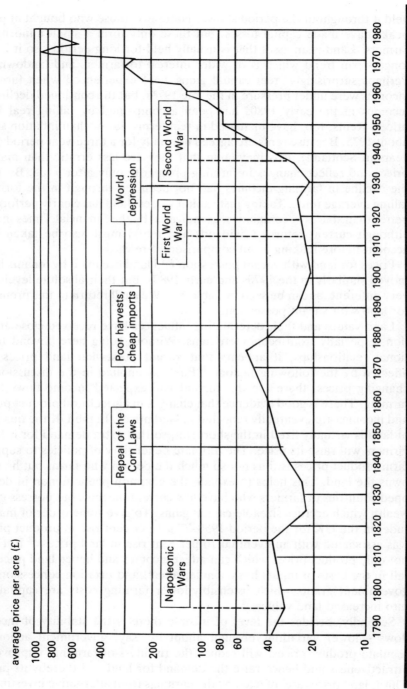

SOURCE  Newby, H. et al. (1978) *Property, Paternalism and Power: Class and Control in Rural England* (London: Hutchinson).

held it throughout the period shown. Naturally, those who bought at price peaks have made capital losses, but these only form a small minority of farmers. Land is an asset that is usually held for long periods, so it is the longer-term trend which is of wider interest to farmers and landowners. Perhaps surprisingly, real capital gains were experienced when farming incomes were under pressure in the late 1970s, but the continued decline in incomes in the early 1980s has been accompanied by falling real land prices. Rents, too, have increased in real terms, faster than inflation since about 1975. Because rents are agreed normally for a three-year period (five years in Scotland), it is inevitable that they are less erratic than market prices and reflect changes in farming prosperity only after a lag. By 1984 the decline in farming incomes had not been fed through in the form of falling average rents. Taking just capital values, land has clearly performed better than shares in companies, as indicated by the Financial Times index, although current returns (dividends and rents) must also be taken into account before making a full comparison of rewards.

Prices for land with vacant possession and land occupied by tenants have moved similarly in the 1970s and early 1980s, but their absolute levels are very different, as can be seen in Table 6.5. We shall return to this premium for land with vacant possession.

Land values and their determining influences have received close attention, especially from econometricians. Without going here beyond functional relationships, it appears that **vacant possession** land prices are affected by the following factors.[26] First, as implied in the discussion of changing prices, there are the current and expected income flows from farming. There is good evidence that changes in agricultural product prices and incomes are eventually reflected in land prices. In the UK the quantity of land is virtually fixed in the short run, so that extra demand for it from farmers will raise its price. The ultimate beneficiary of policies to support farm product prices is thus not so much the person who farms but he who owns the land. This helps to explain the common phenomenon in developed countries of farmers who have low current incomes but possess great wealth which accrues sizeable capital gains. To give some order of magnitude, in the UK for the period 1950–77 a 1 per cent rise in product prices was associated with an eventual 10 per cent rise in land prices.[27] But it is not only product prices which can affect incomes and hence land prices; a fall in the costs of inputs has a similar effect, and taxation concessions or government grants which ostensibly-lower farming costs are capitalised into increased land values.[28]

Secondly, besides the level of income there is the stability of income flows – lower variations brought about by, say, government action to regulate product prices, will reduce the riskiness of farming, improve its attractiveness and hence raise the demand for land and thereby its price. Third, land prices are affected by the earnings from alternative investments

# FIGURE 6.4A
## Land Values and Other Major Economic Indicators – Nominal Values

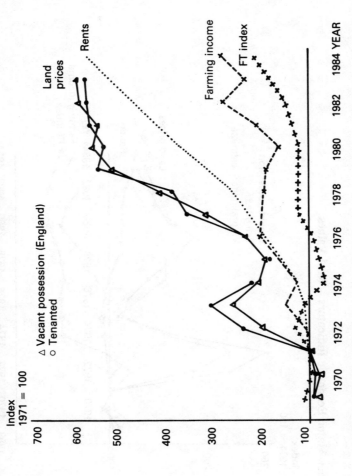

SOURCE   Land prices: MAFF/IR series.
Rents: MAFF rent inquiry.
Farming income: HMSO Annual Review of Agriculture.
FT index: HMSO Annual Abstract of Statistics.

**FIGURE 6.4B**
**Land Values and Other Major Economic Indicators – Real Values**

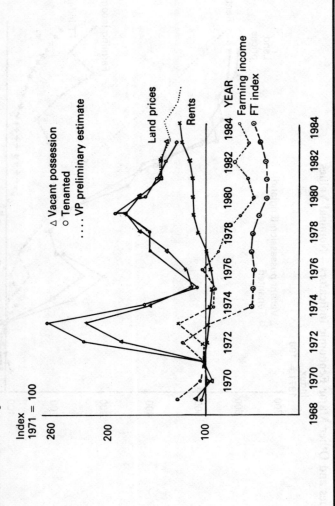

(present and future and their variability). For instance, if the earnings potential of industrial share looks uncertain, as was probably the case at the time of the UK's entry into the European Community, then some investors may switch to land, with repercussions for its price. Fourth, there is the speculative effect which can push up prices of land just like other forms of wealth (gold, antiques, art). To the extent that future price rises can be foreseen, rises will be anticipated by people buying early to gain from the rise, but this will cause demand to expand and prices to rise. Expectation can therefore be partly self-fulfilling; as long as confidence remains that prices will rise, then rise they will; the opposite, expectations of price falls will also tend to 'talk the market down'.

Fifth, land prices reflect in part differences in capital taxation between the way in which land and other assets are treated. It is common among developed countries for lower tax rates to be applied to land than to other wealth, concessions frequently introduced to ease the passing of land between generations of farming families. The outcome has been that non-farmers have been attracted into investment in land, with an effect on price which largely nullifies the lower tax rate.

Sixth, there is the level of interest rates. Lower borrowing rates make the purchase of land more attractive; at low rates the land price which can be justified by a farmer wishing, say, to buy to expand his farm's area is greater. As will be seen in Chapter 8, the way that interest is treated in calculating farmers' liability to income tax, combined with falls in the value of money resulting from inflation, has meant that borrowing for land purchase has been at very low real cost for much of the postwar period in the UK, and frequently the real interest rate has been negative. The implication is that farmers have been encouraged to buy land using credit, a tendency which many have followed with great ultimate benefit as long as their business has been able to bear the burden of interest payments in the

**Table 6.5**
**Average Land Prices in England[1]**

|  | Vacant possession (£/ha) | Index 1971=100 | Tenanted (£/ha) | Index 1971=100 |
|---|---|---|---|---|
| 1969 | 570 | 92 | 379 | 88 |
| 1970 | 526 | 85 | 372 | 86 |
| 1971 | 620 | 100 | 432 | 100 |
| 1972 | 1238 | 200 | 1050 | 243 |
| 1973 | 1643 | 264 | 1354 | 313 |
| 1974 | 1330 | 215 | 961 | 272 |
| 1975 | 1205 | 194 | 797 | 184 |
| 1976 | 1472 | 237 | 1019 | 236 |
| 1977 | 1994 | 322 | 1563 | 362 |
| 1978 | 2602 | 420 | 1687 | 391 |
| 1979 | 3227 | 520 | 2381 | 551 |
| 1980 | 3470 | 560 | 2336 | 541 |
| 1981 | 3418 | 551 | 2450 | 567 |
| 1982 | 3669 | 592 | 2490 | 576 |
| 1983 | 3789 | 600 | 2404 | 556 |

NOTE    1. MAFF/IR series. There is a delay between the date on which a sale is agreed and the date on which it is included in these series. The delay is thought to average 6–9 months. The figures shown in the table for each calendar year relate to sales included in the series in the years ending the following September.

SOURCE    MAFF Press Notices.

first few years following the land purchase. One factor contributing to the decline in agricultural land prices in the UK during the 1980s has been the exceptionally high real rate of interest that farmers have faced, itself in part a product of a lower rate of inflation.

Finally, land prices have risen to reflect the attitude which buyers and sellers have towards land as a source of political power, social prestige, domestic privacy and opportunity for controlling one's own environment. While the extra-economic aspects of land may have changed since the hey-day of the great estates, they remain in different forms. An increasing proportion of farms are part-time, probably a third but more so among small farms, and these families by and large regard the income-generating potential of their land as of secondary importance. High prices paid for small farms are in part a reflection of the demand coming from people with outside wealth and income wishing to enjoy the domestic and environmental benefits derivable from land.

### 6.6.3 The value of tenanted land

A closely similar list to the above could be built up of the determinants of tenanted land prices, the main difference being that the income stream comes not directly from farming but from the rent paid by tenants, although this of course is related to the profitability of farming. Land let to a tenant farmer can be viewed as an income-generating asset, and an investor will be willing to pay for land a sum dependent on the rent stream it yields. However, as with vacant possession land, taxation treatment and anticipated changes in capital values will be reflected in the price an investor is willing to pay. A specific case of this is when land is bought with a sitting tenant who seems likely to give up his tenancy in the near future, allowing the land to be sold again on the higher-priced vacant possession market.

To a potential purchaser such as a pension fund, tenanted agricultural land must be assessed by similar financial yardsticks as other investments. Land will not be bought if clearly better rewards are available elsewhere. Judged by average rents (after allowing for expenses) and average tenanted land prices, the overall yield in 1981 was somewhat less than 2 per cent, as calculated by the formula below,[29] although the rate of return on individual farms with recently-negotiated tenancies could be much higher

$$\text{Value of an income yielding asset} = \frac{\text{annual rent}}{\text{rate of return}}$$

To this current return must be added any rise in the capital value of the land. Taking both forms of return together, the rewards provided by tenanted land do not seem to have been greatly dissimilar from alternative investments of comparable risk and resistance to erosion by inflation; long-dated government index-linked securities were typically yielding 3–3.5 per cent in the early 1980s. However, it is difficult to be categorical about the average returns to owning tenanted land because of the uncertainty which surrounds land prices; the quantity of tenanted land being sold each year has been very small in the recent past, so that for some years it has not been possible to put much reliance on the average land-price figures as indicating general trends in market values.

### 6.6.4  The vacant possession premium

A glance back at Table 6.4 will show that there is a substantial difference between the price of land which is already occupied by a tenant (and so yields a rent to its owner as landlord) and that sold with vacant possession and which can therefore be farmed by its owner and yield him an income stream direct. Before the Second World War there was little difference between the prices of tenanted and owner-occupied land.[30] With postwar rises in agricultural prosperity the demand for land was strong, but rent arbitration legislation prevented much of a rise in rent levels, and hence the price of tenanted land was kept down. There were no such restricting influences on the price of vacant possession land, which responded to rises in farm profits. The price premium for vacant possession was over 50 per cent in the late 1940s and has varied considerably since, touching a low of 13 per cent in 1972 (England and Wales). Because of the more rapid response of vacant land prices, the premium has been seen as a barometer of farming's prosperity.

The existence of the premium for vacant possession also has held major implications for the UK's tenure pattern. Buying as a sitting tenant gives the opportunity for very large capital gains, since the tenant can immediately re-sell his land at a much higher price. If the tenant buys using a loan which is beyond his ability to service, by selling only a part of his newly acquired land the debt burden could probably be reduced to manageable proportions. Any farming advantages of remaining a tenant pale into insignificance when compared with the potential capital gain, so capital reasons alone could be expected to propel an expanding owner-occupation sector in British agriculture.


## 6.7  Conclusions

Agricultural land is far more than just a factor of agricultural production. While its extensive and intimate use is the feature which most distinguishes the farming industry, land has the ability to contribute to other economic activities at the same time as being used for farming, such as water catchment and tourism, and to generate utility for walkers, motorists, birdwatchers and a host of people who in diverse ways take enjoyment from the countryside. These non-farming attributes generally imply that land is in part a public good, and society is increasingly watchful for any threat which erodes its long-customary rights to enjoy the coutryside and what it contains. Formerly the threat was thought to come from urban growth; an unhindered price system has been shown to be unsatisfactory in handling the transfer of land from agricultural to urban use, and a planning system has been established to control such development. More recently farmers themselves and their new agricultural practices have been seen as a source of encroachment; there are moves to restrict the freedom of farmers to use land as they chose through some extension of the planning system.

It follows that the ownership of land involves much more than rights over a productive asset. Owners have power to shape the countryside, generally without the need to consult or heed other members of society. The market value of their land is essentially determined by the narrow considerations of farming and investment appraisal; others in society might well attach different values depending on the relative weights given to wildlife, visual appearance, accessibility and so on. Economics can help explain the nature of these non-farming characteristics of land and how they might be made to bear on the actions of farmers, but the decision on the balance of interest between farmers and other groups is essentially a political one.

Power, politics and land have a long association. Large landowners perhaps no longer dominate rural societies and economies in the way they once did, although the overwhelming majority of UK farmland is clearly still in private hands. Nevertheless land ownership is a major but largely unrecognised power influencing not only land use and countryside appearance, but also the rates of change in farm sizes and numbers, the distribution of benefit from rising land prices brought about largely through public support of farming, and, perhaps most significantly, who can become a farmer. Increasingly, entry to farming is being restricted to those whose families own or control land, a situation not of any deliberate design but brought about by the nature of land and the rights attached to it.

# Appendix: Major Legislation Affecting Agricultural Land

The position of agricultural land in the UK cannot be understood without taking into account the legal framework, since many changes in tenure, ownership and use have their genesis in changes in the law.

## 1 Landlord and tenant legislation

The major legislation under which tenancies of agricultural land are regulated is the 1948 Agricultural Holdings Act. Modifications have been introduced, notably to rent arbitration and to the rights of family members to succeed to a tenancy (see below) but the framework remains that of the 1948 Act. In non-technical terms, the legislation provides:

a. lifetime security for the tenant, subject to a range of conditions the most significant of which is the regular payment of rent.
b. freedom of cropping, within the 'rules of good husbandry'.
c. an arbitration procedure (before the Agricultural Land Tribunal) for disputes between landlord and tenant on matters related to the conditions laid down in the tenancy agreement (such as the responsibility for the repair of buildings).
d. compensation procedure at the end of a tenancy for tenants who have improved the farm, such as by investing in long-life buildings (normally with the agreement of the landlord) or for landlords if the tenant has breached part of the terms of the tenancy and lowered the value of the farm, such as by a failure to maintain ditches.
e. since the 1976 Agriculture (Miscellaneous Provisions) Act, for certain member of a tenant's family (child, spouse, brother or sister) to claim a new tenancy on the death of the tenant, providing certain tests of 'eligibility' are satisfied. This right of succession does not apply to tenancies created since the passing of the Agricultural Holdings Act of 1984, but leaves existing tenants unaffected; succession on the retirement of the tenant was made possible by this Act.

As might be expected, the 1948 Act resulted in some unforeseen imbalances within the industry, and the framework required some modification. The original procedure for rent arbitration caused disputed rents to be fixed on the basis of existing rents in the area: this resulted in rents being kept down during a period of farming prosperity. This in turn created a large premium for land with vacant possession and encouraged the sale by landlords primarily to sitting tenants, producing a contraction in the tenanted area. Modification to the arbitration procedure was made in the 1958 Agriculture Act, changing the arbitrator's yardstick to what had been paid for open market *new* lettings in the area, allowing rents to rise and redressing the balance to some extent between the interest of landlord and tenant. The 1976 legislation, which granted rights of succession to the families of existing tenants, thereby easing their plight if the tenant died, greatly reduced the number of farms coming available for renting by others and aggravated landlords' reluctance to re-let those farms which came to hand.

With very few farms being newly let on the free market from a willing landlord to a willing tenant, the reference point for arbitration in the case of sitting tenants was effectively removed; those few that were newly let tended to be unrepresentative and, it was felt, secured atypically high rents because of the scarcity of available farms and the tendency for new tenants to be taken by expanding established farmers able to outbid others because of a potential to spread fixed costs. The 1984 Act contained a new rent formula, based on the land's productivity, with instructions to arbitrators to disregard the 'scarcity element' in the rents being paid for comparable holdings.

From other quarters comes the suggestion that only radical departures from the established pattern of lifetime tenancies can breathe life into the rented land sector, with the calling for fixed-term tenancies of 10 or 20 years to give a balance of reasonable security for tenants coupled with some improvement in land mobility, but these suggestions have not been incorporated in the 1984 Act (for further discussion see Berkeley Hill[31]).

## Taxation

The other main legislation of special relevance to land is that on the taxation of capital, the two most important being tax levied on transfers of capital and tax on increases in the value of capital. Capital Transfer Tax (CTT) as the name implies, was levied on the death of a person when his assets were transferred to inheritors and on gifts made during lifetime. It replaced Estate Duty in 1975 and, unlike its predecessor, covered gifts made during lifetime. In the 1986 Budget, CTT was itself replaced by Inheritance Tax which reverted to the earlier system of only taxing capital passing at its owner's death (there are exceptions, however, such as gifts to trusts or involving companies); the revenue from lifetime capital transfers under CTT had always been small, and never more than 10 per cent of its total. Under all three taxes the rates have been progressive, that is higher rates are charged on larger amounts of capital. There were many reliefs and concessions under CTT, important among which was the basis of valuing agricultural land; for settling the rate to be paid land was valued at 50 per cent of its market price and the tenanted land at 70 per cent of its (lower) market price. Similar concessions are expected under Inheritance Tax.[32]

The range of concessions make the ownership of agricultural land attractive to persons who might otherwise hold shares in public companies or other forms of wealth, although the worst former excesses of 'death bed' purchases are prevented under present legislation – the emergency selling of other assets and buying land within days or hours of imminent death in order to reduce the amount of tax payable.

Capital Gains Tax (CGT) is a tax applied to increases in the market value of assets such as land normally on the change in nominal values but from 1983 in part calculated to only tax real increases (i.e., if land prices rise faster than inflation). Again there are important concessions such as exemption of large amounts of gain[33] for farmers retiring at 65 or over and selling or giving away their farms. The rate of tax on gains is 30 per cent; this is generally lower than the marginal rates which farmers would expect to pay on income, so there is an incentive to switch income from current to capital gain form. This might be done by a farmer buying extra land by borrowing and paying interest out of current income (reducing current spendable income, but by less than the amount of interest as this would be treated as a cost in calculating income tax); at the end of a given period the farmer would have an asset which history suggests will have risen in value at least as fast as the rate of inflation. If the farmer then sells, the gain will be taxed at 30 per cent; but if he does not dispose of his assets, or if he reinvests the proceeds in more land, payment of the tax is deferred until he does so, which may be never as this tax is no longer payable when the owner dies.

Some important concessions to taxes on both income and capital have been available if the landowner could be classed as a full-time working farmer, a situation which caused some landowners to seek legal arrangements (such as partnerships with tenant farmers) which give them this status. This was also a way of avoiding the legislation which gave rights of succession to the family of tenant farmers. Even now the existence of many concessions in the taxes on capital, and those applying to taxes on income too, have made tax planning a highly lucrative activity both for agriculturalists and tax accountants. Indeed, in the USA economists frequently recognise that in agriculture three sources of reward exist: 1) from farming; 2) from non-farm business activities that farmers engage in, and 3) from tax planning. Certainly, in the UK landownership, the rewards from agriculture, investment policy and incomes must all recognise the importance of taxation in determining present patterns and the changes which are occurring.


# Notes

1. EC (1981) *Factors influencing ownership, tenancy, mobility and use of farmland in the United Kingdom* Information on Agriculture No. 74 (Luxembourg).
2. Best, R. H.(1983) 'Urban growth and agriculture', *British Association for the Advancement of Science*, Annual Meeting.
3. Best, R. H. and Ward, J. T. *The Garden Controversy*, referred to in Best, R. H. (1981) *Land Use and Living Space* (London: Methuen).
4. There were earlier Acts, but the 1947 Town and Country Planning Act is usually regarded as marking the start of effective control: see Blunden, J. and Curry, N. (eds) (1985) *The Countryside Handbook* (London: Croom Helm/ Open University).
5. Raup, P. M. (1982) 'An agricultural critique of the National Agricultural Lands Study', *Land Economics* **58**:2, 260–74.
6. Shoard, M. (1980) *The Theft of the Countryside* (London: Temple Smith).
7. Newby, H. (1985) *Green and Pleasant Land?* (Aldershot: Gower).
8. The Land Registry in England and Wales is incapable of supplying useful information because of its incompleteness and the nature of the information collected; a broadly similar situation exists in Scotland. The Northfield Committee recommended that a more suitable and effective land registry be established. Northfield (1979), *Report of the Committee of Inquiry into the Aquisition and Occupancy of Agricultural Land* Cmnd. 7599 (London: HMSO).
9. Bateman, J. (1883) *The Great Landowners of Great Britain and Ireland* (4th edn) (London: Harrison) reprinted 1971 (Leicester, Leicester University Press).
10. Northfield (1979) op. cit., and Jones Lang Wootton (1983) *The Agricultural Land Market in Britain* (London: Jones Lang Wootton).
11. In the European Community, different histories, social conventions and legal institutions have produced a wide range in the degree of owner-occupation. In 1973 the percentage of land owner-occupied was Ireland (92), Denmark (90), West Germany (78), Italy (70), Luxembourg (65), UK (53), France (52), Netherlands (52), Belgium (29).
12. For a detailed description of these minor forms see EC (1981) op. cit.
13. The equivalent figure for the whole UK was 38 per cent.
14. Statistics in Scotland are gathered on these sorts of tenancies, termed 'less than arm's length'. About 8 per cent of all land there falls into this category.
15. In 1979. Northfield (1979) put the share of land let commercially at 35–40 per cent when the 'official' figure was 43 per cent.
16. Massey, P. and Catalano, A. (1978) *Capital and Land: Landownership by Capital in Great Britain* (London: Arnold).

17. Hill, Berkeley (1985) 'Farm Tenancy in the United Kingdom', *Agricultural Administration* **19**:4, 189–207.
18. Britton D. K. and Hill, Berkeley (1978) *Farm Tenure, Size and Efficiency* (Ashford: Wye College).
19. Gasson, R. and Hill, Berkeley (1984) *Farm Tenure and Performance* (Ashford: Wye College).
20. Harrison, A. (1975) *Farmers and farm businesses in England* Miscellaneous Studies No. 62 (University of Reading: Department of Agricultural Economics and Management).
21. National parks occupy some 9 per cent of the total area of England and Wales, AONB 10 per cent: the SSSI form 5.3 per cent of GB, but these sites are in part located in NPs and AONB. Taking into account other areas where restrictions apply (e.g. MOD land) it has been estimated that effective control applies to 17 per cent of GB (Green, B. H. (1986) 'Controlling ecosystems for amenity' in Bradshaw, A. D. Goode, D. A. and Thorpe, E. (eds) *Ecology and Design in Landscape* (Oxford: Blackwell).
22. Farm buildings need planning permission only if they exceed 12m. in height, or 465 sq. m. floor area or are to be sited within 25m. of a trunk or classified road (special controls apply in National Parks, AONB, SSSI, etc.).
23. Thomas, Edgar (1949) *An Introduction to Agricultural Economics* (London: Thomas Nelson).
24. Newby, H., Bell, C., Rose, D. and Saunders, P. (1978) *Property, Paternalism and Power: Class and Control in Rural England* (London: Hutchinson).
25. Burrell. A., Hill, Berkeley and Medland, J. (1984) *Statistical Handbook of UK Agriculture* (London: Macmillan).
26. Based on Higgins, J. (1979) *Irish Journal of Agricultural Economics and Rural Sociology* **7**:2, 127–148.
27. Trail, B. (1982) 'The effect of price support policies on agricultural investment, employment, farm incomes and land values in the UK', *J. Ag. Econ.* **33**: 3, 369–85.
28. Centre for Agricultural Strategy (1978) *Capital for Agriculture*, CAS Report No.3 (University of Reading).
29. This formula can also be used to calculate the maximum price an investor could afford to pay for a piece of land, given the amount of rent it earns and the rate of return he requires on his investment. Part of the rent would be absorbed by the costs of estate administration (maintenance of buildings, insurance, management and statutory charges). These costs are thought typically to consume between 30 and 50 per cent of the gross rent ADAS (1976) *Expenses of agricultural land ownership in England and Wales*, Technical Report 20/9 (Pinner: MAFF).
30. In 1937–9 the vacant possession price was £25 per acre and the tenanted land price was £23. Gross rents, at £1.2 per acre, represented a 5 per cent yield on tenant land.
31. Hill, B. (1985) op. cit.
32. For latest details of taxation consult the most recent issue of Nix, J., *Farm Management Pocketbook* (Ashford: Wye College).
33. For example, under the regulations applying in 1979 a farmer with a wife and wishing to pass his £200 000 farm with £50 000 worth of stock and machinery to his son upon death could find his estate paying at worst £108 750 of Capital Transfer Tax. Taking advantage of all concessions could reduce this to £5750 (EC, 1981, op. cit.).

# Resources in Agriculture: Labour 7

## 7.1 Numbers of People Engaged in Agriculture

In the United Kingdom about 700 000 people, a little less than 3 per cent of the working population, are currently directly engaged in agriculture. This figure refers to those who work on farms and horticultural holdings either full- or part-time.[1] Taking a broader view of what constitutes agriculture by including the service industries which depend directly on British farming (veterinary services, animal food suppliers, etc.) adds a further 246 000 (or 1 per cent of the working population). To go further and embrace the food distribution industry and tourism which are partly dependent on the activities of farms raises extreme problems of definition. Virtually all sectors of the economy are interrelated, and segmentation is not always practical for descriptive purposes nor necessary in explaining changes. Nevertheless, of the 600 000 engaged in food processing (March 1980), food distribution and retailing (187 000), machinery manufacture (60 000) and chemicals (12 000 for fertiliser alone), it has been estimated that 215 000 jobs are directly dependent on UK agriculture.[2] Together this agricultural and related employment represents about 1 job in 20 for the economy as a whole.

This chapter will confine itself to the labour directly engaged in farming. For at least 100 years, employment in agriculture has been shrinking both in terms of absolute numbers of people and in relation to the rest of the economy (see Table 7.1). Since the beginning of the twentieth century the number of workers in agriculture, horticulture and forestry has halved, a period in which the total working population has increased by between three and four times.

Although for the UK as a whole agriculture accounts for less than 3 per cent of employment, there is considerable regional variation in this figure.

**Table 7.1**
**Percentage of British Working Population Engaged in Agriculture: a Historical Perspective**

|  | A<br>Workers in<br>agriculture<br>(millions) | B<br>Total working<br>population<br>(millions) | A as % B |
|---|---|---|---|
| 1851 | 1.8 | 7.6 | 23 |
| 1901 | 1.3 | 13.8 | 9 |
| 1951 | 1.1 | 17.9 | 6 |
| 1971 | 0.7 | 24.0 | 3 |
|  | ('000s) |  |  |
| 1981 | 635 | 26.7 | 2.4 |
| 1982 | 632 | 26.7 | 2.4 |
| 1983 | 624 | 26.6 | 2.3 |
| 1984 | 618 | 27.0 | 2.3 |
| 1985 (provisional) | 616 |  |  |

NOTE    'Total working population' is greater than 'Total civilian employed labour force'. For 1983 the respective figures were 26.77 m and 23.47 m. The *Annual Review of Agriculture* expressed the agricultural labour force as a percentage of the latter and hence achieved a larger percentage figure (2.7 per cent in 1983).
SOURCES    Historical Abstract *British Labour Statistics*. MAFF *Annual Review of Agriculture*. CSO, *Annual Abstract of Statistics* (London: HMSO).

Agriculture is of particular importance in Northern Ireland, where about 9 per cent of the population is so engaged. In areas of Britain defined as rural in the 1981 population census, 14 per cent of the population was engaged in agriculture.[3]

Detailed labour statistics are collected annually in the June census of agriculture, figures from which are given in Table 7.2. From these a number of significant features emerge. First, while the total number of persons has been declining, the various constituents of the labour force have been changing at different rates. Over the 10 years 1975–85, the number of regular full-time (also called whole-time) hired workers fell by about one third, whereas the number of farmers increased slightly. Farmers returning themselves as whole-time fell a little but part-time farmers

Table 7.2
**Numbers of Persons Engaged in Agriculture, Selected Years, 1960–83 (thousands)**

| Workers | 1960 | 1970 | 1975 | 1980 | 1984 | 1985 (provisional) |
|---|---|---|---|---|---|---|
| *Regular whole-time* | 505 | 269 | 222 | 180 | 161 | 157 |
| of which Hired:  male | n.a. | 186 | 157 | 133 | 116 | 111 |
| : female | n.a. | 16 | 15 | 12 | 10 | 10 |
| Family: male | n.a. | 53 | 37 | 30 | 30 | 31 |
| : female | n.a. | 14 | 13 | 5 | 5 | 5 |
| All male | 462 | 239 | 194 | 163 | 146 | 141 |
| All female | 43 | 30 | 28 | 17 | 15 | 15 |
| *Regular part-time* | 96 | 80 | 80 | 64 | 60 | 61 |
| of which Hired:  male | n.a. | n.a. | 22 | 19 | 19 | 19 |
| : female | n.a. | n.a. | 26 | 25 | 23 | 22 |
| Family: male | n.a. | n.a. | 15 | 13 | 12 | 13 |
| : female | n.a. | n.a. | 18 | 7 | 7 | 7 |
| All male | 59 | 41 | 36 | 32 | 31 | 32 |
| All female | 37 | 39 | 44 | 32 | 29 | 30 |
| *Seasonal or casual* | 92 | 76 | 73 | 101 | 96 | 99 |
| of which All male | 52 | 39 | 41 | 57 | 57 | 59 |
| All female | 40 | 37 | 32 | 43 | 39 | 40 |
| *Salaried managers* | n.a. | n.a. | 7 | 8 | 8 | 8 |
| Total employed | 693 | 425 | 382 | 353 | 325 | 325 |
| *Farmers, partners and directors* | (344) | (325) | 280 | 298 | 293 | 291 |
| of which whole-time | n.a. | n.a. | 212 | 208 | 202 | 199 |
| part-time | n.a. | n.a. | 68 | 90 | 91 | 92 |
| Total persons engaged | 1037 | 750 | 662 | 651 | 618 | 616 |
| Spouses of farmers, partners and directors (engaged in farm work) | n.a. | n.a. | n.a. | 75 | 75 | 77 |
| Man equivalents | n.a. | 687 | 629 | 576 | 567 | n.a. |

SOURCE  MAFF *Annual Review of Agriculture, Agricultural Labour in England and Wales* (figures for UK) (London: HMSO).

rose both as a proportion of all farmers and in absolute terms. The number of people engaged as seasonal or casual labour also increased.

Second, the rate of decline in the labour force has slackened. During the 1960s the number of people engaged in agriculture fell by 28 per cent but over the 1970s by only 13 per cent and by an even smaller amount in the 1980s. This slow-down is a reflection not only of the increasing difficulty of releasing labour once the numbers have fallen to low levels, but of the lack of opportunity for employment elsewhere in the economy, which was in decline towards the end of the period.

Third, the composition of the labour force has changed, with farmers and their families forming an increasing part of the labour force. In 1960 the ratio of all workers to farmers was 2:1; by 1985 it had fallen to 1.1:1. In the 1980s the number of farmers, partners and directors, together with their spouses engaged in farm work, have comfortably exceeded the number of employed workers. If employed family workers were separated from other hired (non family) workers, then on a simple head count no less than 62 per cent of the total labour force consisted of farm family labour (farmers and other family members). This is reflected in the proportion of the total labour input coming from family labour, estimated to be 60 per cent in 1975.

The domination of farm families in the labour force, while high in the UK, is even more marked in the European Community as a whole. In 1975 there were in the 9 member states 5.8 million farm 'heads', 6.0m. family member workers but only 1.0m. regularly employed non-family workers. When part-time workers and farmers were expressed in full-time equivalents (Annual Labour Units), the family was found to provide 82 per cent of the total EC farm labour force (see Table 7.3). In the UK the proportion of non-family labour (40 per cent) was more than double the EC (EUR 9) average of 18 per cent. The close identification of families with farms has many implications, which will be returned to.

## 7.2  Characteristics of Farmers

We shall consider the labour used by agriculture in two sections. First the farmers, who are typically self-employed business operators and whose rewards come from entrepreneurial activities, and second the hired workers who receive a wage and who do not bear directly the risks of the farming business.

**Table 7.3**
**Family Composition of the Farm Labour Force, 1975**

|  | UK | EC |
|---|---|---|
| Total annual labour units | 626 000 | 7 543 000 |
| 1  holders | 36 | 46 |
| 2  family | 24 | 36 |
| (1 + 2) | (60) | (82) |
| Non family regular | 33 | 11 |
| Non family non-regular | 7 | 7 |
|  | 100 | 100 |

SOURCE   EC (1980) *The Agricultural Situation in the Community* (Brussels).

### 7.2.1 Sex and age

The personal characteristics of farmers, such as their ages, family status and education, have great relevance to the pattern of agriculture, and how farming responds to change. Farming is overwhelmingly a male-dominated occupation with a bias towards the middle-aged and elderly; in terms of age structure and sex composition farm occupiers differ markedly from farm workers, and even from the rest of the self-employed sector. In the 1981 population census, males classed as farmers, farm managers and horticulturalists in Great Britain outnumbered females by over 8:1, compared with 4:1 for all self-employed people. The ratio for agricultural workers was over 4:1 whereas for the economically active population as a whole it was less than 2:1. The census also showed that farmers, managers and horticulturalists as a group were much older than agricultural workers and the economically active population in general, with 19 per cent of farmers being aged 60 years and over, compared with 12 per cent for all self-employed persons, 10 per cent for farm workers and 7 per cent for the economically active (i.e., in some occupation) population generally. The comparison is illustrated in Figure 7.1. It is known that hired farm managers tend to be younger than independent farmers, but they constitute only about 3 per cent of the total number of farmers.

**FIGURE 7.1**
**Age Distributions from the 1981 Population Census (GB)**

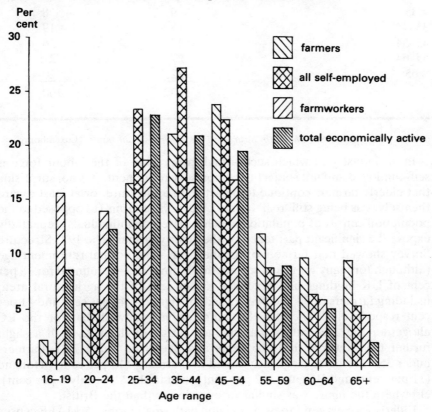

**Table 7.4**
**Distribution of Holdings by Age of Holder, 1983**

| Age of holder | EUR (10) (%) | UK (%) |
|---|---|---|
| <35 | 9 | 8 |
| 35–44 | 15 | 19 |
| 45–54 | 27 | 26 |
| 55–64 | 27 | 27 |
| ≥65 | 22 | 20 |
| | 100 | 100 |

SOURCE Eurostat (1986) *The Farm Structure 1983 Survey – main results* (Luxembourg).

In an industry in which such a high proportion of the labour force is self-employed and not subject to compulsory retirement, it is not surprising that elderly farmers continue to be active in agriculture, or at least regard themselves as being still 'real' farmers. Surveys of farms (as opposed to the population census of population which is based on individuals) repeatedly expose the significant part taken by the elderly farmer. The 1975 Structure Survey showed that farmers of 65 and over – the normal retirement age (although for many groups in society this is now 60), – accounted for 18 per cent of UK holdings and occupied 15 per cent of the agricultural area; including farmers of 55 and over raised the figure to 44 per cent and 41 per cent respectively.[4] The 1983 Structure Survey showed that for the EC altogether the age pattern was similar to that of the UK, with a slight further bias towards those 65 years and over (see Table 7.4). Farmers under 35 were more frequently found in Germany (15 per cent), Holland (11 per cent) and Belgium (12 per cent) than in the UK (only 8 per cent); elsewhere the figure was similar or even lower than the British.

Elderly farmers tend to be associated with small farms. A MAFF census in December 1974 found that in England and Wales the proportion of whole-time farmers, partners and directors over 55 decreased with increasing farm size up to about 120 ha (see Table 7.5). This was mirrored by an almost equal rise in the proportion of farmers under 35. Although there is no direct evidence, many of these young farmers, and particularly those found on the largest farms, were probably junior partners or directors farming with more elderly relatives. Results from the 1975 Structure Survey give a similar picture for the whole UK, with elderly farmers (65

**Table 7.5**
**Percentage Distribution of Whole-time Farmers, Partners and Directors by Age Group and Holding Area Size Group, England and Wales, December 1974**

| Holding size group, (ha) approx. | Age group (years) | | | |
|---|---|---|---|---|
| | <35 | 35–44 | 45–54 | >55 |
| <20 | 14.9 | 20.7 | 24.5 | 39.8 |
| 20–39 | 19.3 | 22.8 | 25.8 | 32.1 |
| 40–59 | 22.9 | 21.1 | 25.5 | 30.6 |
| 60–119 | 25.7 | 21.4 | 24.8 | 28.3 |
| 120–199 | 26.7 | 21.4 | 25.1 | 26.7 |
| 200–279 | 28.2 | 22.2 | 23.3 | 27.3 |
| 280–399 | 26.3 | 23.6 | 23.9 | 26.3 |
| >400 | 25.1 | 24.1 | 24.3 | 26.5 |

SOURCE   MAFF Census data given in a paper by Field, M. J. (1979) *Young Entrants to Agriculture* (Ashford, Wye College: Centre for European Agricultural Studies).

and over) being associated with a smaller than average farm area. In addition, they appeared to be less intensive users of land and labour, indicated by lower livestock units per ha and high labour units per 100 ha. In contrast, farmers in the 35–44 age group had the largest average farm size in ha, carried more livestock and had the lowest man:land ratio. In the EC as a whole this linkage of age and farm size is found too: 67 per cent of the farms occupied by heads aged 65 and over were of less than 5 ha, as opposed to 41 per cent of farm heads of less than 55.

In view of the ageing of the population in general, a somewhat surprising feature of statistics is that farmers in England and Wales seem to be on average younger than they were in the inter-war period. This paradox can probably be explained by changes in the way in which family workers are returned in censuses, the sort of statistical artefact that is an ever-present problem when attempting to trace the development of agriculture. Since the Second World War not only has there been a decay in the patriarchal nature of rural society so that young people demand and get a bigger say in the running of the family farm, but also a greater awareness of the fiscal advantages of taking a son into partnership may well have brought younger adults into the 'farmers, partners and directors' category who formerly would have remained nominally as family workers. This change of status has also been credited with improving the management of farms as new brains are allowed to influence decisions, and with training the next generation of farmers in a better way by easing them into financial responsibility under parental control.[5]

### 7.2.2 Education and training

Education, it is commonly supposed, is linked with productivity and with the ability to adapt to new conditions. Education is in part an investment in people, a form of human capital. Formal education and professional training are not strong features of British farmers, in part a reflection of the age structure of the present farmer population and the educational opportunities existing when they were young. We should not jump to the conclusion that this has led to the industry being hampered in its ability to use the other resources at its disposal or to adjust to changing economic or technical conditions, although circumstantial evidence indicates that there may be some truth in such a suggestion. The most recent large-scale study of farmers' education, by the Economic Development Committee for Agriculture,[6] found that in 1970 only about one-fifth of farmers held some secondary education qualification (18.1 per cent in England and Wales); this varied with the present age of farmer, being highest with the young farmers and declining with age (see Table 7.6). Younger farmers were found to have left school later and were more likely to have been to agricultural college. However, overall only 1 farmer in 10 had studied for some specific agricultural qualification.

Most farmers have little experience outside agriculture. In 1970 the majority had entered the industry (though not as independent farmers) as soon as they had reached the statutory school leaving age. Only 1 farmer in 7 had experience in an other industry (entering farming aged 28 on average) and only 1 in 12 had non-farming qualifications or had formally studied a non-agricultural subject. Recently the position seems to have improved; the Northfield Committee (1979)[7] concluded that attitudes to training off the farm are changing and that over 30 per cent of farmers' sons succeeding their fathers in the recent past have had full-time agricultural education, although this is much higher than the 10 per cent estimated by the Agriculture EDC. The difference between these two figures should cause both to be treated with caution.

There is a clear link between farm size and the amount of education the farmer has received. The EDC study found that farmers on larger farms tended to have left school later than the occupiers of smaller units and there was a noticeably greater amount of specific agricultural education among the larger farmers, and on farms of 500 acres and over (202 ha)

**Table 7.6**
**The Education of Farmers**

|  | Age of Farmer | | | | | |
|---|---|---|---|---|---|---|
|  | 25 and under | 26–34 | 35–44 | 45–54 | 55–64 | 65+ |
| Proportion with one or more types of secondary education (%) | 35 | 29 | 20.9 | 14.1 | 10.6 | 7.8 |

|  | Size of farm group (acres) | | | | | |
|---|---|---|---|---|---|---|
|  | 1–24 | 25–49 | 50–99 | 100–299 | 300–499 | 500+ |
| Proportion of farmers who have studied for agricultural qualifications (%) | 10 | 6.1 | 6.2 | 8.9 | 16.2 | 24.0 |

SOURCE   EDC for Agriculture (1972) *Agricultural Labour in England and Wales* (London: NEDC).

almost a quarter had studied for agricultural qualifications (see Table 7.6). This picture is supported by work in East Anglia which found in the mid-1970s that on farms of 1000 acres and over almost half the farmers (45 per cent) held an agricultural diploma or degree compared with 16 per cent for a sample of farms of all sizes (see Table 7.7).[8] Furthermore, 9 out of 10 of the large farmers had attended grammar or fee-paying schools as opposed to just over half (56 per cent) of the all-size sample. Among farmers, as elsewhere, education seems to reflect not only ability and interest but also social background and economic status; large farms can generate the large incomes necessary to finance private education.

The schooling and training which most of the present generation of farmers has received does not, at least on first impression, equip it well for change, particularly for transfer to other sectors of the economy. In conventional economic terminology it would be labelled 'specific' to farming, with low transfer earnings (ie what it could command in other

Table 7.7
**Characteristics of Occupiers of Farms of 1000 Acres and Over Compared with an All-size Sample, East Anglia**

|  | 1000+ acres (%) | 44-parish sample (%) |
|---|---|---|
| Inheritance of land as a source of initial capital (non-owners excluded) | 69.5 | 60.4 |
| Father was farmer | 80 | 72 |
| Brought up on present farm | 47.5 | 33.3 |
| Education: | | |
| Grammar or fee-paying | 90.2 | 56.1 |
| Agricultural diploma/degree holding | 45.1 | 15.8 |

SOURCE   Tables in Newby, H. *et al.* (1978) *Property, Paternalism and Power: Class and Control in Rural England* (London: Hutchinson).

industries). As will be shown later, its inability or unwillingness to move to other sectors – in other words its occupational immobility – is a major factor underlying government policy towards agriculture.

### 7.2.3   *Geographical mobility of farmers*

One of the more striking features of the UK farming's management is the very limited geographical experience it has encountered, but again the quantitative information is limited. In 1969 a survey found that 96.8 per cent of English farmers (taking the oldest business principal as the farmer) had made no more than one move in their careers and 78.9 per cent were on the same farm as the one on which they had begun their careers.[9] Only 2.4 per cent had farmed on their own account more than 100 miles from their present farm, and many of these larger relocations were associated with the period of agricultural depression in the 1930s. Farmers using a mix of tenanted and owned land, noted in Chapter 6 as an efficient and dynamic sector of the industry, were more likely to have moved, and this ties in with their tendency to be a growth point in farming.

A study in the 1960s of an upland farming area dominated by small businesses found that 70 per cent of farmers had not moved during their careers.[10] At the other end of the farm size spectrum, a survey of farmers in East Anglia during the mid-1970s showed a broadly similar picture of low mobility, with almost half of farmers with 1000 acres and over having been brought up on their present farms; in a sample of all sizes of farm in the same area the corresponding figure was one-third, suggesting a lower geographical mobility by large farmers.[11] However, both samples were similar in that just over half the farmers in each had been brought up in their present or neigbouring parish. Only about a quarter of the farmers, big or small, had been brought up more than 50 miles away from their present farms. Very few were born in urban areas and most of those who did not have a lifetime's association with their farm were at the very least born in East Anglia. This low geographical mobility, however, must be viewed against the legal and economic conditions prevailing over their total period as farmers and which need not necessarily apply in the 1980s.

The growth of owner-occupation, described in the previous chapter, has probably caused farmers to become increasingly reluctant to make whole-farm changes – although perhaps not acreage changes – as part of the continuing process by which they adjust their scales of business operations to changing needs and opportunites. Until recently (1976) there was no legislation providing for close relations to inherit tenancies in England and Wales; this Act has probably further reduced the already limited geographical mobility of farmers, especially tenants.

### 7.2.4  How farmers enter farming

There is no shortage of people from many walks of life who would like to become farmers, but economic conditions have produced a situation where the industry is virtually closed to most of them. Without farming parents of other near relative, or considerable family wealth, opportunities to farm are severely limited.

Farmers in the UK come very largely from farming families. The 1970 EDC study referred to above found that 76 per cent of farmers had been trained on the family farm, and over 83 per cent of farmers in England in 1969 had social origins in the farming community. More recent work[12] found this to apply to a marked extent among occupiers of very large farms in East Anglia; almost 80 per cent of farmers of 1000 acres and over had fathers who were farmers themselves, rather more than the 72 per cent for farmers in general in the area. In Scotland the picture is similar, with about three-quarters of new occupiers coming from within the agricultural industry.[13]

From this evidence clearly it would be wrong to infer that the tradition of passing a farming career from father to son on the same farm is the exclusive characteristic of small farmers with limited horizons. In the East Anglian work cited above, compared with a random sample of farms, more farmers on farms of 1000 acres had inherited their farms; more were sons of farmers; more were born and/or brought up on their present farms; and far more had experience on only one farm – their own and their father's before them. Despite the influence of a more business-oriented approach to farming which is supposed to apply among larger farms and the publicity which a few rapidly expanding agri-businesses have attracted, the occupational (and geographical) immobility which can be shown to characterise UK farmers applies to most of the farmers at the top of the size spectrum, and perhaps in a more rather than less marked form. The explanation seems to be related to the problem of the acquisition and control of land.

If farmers come largely from farming backgrounds, and some two-thirds of land is owner-occupied, then one might expect inheritance to play a significant part in the transfer of land to new farmers from their elder relations. Inheritance of land was the way in which 6 out of 10 farmers in the East Anglia of the mid-1970s acquired their initial capital, and this was of even greater importance among large farms. Other small-scale studies have found about 70 per cent of full-time farmers to have taken over occupancies from close relatives.

Information on how those farmers enter the industry who do *not* inherit their land is even less readily available. Statistical detective work and guesstimates have to be employed. In the late 1970s some 3500 to 4500 opportunities appeared to arise annually for new farmers to enter agriculture.[14] On the informed guess that about half of farmers had a natural successor (taking 2500–3000 of the available new occupancies), there remained some 1000–1500 openings for new entrants, whether as tenants, purchasers or managers, plus those already farming elsewhere but wishing to move 'up the ladder'. As a check on this guesstimate, other sources suggest that about 500 new purchases are made annually and about 500 holdings have been re-let in England and Wales,[15] more if Scotland is included; the total is 1200, not dissimilar from the result of the earlier estimate, but these figures may be wide of the mark. The only reasonably reliable component is the 150 or so new lettings of smallholdings made annually by county councils. Accepting the figures as the best available, it is clear that only a third, and probably less, of new farmers are not parts of farmers' families. In an industry involving about 300 000 farmers, partners and directors in the UK, the 1500 or so opportunities for 'outsiders' to enter annually represents a very low rate of infusion.

Given that many people from non-farming backgrounds would dearly like to get into this industry, what are the characteristics of the relatively few who manage the task? Unfortunately, the social origins of those 'outsiders' who establish themselves as independent farmers, and the economic resources they utilise to get in, remain tantalisingly obscure. A number of small studies suggest that part-time farming is important for these new entrants, that is they have another source of income, and that they command resources outside farming, such as wealth created in some other line of business, to buy their way in as owner-occupiers. Predominantly, however, entry to farming is restricted to those who already have strong, tangible and usually family links with the industry, or to the holders of other forms of wealth.

### 7.2.5 Exit from farming

If farmers, particularly those on small acreages, were more willing to quit agriculture when rewards fell, and history shows a strong trend for small farms to become increasingly less viable, then much of the need for government involvement with this industry would disappear. Governments, seeking a less expensive and more permanent alternative policy to the support of farm product prices as a means of avoiding low farm incomes, have turned on a number of occasions to 'structural' measures, in the form of easing the exit of farmers from the industry by offering them pensions or lump sum payments. The assumption is that once the 'problem' farmers have been adequately taken care of, the others can be viable at prices which would be dictated by the freer interplay of demand and supply.

Unfortunately, the process of leaving farming has not been well studied, so it is difficult to lubricate it by policy measures. Research tends to have been small scale. One estimate[16] suggests that about 3 per cent of farmers leave each year by death or retirement (of which half are replaced by new entrants); death could alone be responsible for half the leavings.[17] A study in Scotland[18] found that 60 per cent of outgoers retired, that is took no other job; retirement was universal for those aged over 65, and for those below this age only 13 per cent took a non-agricultural job. Judged on what little evidence there is available, it seems that farmers who stop farming contribute little to other sectors of the economy in terms of their labour or management ability.

Because farming is a family-dominated activity, with sons (more rarely daughters) succeeding parents on the majority of viable farms, there is rarely a single date on which business control is handed in one single and complete step to the next generation. In this context '**succession**' should be distinguished from '**inheritance**'; the latter by contrast is usually associated with a single event, the death of a parent or the legal transfer by gift sometime earlier. Succession is a gradual process in which responsibility for the farming business passes, activity by activity, from one generation to the next. Starting typically with day to day technical management, sons progress to involvement with longer-term business management decisions, frequently assuming wider responsibilities for one aspect of the farm, such as a single enterprise or the farm's machinery. Involvement with cash control or capital management is likely to be delayed until the father's health and age force him to relinquish control. This process of succession will vary from case to case, but its speed seems to be related to the farmer's age, son's ability, size of farm (with the larger farms offering the greater opportunities for early delegation) and possibly the father's own values concerned with independence, achievement and enjoyment of being in control.[19] The practice of fathers and sons forming farming partnerships with the younger man gradually assuming greater control makes for difficulty in deciding when formally to classify the elder as 'retired'. This is reflected in the age structure of men classes as active farmers, described earlier. Given this classification difficulty, the idea of 'conventional' retirement rather than continuing in farming until death seems to have been on the increase,[20] although this is probably a formalisation (for tax purposes) of the handing-over process between generations which has always gone on.

Despite the retirements which take place, the rate of farmers leaving, is too slow; the basic agricultural policy problem faced by the UK and most industrialised countries is a surplus of farmers. The low ex-migration can be explained by factors such as the inertia resulting from advancing age, lack of formal qualifications and experience in other fields, ignorance of conditions and opportunities elsewhere, the non-monetary advantages of an independent rural life and government support of farm incomes which has blunted the economic goad to quit. With owner-occupiers, the security of holding an asset which until recently was appreciating in real terms is probably an important reason why farmers have remained in the industry. Barriers to off-farm occupations increase substantially among small farmers over the age of 50.[21] Lack of suitable employment alternatives hamper out-flow, and areas with high proportions of marginal farms (mid Wales, Northern Ireland) also suffer from high levels of general unemployment. Frustratingly for policymakers, where there is the greatest pressure for farmers to leave – those older men on small farms and without successor – there exist the most factors militating against the change.

### 7.2.6   Full-time and part-time farmers

Being a farmer does not necessarily imply being a farmer all the time. An important minority of farmers manage to combine agriculture with some other occupation, the two (or more) jobs together absorbing their time and other resources at their command (particularly their capital) and in turn together providing their income and other rewards. Semantic problems then arise. What is a farmer? Is a farmer simply a person who engages in farming-type activities (which would include the owner of a paddock growing grass for his own horses)? Has he to have more than a given land area? Is he a farmer only if he sells his output (which might exclude some subsistence producers)? Is he a farmer only if most of his income comes from farming? Whatever the preferred definition of a farmer it is clear that many people exist who could reasonably be described as being independent agricultural producers but who are not solely dependent on farming for their livelihood. They cannot be ignored when considering the level of national agricultural output, or the question of entry to the industry or, especially, the level of incomes earned by farmers, with its important link to agricultural policy.

'Part-time' also has its definitional problems. A simple form of part-time farming is where a farmer has another job or business which includes travel from the farm to the other place of work, and where the other occupation is in no way connected with agriculture. In reality much more complex situations are encountered. A farmer may be full-time in agriculture but his (or her) spouse may have some other occupation, so that the couple are part time in that they are not solely dependent on farming. Husband and wife usually form a single unit as far as expenditure for living is concerned, so that it is more sensible to refer to the couple's combined sources of income rather than confine interest to just one of them, although available statistics do not always make this possible. There is also the problem of deciding what constitutes a non-farming activity. Some farmers do agricultural contract work (ploughing, combining, etc.), and this can form a major source of their income, but it is essentially not part of their farming business. Others have non-agricultural activities taking place on their farm premises; bed-and-breakfast accommodation and caravans are clearly not part of normal farming activities, but farm shops are less clearly categorised.

According to EC statistics for 1983, on almost a third of all UK holdings the farmer or spouse has some other form of earned income (31 per cent excluding farms arranged as companies), and this is probably an understatement of the real position.[22] Second earned incomes occur in other EC countries with similar or greater frequency than in the UK, with 43 per cent of German farmers and 30 per cent of Italian farmers having other jobs in 1975. The figure for the community as a whole (excluding Greece) was 27 per cent and for the UK calculated at the same time and in the same way 23 per cent; this UK estimate for 1975 is widely regarded as a serious understatement.

There seems to be almost no job which cannot be found combined in some way with farming, and examples found in surveys of the UK's part-time farmers range from cabinet ministers, pop stars, solicitors, shopkeepers through to postmen and road menders. In the UK non-farming incomes in the main come from some other business which the farmer owns, usually off the site of the farm, or from a profession or from the higher levels of management, and only a relatively small proportion of part-time farmers are employees in manual or lower white-collar jobs.[23,24] This contrasts with the situation in other member States of the EC; in Germany, for example, off-farm hired employment predominates. While many of the second businesses have associations with farming (a wide range including butchers and greengrocers shops, seeds and feeds merchants, food processing and marketing), almost half have no obvious agriculture connections. Self-employment in both farm and non-farm occupations could be expected to permit considerable flexibility in allocating time according to each business's need, a facility which might not be available to most manual or lower white-collar employees. Parallel proprietorship of a farm and non-farm business also means that there must be competition for capital funds, and that bank loans ostensibly intended to support activities in one sector are likely to have some impact in the other business under the farmer's control, although the extent to which loans to farmers have been used to develop other activities, and vice versa, is difficult to gauge.

Part-time farming is not restricted to the smallest sizes of farm; an element can be found throughout the size spectrum. The Structure Survey of 1983 found that the proportion was greater among small farms (about 40 per cent on holdings of under 10 ha) and declined to about 11 per cent on those above 50 ha. However, other evidence suggests that its incidence rises again with the largest farms. In England in 1969 among farms of over 202 ha there was well in excess of one part-time farm for every two full-time (see Table 7.8). In absolute terms part-time farming is most usually found among small farms (in 1983 almost half the total number of part-time farms were of less than 10 ha), where farm earnings are likely to be lowest. The incomes of these farms families are clearly greater than those from their farms alone; indeed, farm incomes may be deliberately kept low for tax reasons. The main purposes for being in farming may be environmental, social, domestic and for investment rather than as a source of current income. Surveys covering part-time farmers in Britain have shown that most do not rely on their farms for their main source of income; typically in 2 cases out of 3 their non-farm income exceeds that arising from the farm.

Part-time farming is by no means a recent phenomenon. In the UK farmers with additional incomes from other gainful activities have been recognised for at least one hundred years; in the nineteenth century two Royal Commissions appointed to examine the great hardship caused to certain large sections of British agriculture in the late 1870s to the late 1890s found a wide range of occupations combined with farming (fishing, retailing, road haulage, wholesale distribution, factory work, banking and

**Table 7.8**
**Composition of Farming by Farm Size and Full- or Part-time Status, England, 1969\***

| Size group (acres) | Percentage of Total numbers | | |
| | Full-time farms | Part-time farms | Total |
| --- | --- | --- | --- |
| Under 50 | 19.4 | 15.3 | 34.7 |
| 50–499 | 47.5 | 13.0 | 60.5 |
| 500 and over | 3.1 | 1.8 | 4.9 |
| | 69.9 | 30.1 | 100.0 |

NOTE  *Raised results from a sample of holdings of five acres and over.
SOURCE  Derived from Harrison, A. (1975) *Farmers and Farm Business in England*, Miscellaneous Studies No. 62 (University of Reading: Department of Agricultural Economics and Management).

agricultural work on other farms), much in permanent combination, the symbiosis resulting in increased security and not uncommonly proving extremely profitable. With the addition of a few other categories such as tourism, middle and upper management and the professions, the list of non-farm activities would serve to describe the situation found in the 1980s.

Nor should part-time farming be regarded as only a transitionary stage for farmers who are in the process of leaving, or for new entrants who will eventually become full-time farmers. While it is true that some farmers are forced by circumstances to seek an outside additional source of livelihood and eventually leave agriculture, these are by no means typical of UK part-time farmers. Probably only one-fifth of part-time farmers have arisen this way. Most have bought or inherited their farms (8 out of 10) and three-quarters continued with their outside job when they acquired their land.[25] On balance, people who have set up as farmers tend more to be part-time than those they replace. Although part-time farming seems to be in a greater state of flux than full-time farming, once established, part-time farms tend to be permanent arrangements which suit the abilities, interests and wealth holdings of their occupiers.

Part-time farmers are thus a very heterogeneous group, exhibiting a diverse range of motives for including farming as one of their activities and wide differences between the relative importance of farming to their livelihoods. One feature of this phenomenon is, however, that in those countries where data are available, part-time farming is clearly growing in relative importance, particularly that sector of part-time farmers who are primarily dependent on *off*-farm occupations, and this increase has occurred through the farm size spectrum.[26] This is of major importance to governments which have policies of supporting the incomes of farmers; many of those with low farm incomes may already enjoy adequate total incomes because of their other occupations. In addition, the output from part-time farms is likely to be less sensitive to price changes than from full-time farms, blunting the effectiveness of falling prices as a curb to agricultural surpluses.

## 7.3  The Hired Labour Force

Since there are so many farmers relative to hired workers, and since farmers usually work manually alongside their employees, at least for part of their time, for many purposes it is convenient to include both farmers and workers when describing the agricultural labour force. However, fundamental difference exists between them which cannot be ignored.

First, hired workers usually have a pre-determined income, whereas the farmer's income from farming is a residual, left over from the farm business after other costs have been deducted from sales, and this residual will fluctuate in a way which is not precisely predictable.

Second, the wage of the hired worker is the price paid for his labour, whereas the reward the farmer receives is the return to his manual labour (assuming that he does some), his managerial organisation and his compensation for taking the risks associated with farming, and a return on the capital locked up in the business. Landowning farmers will also receive part of their reward in the form of capital gain.

Third, the farmer is free to change the pattern and size of his farming activities whereas the hired worker generally has to operate within the system determined for him; in conventional terms of employment the worker has pre-determined periods of work, whereas the independent farmer can put in as many or few hours as he chooses.

Fourth, farmers frequently have the right to pass the occupancy of their farm to their chosen successor, whereas employees do not have such rights over who succeeds them to their jobs.

Fifth, frequently there is a large absolute difference between the financial rewards enjoyed by farmers and those of hired workers. Accompanying disparities in spending power there are usually differences between the housing conditions of the two groups.

Such differences, and in particular the consequences of different levels of economic status, can produce a dual society within the agricultural community. But, as Newby has pointed out,[27] (hired) agricultural workers are not a group of whom most people have much knowledge or understanding. To the public consciousness the farm worker is 'socially invisible', the work which he does being commonly attributed solely to 'farmers' and the conditions of his employment not raising much concern.

### 7.3.1 Numbers and distribution of hired workers

Reference has already been made to the falling numbers of regular full-time hired workers in UK agriculture. In 1984 there were some 161 000 but this was only a third of the total on farms in 1960.

Most agricultural holdings in England and Wales do not employ any whole-time hired labour, the proportion rising from 65 per cent in 1971 to 71 per cent in 1983 (changes in June census coverage require these figures to be treated with caution). Another 21 per cent employed only one or two whole-time workers. Many of the holdings with no hired workers are very small and not what might be regarded as commercial farms; in 1976 beyond 50 ha more than half the holdings had hired workers, the percentage rising with holding size.

Over a third (38 per cent) of all workers in 1982 were either the sole hired worker on the holding or one of two. Over half (58 per cent) found themselves in groups of 4 or less; at the other extreme almost one quarter (22 per cent) of the total were in groups of 10 or more. Consequently, while much of the hired full-time labour force is thinly spread, a small number of relatively large holdings account for a substantial minority; in 1983 half of the total number of workers was found on only 9000 holdings, 5 per cent of the total holding numbers. Such a dispersion makes the organisation of trade unions difficult; in the mid 1970s only about 40 per cent of the regular full-time hired workers were members of the National Union of Agricultural and Allied Workers (now joined with the TGWU). Unionisation was strongest in those counties with large arable farms and the greatest concentration of workers (Norfolk, Lincolnshire and Huntigdonshire) and lowest in Wales.[28]

### 7.3.2 Ages of hired workers, and skills

We have already seen that hired agricultural workers tend to be younger than farmers. Farming has traditionally attracted a relatively high proportion of school leavers, of whom some leave after a year or two for other jobs or further education; the age distribution of youths suggests that a number are attracted in at ages 17 and 18, only to leave at 19. The extent of this 'wastage' has probably diminished in the UK as a whole, as it clearly has over the last 20 years in Scotland.[29]

During the 1970s it appears that the hired labour force became slightly younger, with the proportion of men aged 20 to 34 years rising from 32 to 39 per cent between 1971 and 1980 and those aged over 54 declining from 22 to 19 per cent. The figures for 1982 shown in Table 7.9 indicate that almost two-thirds of the hired regular whole-time males are now under 36. As would be expected, foremen tend to be drawn more from the older groups, while dairy cowmen and other stockmen tend to be younger than average.

In terms of types of job undertaken, 'general farm worker' and 'tractor drivers' (groups that are difficult to separate clearly) together account for two-thirds of all workers. Despite the marked contraction in overall numbers (26 per cent in the number of whole-time male employees over the 10 years 1972–82) the overall composition has remained remarkably stable, suggesting similar rates of decline among the various skills except for foremen whose numbers have fallen less and workers in horticulture where numbers fell faster, reflecting the relative decline of this sector of the agricultural industry. Quotas on milk production, introduced in 1984, could result in a more rapid fall in dairymen from now on.

### 7.3.3 Wages, housing and conditions of employment

It is well established that farm workers are far down the scale of wage earners, and a number of studies have had little difficulty in showing some farm workers' families as suffering poverty and real deprivation. Minimum wages are set by the Agricultural Wages Board; these are widely interpreted by farmers as the effective rate (plus perhaps a small supplement) rather than merely as a safety net for those workers least able to protect their own interests.

In 1949, farm workers received 69 per cent of manual industrial earning but this dropped to 65 per cent in 1960. Agriculture earnings rose faster after 1972 than those in other industries and by 1976 they reached 72 per cent of non-agriculture earnings. This relative position has remained broadly unchanged (see Table 7.10).

It must be remembered, however, that the type of employee cited in most comparisons – the hired, whole-time adult male worker – is a fairly restricted labour group in agriculture. These workers provided less than a quarter of the man-equivalents used in farming in 1982. Rates of pay for other groups of employees (part-time and girls and male youths) will vary a great deal from the earnings shown in the table. Therefore, the gap between the reward for an hour's work in agriculture and an hour's work in

**Table 7.9**
**Percentage Distribution of Hired Regular Whole-time Men and Youths by Age Group, England and Wales, Year ended 31 December 1982**

| Occupation | Youths aged 15 and under 20 (%) | Men aged 20 and under 36 (%) | Men aged 36 and under 51 (%) | Men aged 51 and under 65 (%) | Men aged 65 and over (%) | All Hired Men & Youths (%) | Numbers of whole-time men workers aged 20 and over* ('000s) | (%) |
|---|---|---|---|---|---|---|---|---|
| Foremen | – | 49.1 | 30.3 | 19.4 | 1.2 | 100.0 | 7.6 | 8 |
| Dairy cowmen | 7.0 | 60.3 | 20.6 | 11.6 | 0.5 | 100.0 | 6.7 | 7 |
| All other stockmen | 14.4 | 55.8 | 15.7 | 13.6 | 0.4 | 100.0 | 10.7 | 11 |
| Tractor drivers | 3.6 | 53.1 | 24.8 | 17.6 | 0.8 | 100.0 | 21.0 | 22 |
| General farm workers | 20.0 | 48.1 | 16.6 | 14.6 | 0.7 | 100.0 | 41.0 | 44 |
| Horticultural workers | 21.2 | 53.0 | 14.7 | 10.8 | 0.2 | 100.0 | 5.8 | 6 |
| Other farm workers | 11.5 | 47.0 | 22.5 | 18.3 | 0.7 | 100.0 | 1.5 | 2 |
| All hired males | 13.8 | 51.2 | 19.3 | 15.0 | 0.7 | 100.0 | 94.2 | 100 |

NOTE  * There were also 15 100 youths (under 20 years old) and 10 100 females.
SOURCE  MAFF, *Agricultural Labour in England and Wales* (London: HMSO).

**Table 7.10**
**Average Earnings of Hired, Whole-time Adult Manual Male Workers**

| Year | Agriculture[1] Weekly earnings (£) | Earnings per hour (A) (p.) | Industry[1] Weekly earnings (£) | Earnings per hour (B) (p.) | A as percentage of B |
|------|------|------|------|------|------|
| 1974 | 34.52 | 74.4 | 48.63 | 107.8 | 69 |
| 1976 | 50.50 | 109.8 | 66.97 | 152.2 | 72 |
| 1978 | 61.80 | 133.5 | 83.50 | 188.9 | 71 |
| 1980 | 86.48 | 187.2 | 113.06 | 262.9 | 71 |
| 1982 | 106.65 | 228.9 | 137.06 | 319.5 | 72 |
| 1983 | 117.93 | 250.4 | 149.13 | 344.4 | 73 |
| 1984 | 124.16 | 266.4 | – | – | – |

NOTE 1. Agriculture: pre-1978, UK; post-1978, England and Wales. Industry, UK all years.
SOURCE MAFF, *Agricultural Labour in England and Wales*, CSO, *Annual Abstract of Statistics*, CSO *Monthly Digest of Statistics*. (London: HMSO).

other industries may, in fact, differ considerably from that presented here. Even so, it seems safe to conclude that in terms of earnings, farm workers are a low paid section of the UK labour force. According to the Low Pay Unit, agricultural workers came fourth from bottom in the nation's wage league and in 1982 38 per cent of its adult male workers were below the official poverty line, taken to be the thresh-hold for Supplementary Benefit entitlement.[30]

The disparity between youths' earnings inside and outside farming since the Second World War has been noticeably less than among older workers.[31] The drain of workers in the 21–45 age group from the industry to some extent marks the exodus of those who entered agriculture in search of economic rewards, lured by an outdoor life and the attraction of machinery, and who left when faced with a deteriorating position vis-à-vis industrial workers.

The reasons why farm workers are poorly paid are complex, but together they place farm workers in a weak market situation and the union acting on their behalf in a difficult position. They include the scattered nature of the workforce, the smallness of work groups, the problems of organising effective strike action and the reluctance of livestock workers to risk the welfare of their charges.[32] Of particular importance seems the fact that many farm workers find themselves frequently working alongside their employers. This has several important consequences: first, the close employer/employee relationship and small total work force means that there are few if any formalised work procedures so that a 'work to rule' is meaningless; second, any withdrawal of labour is likely to be offset by the farmer or his family working longer hours (possibly with extra machinery); third, the 'face-to-face' relationship between farmer and employee inhibits any idea of withdrawal of labour and undermines the whole basis on which normal industrial trade unions operate – 'Trade unionism is essentially concerned with the erection and maintenance of that set of formalised rules between employers and employees which is inoperable in the close personal relationships of agriculture' (Newby[33]). While the closeness may permit the easy negotiation of minor supplementary payments above the legal minimum wages, it militates against more radical improvements.

Compared with most manual workers in industry, farming seems to offer greater non-economic rewards in terms of job interest, diversity, challenge and the exercising of responsibility and control. This must add to the explanation of why farm workers are willing to accept low wages. However, the value of those traditional perquisites of farm workers (free milk, fuel, etc.) has, according to the Low Pay Unit, been greatly exaggerated.[34] And working on farms is no longer an idyllic occupation, if ever it was; agriculture is dangerous and shares with coal mining the undesired distinction of being at the top of the industrial risk ladder. Changes in the industry have exposed the farm employee to more mechanisation and more chemicals, and fewer workers and larger machines have tended to make him more isolated. A combination of these factors, together with housing problems, difficult social conditions and low wages means that, according to one study in Suffolk, three-quarters of farm workers would not recommend a job on the land to their sons.[35]

Certainly the profitability of farming seems to bear little on the level of wages, which effectively denies the old saw that workers' wages were low simply because farmers could not afford more. More important to their level seems to be the state of the local labour market, and whether there are alternative jobs available. Critical to the effectiveness of this market is the geographical closeness of the jobs (travel is disproportionately expensive to the low paid), their education requirements (farm workers, like farmers, frequently lack formal, therefore easily transferable, qualifications) and the matter of accommodation. Even if suitable housing could be found elsewhere the cost of moving could prove a major barrier.

Farm houses which go with the job are undoubtedly an attraction for younger people. Housing is an essential provision for workers in areas where farms are remote and/or council houses are unobtainable. The proportion of whole-item workers receiving a house rose from 34 per cent to 55 per cent between 1948 and 1975 and thereafter declined to 47 per cent in 1982. The proportion varies with the type of workers, from 68 per cent of dairymen to 14 per cent of horticultural hired whole-time men in 1982, with only 12 per cent of women being provided with a house. Youths and girls are more likely to receive board and lodgings than a house, though three-quarters of boys and even more of the girls receive nothing in the way of accommodation. As the agricultural labour force has declined, many farm cottages previously used for workers have been sold or rented, often to people otherwise unconnected with the rural economy. The Rent (Agriculture) Act in 1976 partially 'untied' cottages in England and Wales, and it seems likely that this will reduce the long-term supply of farm cottages for farmworkers.

Housing is particularly critical to mobility. Because of low wages farm workers are largely dependent on tied cottages or local authority rented housing in rural areas. Farm workers exhibit a strong preference for council housing because of the increased range of employment opportunities they imply.[36] But in many rural areas to take a non-agricultural job probably means moving, losing a worker both his existing accomodation and his right to council housing priority. Effectively excluded from the owner-occupier housing market, his ability to leave agriculture is constrained, though possibly not his desire. 'So he stays where he is, still fed up, but apparently the epitome of the loyal, long-serving contented farmworker.'[37]

## 7.4 Motives and Goals

To conclude this section on labour we look at the evidence on declared motives and goals, turning first to consider farmers. One of the most important aspects of this group of businessmen, yet a sadly neglected subject among academic economists, is the nature of the driving forces which cause them to farm and which keep them in farming. As will be seen in those sections in this book concerned with the farm as a business, a broad understanding of the motives behind the organisation of production – the values held by farmers and the goals and objectives behind decision taking – is essential in explaining why the agriculture industry arranges itself in the manner that it does or responds to stimuli (grant aid or product price signals and so on) in the observed manner. If ever there was a case of over-simplification leading to a misunderstanding of function and the bad prediction of behaviour, it would be if farmers were assumed to be simple profit-maximising businessmen. Because for the vast majority of farmers their private and business lives are inextricably mixed, it is impossible to explain farming decisions (e.g. to invest in a new building or extra land) without considering the background against which they take place – the personal prejudices and preferences of the farmer, the size of his family and its composition, the presence or absence of successors, etc. Many apparently illogical actions by farmers are shown to be rational and more predictable once the simple profit-maximising model is augmented by the additonal influences of personal circumstance and motivation. To this end the motives, goals and values of farmers and the attitudes they bring to their farming must be examined.

First it is necessary to clear some terminology. Personal 'values' are concepts like honesty, the preference for independence, freedom and success. Each value is 'a conception of the desirable referring to any aspect of a situation, object or event that has a preferential implication of being good or bad, right or wrong'.[38] Values tend to be the permanent property of an individual and unlikely to change. 'Goals' on the other hand are more clearly defined events, such as owning more land, or becoming technically up-to-date in farm machinery. One goal may lead on to another; buying a neighbouring farm may be just a stepping stone to setting up a son as an independent farmer. More formally, goals are 'ends or states in which the individual desires to be or things he wishes to accomplish'.[39]

Individual farmers will have their own sets of values and goals which will determine the attitudes they take towards their farms; this attitude is sometimes called 'orientation'. Each person will bring a number of values, and the farm will be able to provide simultaneously for a number of these to various extents. A list of values and goals associated with farming are given in Table 7.11. They are classified under four headings which indicate the attitude (or orientation) they imply towards farming. An **instrumental** orientation implies that farming is viewed purely as a means of obtaining something else, such as an income or security with pleasant working conditions. Farming is therefore just a means to an end, an instrument in obtaining an objective. Farmers with a predominantly **social** orientation are farming for the sake of interpersonal relationships in work. **Expressive** values suggest that farming is a means of self-expression or personal fulfilment, while an **intrinsic** orientation means that farming is valued as an activity in its own right.

The classification does not attempt to be absolute. For example, working with family members, given under the 'social' heading, can also be a means of ensuring family income for the future (instrumental) and of the early buying of land to give pride of ownership (expressive).[40] Also, an individual farmer is likely to view his farm as performing a role in all four groups simultaneously. Few will see it as solely a means of making money or solely as a way of enjoying open-air work conditions; even the head of a

**Table 7.11**
**Values and Goals Associated with Farming**

*Instrumental*
making maximum income
making a satisfactory income
safeguarding income for the future
expanding the business
providing congenial working conditions: hours, security, surroundings

*Social*
gaining recognition, prestige as a farmer
belonging to the farming community
continuing the family tradition
working with other members of the family
maintaining good relations with workers

*Expressive*
feeling pride of ownership
gaining self-respect for doing a worthwhile job
exercising special abilities and aptitudes
chance to be creative and original
meeting a challenge, achieving an objective, personal growth of
character

*Intrinsic*
enjoyment of work tasks
preference for a healthy, outdoor, farming life
purposeful activity, value in hard work
independence – freedom from supervision and to organise time
   control in a variety of situations

SOURCE    Gasson, R. (1973) 'Goals and values of farmers', *J. Agric. Econ.* 18:3, 521–42.

family whose source of livelihood is some other form of business and who
buys a small farm in order to live in a pleasant environment will not be
immune from the social prestige of landownership. For all farmers the four
basic orientations will be in some form of balance which will govern their
overall approach to their farms, a balance which is likely to change over

time as they become older, move through their family cycle, and as their financial conditions alter.

A narrow 'farm-is-a-business' approach puts major stress on the instrumental orientation to farming. Yet when farmers are questioned about the strengths of their feeling towards those aspects of farming listed in Table 7.11 the evidence suggests that, overall, it is the intrinsic and expressional motivations which are dominant among farmers – the enjoyment of the very process of farming and of the associated independence, and the challenge it presents.[41] Instrumental and social motivations are of a lower order of importance (see Table 7.12). Differences emerge between small and large scale farmers, with the smaller placing even more emphasis on intrinsic aspects, particularly on the independence they enjoy, whereas large farmers, although still basically intrinsically orientated, attach more importance to the commercial aspects of farming than the small farmers. The pattern is not simple; some social values are rated more highly by small farmers (working close to home and family) and others by larger farmers (belonging to the farming community). Small farmers above certain income levels appear to be interested in maximising satisfaction rather than money income. There may be an element of chicken-and-egg causality in their behaviour; the low rating of high incomes by small farmers may be in part a reflection of their inability to generate them, so high incomes are disparaged.

However, the evidence clearly shows that motives for farming other than commercial ones are important, particularly among small farms. This helps to explain a number of observed phenomena. For example, official schemes to encourage small farmers to leave the industry by offering pensions, to compensate them for the loss of income have failed because, by taking a commercial attitude to profits foregone in calculating the size of compensation payments, the schemes in no way provided a substitute for the major loss farmers would have to face – the loss of enjoyment of the farming process itself. Inadequate compensation and rising land values if they held on to their farms together rendered ineffectual both the European Community retirement scheme as applied in the UK and the earlier national measure (see Chapter 14 for an account of the history of retirement schemes).

A second example is the alacrity with which farmers responded to the first Farm Improvement Scheme (1957) which gave grants for new buildings and other capital improvements, a response which took MAFF by surprise. Investment decisions by farmers during that period were heavily influenced by the desire to be technically up-to-date in machinery and buildings, with financial viability low on the order of priorities,[42] an attitude which persists today, encouraged by grant schemes and advantageous tax allowances.[43] A willingness to invest for technical reasons reflects the basically intrinsic attitude which farmers take. Even farmstead

**Table 7.12**
**The Importance of Attributes of Farming to Cambridgeshire Farmers by Size of Farm Business**

| | Size of farm business | |
| --- | --- | --- |
| | Smaller (SMD 600–950) | Larger (SMD 1300–) |
| Number of farmers | 38 | 38 |
| *Attribute* | *Score* | |
| Intrinsic | 59 | 59 |
| Expressive | 35 | 50 |
| Instrumental: income | 38 | 46 |
| conditions | 22 | 33 |
| Social: belonging | 39 | 23 |
| prestige | 23 | 32 |

NOTES   Scores for smaller farmers differ significantly from larger (p.<0.01) Responses have been scored as follows:

| | |
| --- | --- |
| Most important | 3 |
| Very important | 2 |
| Important | 1 |
| Not important | −1 |
| Not relevant or no reply | 0 |

SOURCE   Gasson, R. (1973) 'Goals and values of Farmers', *J. Agric. Econ.* 18:3, 521–42.

tidiness is given as a good enough reason for major expenditure, as are semi-commercial factors such as improved working conditions. And, finally, the presence of a successor has been shown to be of major importance in determining many aspects of farm practice, especially investment in land improvements such as drainage and buildings.[44] To pass a farm on to the next generation, preferably in a better state than it was acquired, is a fundamental and powerful instinct of great practical relevance. To ignore such influences is to omit major elements in the explanation of the present pattern of farming and of the changes which are occuring.

## 7.4.2 Motives and goals of farmworkers

Farmworkers, like farmers, seem to derive much satisfaction from the nature of agricultural processes themselves rather than from the monetary rewards to be gained, thereby exhibiting an intrinsic attitude to their work. A survey of Suffolk workers found that the variety of the job was overwhelmingly the major source of satisfaction, together with the pleasure coming from seeing crops and animals grow together accounting for almost 9 out of 10 replies.[45] Work with modern machinery was seen as a particularly rewarding task, a badge of the workers' standing and frequently producing a strong air of proprietorship in its operators.

Evidence on overall job satisfaction, absenteeism and turnover rates among agricultural workers is not plentiful, although what limited data there are available suggest a relatively high degree of job satisfaction in terms of whether they found the job interesting all or most of the time compared with car industry and textile workers, although not higher than some others such as printing.[46]

However, the relatively low concern of farmworkers with money is not one entirely of free choice; as we have already seen, they are restricted in their alternative employment opportunities by their housing, education, immobility and, not least, by the expectations they have learned through growing up usually as part of farming communities. Many do not have the luxury of being able to weigh the various advantages and disadvantages of a range of jobs and of chosing the balance of financial and non-financial rewards best suited to their personalities and expectations. Being trapped in a low-income occupation, with little likelihood of being able to move to other forms of employment, there may be a tendency for the farmworker to accept the positive aspects of his job and ignore those over which he has no effective power. The level of financial rewards seems reflected in the fact that in the Suffolk sample, 55 per cent said that they would not take up farm employment if they had their time over again and 73 per cent would not recommend a job on the land to their sons.

In an effort to probe the goals and motives of farmworkers one recent small scale exercise in Scotland has attempted to identify sources of satisfaction and dissatisfaction among farm workers.[47] Work studies in industrial contexts have shown that certain words ('responsibility', 'supervision', etc.) are associated with satisfaction or dissatisfaction, one important finding being that the removal by management of factors associated with dissatisfaction (such as administration or bad working conditions) does not necessarily result in a more contented workforce if 'satisfiers' are absent. For an improvement, factors capable of allowing workers to achieve targets, gain recognition or take responsibility must be present.

Table 7.13 illustrates the factors claimed as giving a short-or long-term effect in the Scottish context. Clearly an effective agricultural management

**Table 7.13**
**Factors Associated with Job Satisfaction by Farm Workers**

| | |
|---|---|
| Satisfiers – long duration | Good equipment, work itself. Job security, responsibility, salary, advancement. |
| Satisfiers – short duration | Achievement, recognition, farm performance, everything running smoothly, tidiness. |
| Dissatisfiers – long duration | Responsibility, salary, management and organisation, supervision. |
| Dissatisfiers – short duration | Lack of recognition, lateness for meals. Long working hours, observation of unnecessary waste. |

SOURCE  Critchley, R. and Birse, M. (1980) 'Job satisfaction and the farm worker.' *Farm Management Review* 13:19–24 (Aberdeen, North of Scotland School of Agriculture).

should be aware of minimising sources of dissatisfaction and be willing to provide means of satisfaction, in particular the necessity for opportunities to achieve and for achievement to be both obvious to the worker and to be acknowledged by others. Conditions conducive to achievement, such as good equipment enabling more acres per day to be cultivated, are therefore important.

Perhaps surprisingly there is little evidence in the Scottish study for the existence of a relationship between the state of contentedness and the amount of work or quality which workers said they would produce. Most stated that they would not be affected, nor would work contentedness affect their domestic relationships and relationships with other workers. There is a suggestion that unhappiness produces a withdrawal *into* work, that is workers adopt a more intrinsic approach.

In terms of the ultimate goal of many farm workers – to be an independent farmer – while this transition at times in the past has been feasible, it is no longer so unless sizeable family resources can be drawn upon. County council smallholdings, first established at the end of the nineteenth century, partly through concern with the disappearance of small farms, believed to be the first step on the farming ladder, are now beyond the financial resources of most farm workers; in 1983 one council (Devon) was reputedly requiring would-be tenants to show evidence of personal assets of £25 000. Even for those who attain them, smallholdings seem incapable of building much farming capital, indicated by the fact that under the local authority scheme (in one recent year) only 55 out of 15 000 tenants succeeded in moving to larger farms.[48] Entrepreneurial status for young people without inheritance rights to land seems more likely under present conditions through the management of single enterprises which form part of large farming businesses (see Chapter 8), although it is the highly trained specialist rather than the general worker for whom this avenue has opened.

# Notes

1. This figure includes 76 000 spouses of farmers, partners and directors, engaged in farm work, who are not included in Table 7.1. Information on these was not collected separately before 1977.
2. Rickard, S. (1982) *Agriculture and employment*. Paper presented to Royal Society of Arts Conference, 'Town and Country: Home and Work'. (London: National Farmers Union). A similar estimate in Craig, G. M., Jollans, J. L. and Korbey A. (1986) *The Case for Agriculture: An Independent Assessment* CAS Report No. 10 (University of Reading) gives 5.15 per cent of jobs and 4.69 per cent of whole-time equivalents as being in or related to agriculture.
3. Hodge, I. (1984) *Rural economic development and the environment*. Paper to conference of the Agricultural Economics Society, 'UK agriculture and the environment', 23. November 1984.
4. Commission of the EC (1980) *The Agricultural Situation in the Community 1979 Report* (Luxembourg).
5. Hastings, M. R. (1984) 'Succession on farms' *J. Agric. Manpower Society* (Summer). Other work on succession, see Fennell, R. (1980) 'Farm succession in the European Community', *Sociologica Ruralis* **21**:1, pp. 19–42; Thomas, H. A. (1981) 'The need for a business approach to the problems of succession, inheritance and retirement', *Farm Management* **4**:4, 157–62 and Weston, W. C. (1977) 'The problems of succession', *Farm Management* **3**:5, 237–47.
6. EDC for Agriculture (1972) *Agricultural Labour in England and Wales* (London: National Economic Development Office).
7. Northfield (1979) *Report of the Committee of Inquiry into the Acquisition and Occupancy of Agricultural Land* Cmnd. 7599 (London:HMSO).
8. Newby, H. *et al.* (1978) *Property, Paternalism and Power: Class and Control in Rural England* (London: Hutchinson).
9. Harrison, A. (1975) *Farmers and Farm Businesses in England* Miscellaneous Studies No. 62, (University of Reading: Department of Agricultural Economics and Management).
10. Nalson, J. S. (1968) *Mobility of Farm Families* (Manchester University Press).
11. Newby, H. *et al.* (1978) op. cit.
12. Ibid.
13. Rettie, W. J. (1975) 'Scotland's farm occupiers', *Scottish Agricultural Economics* **25**:387–93.

14. Northfield (1979) op. cit.
15. Figures for purchasers of new purchases as opposed to existing farmers adding to their holdings grossed up from loans made by the Agricultural Mortgage Corporation; figures for rented farms taken from the MAFF Rent Enquiry.
16. Harrison, A. (1967) *Farming Change in Buckinghamshire* Miscellaneous Studies 43 (University of Reading: Department of Agricultural Economics).
17. Approximation based on mortality rates and population numbers of males of 15 years and over.
18. Rettie, W. J. (1975) op. cit.
19. Hastings, M. R. (1984) op. cit.
20. Gasson, R. (1969) *Occupational Immobility of Small Farmers*, Occasional Paper No. 13 (University of Cambridge: Department of Land Economy).
21. Ibid.
22. Gasson, R. (1985) *The Nature and Extent of Part-Time Farming in England and Wales*. Paper to conference of the Agricultural Economics Society, 'Farm family occupations and income' 13 December 1985.
23. According to Harrison (see Note 9) about three-quarters of all part-time farmers gain their non-farm income as proprietors of second businesses.
24. Gasson, R. (1983) *Gainful Occupations of Farm Families* (Ashfard: Wye College, School of Rural Economics).
25. Ibid.
26. Organisation for Economic Cooperation and Development (1978) *Part-Time Farming in OECD Countries: General Report* (Paris: OECD).
27. Newby, H. (1977) *The Deferential Worker* (Harmondsworth: Penguin).
28. Ibid.
29. Martin, P. C. (1977) 'The age structure of Scottish farm workers', *J. Scottish Agricultural Economics* **27**, 87–91.
30. Winyard, S. (1982) *Cold Comfort Farm: A Study of Farmworkers and Low Pay* (London: Low Pay Unit).
31. Gasson, R. in Edwards, A. and Rogers, A. (1974) *Agricultural Resources* (London: Faber).
32. Pierson, R. (1978) 'What about the farmworkers? – a trade union view', *J. Agric. Econ.* **29**:3, 235–42.
33. Newby, H. (1972) 'The low earnings of agricultural workers: a sociological approach' *J. Agric. Econ.* **23**:1, 15–24.
34. Winyard, S. (1982) op. cit.
35. Newby, H. (1977) op. cit.
36. Pierson, R. (1978) op. cit.
37. Ibid.
38. Gasson, R. (1973)'Goals and values of farmers', *J. Econ.*, **18**:3, 521–42.
39. Ibid.
40. Similarly, money might be an end in itself to a miser, but is normally desired as a means of consumption, leisure, security, progress or prestige. Income may in fact have different meanings for different classes of society; for the economically deprived it may be valued chiefly as a means of security, in the middle classes for the prestige it confers and among the wealthy as a mark of achievement.
41. Gasson, R. (1973) op. cit.
42. Black, C. J. (1965, 1966) 'Capital deployment on farms in theory and practice', *Farm Economist* **10**:475–84. **11**:10–23.
43. Burrell, A. M., Hill, G. P. and Williams, N. T. (1983) 'Grants and Tax Reliefs as Investment Incentives' *J. Agric. Econ.* **34**:2, 127–38.

44. Harrison, A. (1975) op. cit. and Gasson R. and Hill, Berkeley (1984) *Farm Tenure and Performance* (Ashford: Wye College, Department of Agricultural Economics).
45. Newby, H. (1977) op. cit.
46. Ibid.
47. Critchley, R. and Birse, M. (1980) 'Job satisfaction and the farm worker', *Farm Management Review* **13** Aberdeen: North of Scotland College of Agriculture).
48. Pierson, R. (1978) op. cit.

# Resources in Agriculture: Capital and Finance 8

In Chapter 5 we saw that British agriculture uses a little less land as the years go by and much less labour but manages to produce a rising volume of output. A major part of the explanation for this is the increasing amount of machinery, equipment, buildings, drainage and other land improvements which farmers employ. Because of these changes agriculture is said to be an increasingly capital-intensive industry. In this chapter we examine what such a statement means and the evidence behind it. And because part of this additional capital has been acquired using funds borrowed from outside the farm business, it is convenient in this chapter to review the financial position of UK agriculture.

## 8.1  Investment and Stocks of Capital

In the strict economic sense 'capital' refers to items like tractors, barns and dairy cattle which are used in production, that is, goods which have been created not because consumers want them but because they enable those goods and services which consumers do desire (in agriculture's case mainly food, but also clothing, furniture, etc.) to be produced more effectively. At the national level, the creation of capital goods requires resources which could be used for the more immediate production of consumer goods, but while the diversion of resources to the build-up of the nation's stock of capital means that short-term sacrifices have to be made, in the longer term it is hoped that this investment will be rewarded by much greater levels of production of consumer goods once the capital goods start being used. Hence a sharp eye is kept on how much the agricultural industry is spending on capital goods, any fall-off raising concern over farming's future ability to produce food. At the farm firm level investment again takes the form of putting part of one's income aside, not spending it on living but buying capital items (machinery, etc.) in order, first, to compensate for the wearing-out of the equipment already in used and, second, to build up further the stock of capital employed in the business.

### 8.1.2  Gross and net estimates

At both the national and individual farm level an important distinction must therefore be drawn between gross and net investment. **Gross investment** (or gross capital formation) is the total spending on capital goods. Part of this will be simply to compensate for the wearing out of the existing stock of buildings and machinery – termed capital consumption. (In the individual farm business the sum in the accounts corresponding to this capital consumption is called 'depreciation'.) Any investment in excess of this compensation amount is termed **net investment** (or net capital formation), and this net figure represents the true increase in capital used by agriculture. Similarly, the gross stock of capital refers to all the assets used by farming valued as if they were all new (that is, valued at what they cost) whereas the net stock figure reduces the valuation according to how much working life is left in them. The precise method of calculating capital consumption for UK agriculture is too complex to describe here[1] but it involves knowing how long buildings and machinery take to wear out or become obsolete, or rather it involves **assumptions** about such things. These assumptions can be challenged. The result is that, while reasonable confidence can be ascribed to estimates of gross investment at the national level and, to a lesser extent, gross stocks of capital, net figures should be treated with caution.

UK agriculture currently accounts for between 2 and 3 per cent of the country's total spending on capital goods (see Table 8.1). This proportion was higher in the 1950s when agriculture was a larger sector of the economy and also during the early 1970s when the industry's confidence and prosperity was high as Britain joined the European Economic Community. The volume of gross capital formation (that is in real terms) was noticeably above the longer term upward trend in 1973 and 1974, particularly spending on buildings. Within the last few years the decline in incomes for the industry as a whole has resulted in a fall-off in investment, particularly in 1981 following the particularly poor income of 1980. Spending on machinery was noticeably affected (a fairly reliable barometer of

### Table 8.1
### Gross Fixed Capital Formation in UK Agriculture (£ million)

|  | Buildings and works | Plant, vehicles and machinery | Total | | % of national GFCF |
|---|---|---|---|---|---|
|  | (current £M) | (current £M) | (current £M) | (constant 1980 prices £M) | |
| 1955 | 26 | 78 | 104 | n.a. | 3.8 |
| 1960 | 45 | 96 | 141 | 842 | 3.5 |
| 1965 | 67 | 117 | 184 | 961 | 2.9 |
| 1970 | 117 | 133 | 250 | 1097 | 2.7 |
| 1971 | 133 | 152 | 285 | 1149 | 2.8 |
| 1972 | 159 | 192 | 351 | 1270 | 3.2 |
| 1973 | 209 | 231 | 440 | 1408 | 3.2 |
| 1974 | 250 | 309 | 559 | 1427 | 3.4 |
| 1975 | 240 | 351 | 591 | 1235 | 2.9 |
| 1976 | 225 | 453 | 678 | 1205 | 2.9 |
| 1977 | 250 | 519 | 769 | 1187 | 3.0 |
| 1978 | 331 | 563 | 894 | 1237 | 3.0 |
| 1979 | 397 | 601 | 998 | 1195 | 2.9 |
| 1980 | 557 | 507 | 1064 | 1064 | 2.6 |
| 1981 | 506 | 464 | 970 | 910 | 2.3 |
| 1982 | 610 | 596 | 1206 | 1124 | 2.7 |
| 1983 | 634 | 721 | 1359 | 1245 | 2.8 |
| 1984 | 684 | 678 | 1362 | 1234 | 2.5 |
| 1985 (forecast) | 550 | 715 | 1262 | 1081 | 2.1 |

NOTE 1980 series has been carried back before 1979 by splicing with constant price series using earlier base years.

SOURCE MAFF (various years) *Annual Review of Agriculture* (London: HMSO).

farming prosperity) but expenditure on buildings less so. The poor income of the industry in 1985 and little prospect of much improvement in the foreseeable future, coupled with costs of borrowing which are high by historical standards, have kept recent investment (all assets taken together) at levels similar to those experienced in the late 1960s.

**Table 8.2**
**Estimate of Capital Stocks (Central Statistical Office), UK, 1968–85**

| | Gross capital stock[1] at 1980 replacement cost | Changes in capital stocks at 1980 prices | | |
| --- | --- | --- | --- | --- |
| | | Gross domestic fixed capital formation | Capital consumption | Net domestic fixed capital formation |
| | (£'000m) (Agriculture) | (£m) (Agriculture) | | |
| 1968 | 14.5 | 1196 | 828 | 368 |
| 1969 | 15.0 | 1143 | 848 | 295 |
| 1970 | 15.6 | 1214 | 865 | 349 |
| 1971 | 16.2 | 1247 | 885 | 362 |
| 1972 | 16.8 | 1370 | 909 | 462 |
| 1973 | 17.6 | 1464 | 940 | 523 |
| 1974 | 18.2 | 1263 | 969 | 294 |
| 1975 | 18.5 | 1102 | 987 | 116 |
| 1976 | 18.9 | 1116 | 1001 | 115 |
| 1977 | 19.3 | 1117 | 1017 | 100 |
| 1978 | 19.7 | 1200 | 1035 | 165 |
| 1979 | 20.0 | 1066 | 1048 | 18 |
| 1980 | 20.2 | 949 | 1048 | −99 |
| 1981 | 20.2 | 812 | 1036 | −225 |
| 1982 | 20.3 | 978 | 1023 | −46 |
| 1983 | 20.5 | 1068 | 1016 | 52 |
| 1984 | 20.6 | 991 | 1007 | −17 |
| 1985 | 20.5 | 702 | 986 | −283 |

NOTE 1. Does not include land value.
SOURCE. CSO, given in Johnson, C. (1986) 'The balance sheet of British Agriculture' in *Agriculture and Food Statistics* (London: Statistics Users Council).

### 8.1.3 *Quantity of capital used by agriculture*

The common assertion that agriculture uses an increasing real amount of capital, and has substituted capital items for labour in its mix of inputs, is borne out by Table 8.2. This shows estimates of the overall value of the stock of capital (excluding land) used by agriculture in real terms, that is, after allowing for the falling value of money, and a breakdown of investment into that which compensates for capital consumption resulting from wear and tear and obselescence and that which produces a net addition to capital used by agriculture. Positive figures are shown in the column of net domestic fixed capital formation for most year. Even recently, when gross spending has been below the trend and the high investments of the early 1970s were being felt in the consumption estimates, substantial negative investments have only occurred in 1981–2 and 1985. A similar picture is presented by the rising figures for the gross capital stock, which correspond to the accumulation of past spending on buildings, machinery, etc., less the value of those assets which are assumed to have dropped out of use through age, but before making allowance for depreciation of those assets still in use. Despite the negative investments of 1980–5 the value of the gross stock of capital used by farming has not been eroded, and it is the gross stock which largely determines the productive capacity at farming's command. Combined with a falling labour force, the stock of capital per person rose in real terms by 36 per cent in the period 1972–81, slightly more (40 per cent) if only the gross stock of 'fixed assets' (buildings, machinery, etc.) is considered.

Caution should be exercised in relying too heavily on the estimates since, in the case of buildings, the life assumed for stock estimates (30 years) is much shorter than the period for which most buildings actually last in use, although their current use may be not what they were designed for. It is worth pointing out here that the periods chosen to depreciate assets in individual farm management accounts are usually much shorter; for buildings it is typically only ten years. For tax purposes the period may be even briefer. These are not necessarily incompatible. The estimates of life used in national accounting reflects what actually took place, that is they are retrospective (ex post) estimates. The farmer, however, in deciding how fast to write off his investments, will as part of his management decision-making be looking forward (that is, taking an ex ante view). His chosen depreciation period will reflect not only the expectations of physical life of the asset but also the riskiness of the enterprise with which it is associated, the likelihood of obsolescence brought about by as-yet unforeseen new technology and possible alternative uses for the asset. The higher the degree of risk and the less willing the farmer to accept risk the more rapidly will the asset be written off. Taxation authorities have rules of thumb which apply over broad classes of assets (buildings, machinery, vehicles) but sometimes allow very short depreciation periods if they are acting in accordance with government policy to encourage farmers to invest more heavily.

In view of the problems associated with estimating the actual lives of assets, a more direct approach to measuring the amount of capital used by

**Table 8.3**
**Agricultural Machinery in Use, UK, 1959–83 (thousands)**

|  | 1959 | 1966 | 1970 | 1976 | 1979 | 1983[1] |
|---|---|---|---|---|---|---|
| Tractors: total | 505 | 517 | 511 | 526 | 508 | 545 |
| Tracklayers) 10 hp and | 18 | 18 | 14 | 13 | – | 12 |
| Wheeled ) over | 436 | 448 | 431 | 470 | 486 | 518 |
| Under 10 hp | 50 | 52 | 66 | 43 | 22 | 15 |
| Tractor ploughs | 368 | 323 | 308 | n.a. | 163 | 193 |
| Disc harrows | n.a. | 100 | 111 | n.a. | – | 95 |
| Cultivators: rotary and others | n.a. | 447 | 455 | n.a. | 410 | – |
| Corn drills: cultivation | 136 | 124 | 132 | n.a. | n.a. | 74 |
| Farmyard manure spreaders | 92 | 129 | 129 | n.a. | n.a. | 119 (1982) |
| Mowers | 250 | 216 | 197 | 177 | n.a. | |
| Pick-up balers | 68 | 105 | 105 | n.a. | n.a. | 114 (1980) |
| Combine harvesters | 52 | 67 | 66 | 59 | 58 | 57 (1982) |
| Drying machines | 17 | 39 | 64 | n.a. | n.a. | 40 (1981) |

NOTE 1. Scotland and N. Ireland 1980. England and Wales 1983 unless shown otherwise.
SOURCE CSO, *Annual Abstract of Statistics* (London: HMSO), MAFF, *Agricultural Statistics United Kingdom* (Guildford: MAFF).

agriculture might appear attractive, based on surveys of the numbers of tractors, ploughs, etc., found on farms. This would give some idea of the gross stock, although not the net stock unless data were also collected on the ages and conditions of the machines. In the UK there are periodic surveys of machinery and equipment on farms, reproduced in Table 8.3. Rather surprisingly the pattern is not quite what is expected, with the number of ploughs, for example, showing a decline in numbers since 1959, and the peak number of combine harvesters occurring in the mid-1960s. The total number of tractors was almost the same in 1979 as it had been in 1959. Much of the explanation for this apparent anomaly is, of course, that major changes have occurred over the period in the quality of equipment and each unit's work capacity.

There is no regular enumeration of the quantity of buildings and other immovable items which form the other principal group of wasting assets (that is, those that fall in value over time and usage) used by farming. This deficiency in the statistics seems to have resulted from the tendency to treat the provision of buildings as a landlord responsibility and therefore outside the direct sphere of farming activity, even though owner-occupiers have increased in importance as land users and have been responsible for providing most of the new buildings over the past two decades. The most recent survey of farm buildings in England and Wales,[2] there being no equivalent survey in Scotland, found that in 1973 over one third of the buildings originated from before the First World War, and just under a third in the 17 years following the introduction of the Farm Improvement Scheme in 1957. (An explanation for the industry's rapid response to the encouragement to invest provided by this measure was given in Chapter 7.) Considerable differences were found between tenures and sizes of farm in the array of buildings at their disposal. Owner-occupiers of below 300 acres possessed markedly greater quantities of farm buildings than did tenants, particularly of buildings erected since the introduction of the Improvement Scheme. The largest farms had relatively more newer buildings. Older buildings were a characteristic of tenanted farms, especially of those below 150 acres. The landlord- and- tenant system thus seemed *not* to result in a more ready supply of capital to farming, as some of its advocates suggest, and there is no evidence that owner-occupiers, by shouldering the financial responsibilities of both farmer and landowner, find their resources stretched to the point where they are starved of capital for improvements – rather the reverse, at least among long established and low indebted occupiers. There is some evidence[3] that tenure has an effect on the balance between machinery and building investment, with owner-occupiers preferring relatively larger quantities of buildings and tenants more machinery. This probably reflects the greater security of owner-occupiers, their longer planning horizons and their greater borrowing power based on the collateral of their land.

### 8.1.4  *The motive for investment*

The classic economic view of investment is that the cost of any additional unit of capital will be viewed in relation to the additional revenue the item generates for the business. In the case of purchasing replacement machinery as existing equipment becomes inoperative, the cost has to be viewed against the consequences to the business as a whole of *not* replacing. However, in practice when it comes to investment in farm buildings and machinery, motives other than narrow financial ones seem to play an important part.

The return to (marginal) investments in machinery and buildings on farms is not high on a prima facie examination. Indeed, it is often hard to demonstrate measurable rewards for such investments, which are often aimed not so much to generate additional future income but rather to minimise short-run taxation payments and to keep the farm technically up to date. The so-called **residual investment hypothesis** suggests that the level of spending on capital goods is a function of the margin between farm income and the reasonable living expenses of the farm family; the variability of performance over time which seems a characteristic of UK agriculture means that in some years a relatively large margin is available to be spent on capital goods. Advantageous taxation depreciation allowances, used as a way of encouraging investment and extended in the 1970s to cope with accounting problems during times of rapid inflation, have helped channel these funds principally into on-farm (gross) capital formation rather than into off-farm investments by farmers or to consumption spending.

The stimulation of investment by high incomes is commonly encountered in reviews of farm investment patterns; a limited study by the Centre for Agricultural Strategy[4] of net investment in machinery (gross investment less depreciation plus, in this instance, expenditure on contract work) lends some statistical support to this theory of determination by a residual. The study showed that investment in machinery increased more than proportionately with increases in the farmers' current income and liquidity position, as would be predicted by the hypothesis. It is perhaps not surprising in such circumstances, where investment is triggered off in a relatively short-term planning context by largely unpredictable income fluctuations, that it is difficult to demonstrate attractive returns to marginal investments in farm machinery (although the CAS study found an association between increasing the level of machinery stocks and a rise in production intentions).

The position regarding the returns to investment and the residual nature of spending is even less clear with buildings. Complications arise because it is difficult to distinguish replacement investment from genuine additions to the capital stock. Investment in buildings takes longer than in machinery and, at least for the larger projects and on individual farms, is less of a continuous process. It is less likely, therefore, to be affected by year-to-year variations in net income and more by factors such as longer-term expectations of interest rates on borrowings and trends in profitability.

Earlier work in Yorkshire[5,6] found that, following the introduction of the Farm Improvement Scheme (1957) with grant-aided spending on buildings and works, the heavy investment undertaken by owner-occupiers was not rewarded by a benefit visible in the farm accounts by 1961, the end of the period reviewed. Of much greater importance than this, however, was what was revealed about the reasons behind the investments made by farmers. This research demonstrates the technical motives behind much investment; the firm conclusion drawn from the studies is that both tenants and owner-occupiers place great emphasis on the maintenance of the farm's technical efficiency. This can be interpreted not only as a matter of pride but one of reducing vulnerability to adverse business conditions. The first call on funds available for investment was for re-equipment with field machinery, showing the farmer's direct concern for the future, as well as present, performance of the business. With owner-occupiers, the range of possible investments extends to buildings, opportunities generally less open to tenants. The relatively heavy investment in buildings on owner-occupied farms which has continued since this Yorkshire study can be seen as a reflection of farmers' preoccupation with keeping technically up-to-date, influenced, where a successor is evident, by the desire to pass on a viable farm to the next generation.

## 8.2   Finance and Indebtedness

At the level of individual farms additional capital goods which are felt desirable for the business have to be paid for either out of present profits, saving from past profits, or from funds outside the farm business, such as income from other businesses which the farmer may control, or by borrowing. But it is impracticable to separate those funds drawn on to provide extra capital goods from those used to buy land or to purchase 'working capital' such as fertilisers and seed or to pay labour in advance of selling farm output. Therefore one has to consider the whole financial structure of agriculture and the methods it uses to provide itself with the funds required to maintain its farming activities and make the adjustments farmers feel compelled to carry out.

In describing the financial structure of the farming industry land must be treated in much the same way as those other productive assets which the farmer uses – his machinery, buildings etc. In the narrow economic use of the term 'capital', land would be excluded as it is not the product of past human activity (although land *improvements* such as drainage or acidity correction would qualify). But as far as the individual farmer is concerned land has many of the attributes of capital goods – land is an asset used in the production of his output and land can be bought or sold by him. Admittedly it has the additional qualities of not wearing out in the same way that a tractor does, although without proper fertility 'maintenance' the land's productive capacity can be severely harmed. Also, unlike capital, land is by definition not geographically mobile.

### 8.2.1   *The balance sheet for agriculture*

Land, then, is usually treated together with capital goods as making up agriculture's stock of productive assets and is included in any industry balance sheet at a value calculated by multiplying the area of land by market prices.[7] Farming's assets and liabilities are presented in Table 8.4. Any balance sheet for UK agriculture estimated in this way is dominated by the value of land and its associated fixed equipment (buildings, fences, roads, drainage and so on), which together can be labelled 'real estate'. Sometimes the term 'landlord capital' is also employed for these assets, even where land is owner-occupied, as under the landlord and tenant system these were traditionally what was provided by the landlord, who was also responsible for the repair and maintenance of the farm's buildings and other immovable equipment. The method of value calculation will influence the overall picture to some extent, but the importance of land is obvious; real estate formed 80 per cent of the total value of all assets in 1983. The value of land is not readily realisable, but it can serve as collateral for borrowing and has appreciated markedly since the late 1960s under pressures originating at least in part from outside the farming sector (see Chapter 6).

Despite rising liabilities (at least in current terms) at the industry level, borrowing relative to assets is not high, probably only of the order of 10 per cent, and the ratio was little different in 1983 from its 1970 level. Other non-farm wealth held by farmers and landowners is not usually considered in estimates of agriculture's financial position, although it is often important to their business activities. Similarly, their non-agricultural liabilities are usually ignored, although not infrequently these may be secured using farming assets as collateral.

In view of the importance of part-time farming and, in the UK, its common form of running another business in parallel with the farm which would be expected to share the same capital base and compete for funds, a narrow view of 'agriculture's' balance sheet is probably outmoded and, possibly, misleading. Unfortunately, evidence for a wider approach is not yet available.

Over the period 1970–83, the balance between the value of land and buildings ('landlord capital') and that of machinery, livestock and crops altered markedly from about 2:1 to 4:1. This has resulted largely from land appreciation and means that much more borrowing is required than before if a farm is to expand by purchasing land, the principal way of expansion in the 1970s and 1980s. In practice this is thought to have reduced the rate at which farm enlargement has occurred, as only smaller parcels of land can be financed using farm profits which themselves have declined since the mid-1970s; the slowdown in structural change in British agriculture is examined further in Chapter 9. While since 1979 land prices have declined in real terms, which might suggest greater mobility once more, so too has the income of the industry, constaining the ability to purchase. The net outcome of these changes is as yet unclear, although some disturbance of the pattern of the 1970s seems to have taken place (see Chapter 9).

Another feature discernable from Table 8.4 is that, although the ratio of

214

Table 8.4
Estimated UK Farming Balance Sheet as at June (£ million, current prices)

| | 1970 | (%) | 1978 | (%) | 1983 | (%) |
|---|---|---|---|---|---|---|
| *Assets* | | | | | | |
| Cash | 200 | (2) | 343 | (1) | 534.3 | (1) |
| Debtors | 200 | (2) | 443 | (1) | 701.8 | (1) |
| Stocks on farm | n.a. | | 101 | (0) | 138.2 | (0) |
| Quick assets | 400 | | 887 | | 1 383.3 | |
| Growing crops | 600 | (6) | 737 | (2) | 1 187.7 | (2) |
| Livestock | 1 300 | (12) | 4 215 | (11) | 6 393.3 | (10) |
| Machinery | 800 | (7) | 3 325 | (8) | 3 871.1 | (6) |
| Sub total (tenants capital) | 3 100 | (28) | 9 164 | (23) | 12 835.4 | (20) |
| Land and buildings | 7 800 | (72) | 30 711 | (77) | 51 211.2 | (80) |
| | 10 900 | (100) | 39 875 | (100) | 64 046.8 | (100) |
| *Liabilities* | | | | | | |
| Creditors (trade) | 300 | (3) | 407 | (1) | 478.4 | (1) |
| Bank | 500 | (5) | 1 703 | (4) | 4 696.0 | (7) |
| HP | n.a. | | 136 | (0) | 152.0 | (0) |
| Leasing | n.a. | | 60 | (0) | 362.0 | (0) |
| Current liabilities | 800 | | 2 306 | | 5 688.4 | |
| LIC AMC | 200 | (2) | 349 | (1) | 462.0 | (1) |
| Other long term | | | 81 | (0) | 106.3 | (0) |
| Total liabilities | 1 000 | (9) | 2 736 | (7) | 6 256.7 | (10) |
| Net worth (including family loans) | 9 900 | (91) | 37 139 | (93) | 57 790.1 | (90) |
| | 10 900 | (100) | 39 875 | (100) | 64 046.8 | (100) |

SOURCES   Midland Bank (1970); Barclays Bank (1978, 1983).

liabilities to total assets has, if anything, declined, borrowing has *risen* as a percentage of those assets which could be sold fairly quickly if a loan were called in.

### 8.2.2 Borrowing at the industry level

Historically, agriculture is not an industry which has used large quantities of publicly subscribed capital or bank borrowing. No doubt some borrowing by farmers has always gone on, much of it in the past between members of families. Farming has traditionally relied upon internally generated sources of finance and credit, especially for its tenant capital. The forms of business used in UK agriculture (mainly sole-proprietorships, partnerships and private companies, discussed in Chapter 9) mean that it cannot borrow direct from the public but has to use credit institutions such as banks, finance houses and the specialist Agricultural Mortgage Corporation (for land purchase and improvements).

Between the two wars, Professor Ashby found that the principal sources of credit were: inheritance, gift, marriage, two-thirds; saving out of current income, one-third. This picture was summarised as patrimony, matrimony and parsimony. Borrowing does not even merit a mention in this scheme. More recently (1960s and early 1970s) 70 per cent of funds for land purchase came from farmers' own resources. Even in the early 1980s, extra borrowing was only a minor way of financing new capital.

The total borrowing by UK agriculture at any one time cannot be precisely quantified, because only inexact estimates have been made of the amount of private credit and mortgages, frequently arranged between members of the same farming family, and of trade credit from merchants. It is generally agreed that, in the postwar period, the share accounted for by the banks, together with the specialist agencies for credit for land purchase and improvement, has risen and has been accompanied by a decline in family and trade credit, although the latter has commonly been overstated. Rather surprisingly the magnitude of total debt at constant prices was of a similar level in 1985 as in 1953 (Table 8.5) although it was lower in the 1970s and has risen again in the 1980s.

Within the more readily quantified institutional credit sector, the balance between the banks and the land mortgage corporations (principally the Agricultural Mortgage Corporation (AMC) and the Scottish Agricultural Securities Corporation (SASC)) has altered markedly; during the latter 1960s and up to 1971, the AMC and SASC expanded their lending much more rapidly than did the banks, a pattern associated with rising land prices of the period. The new balance was approximately maintained until the late 1970s when the increase in bank credit outstripped that of the other institutions (Table 8.6). Over the 1970s land purchase accounted for a

**Table 8.5**
**Loans to Agriculture (Borrowing by UK Agriculture)**

| | 1953 (£M) | (%) | 1963 (£M) | (%) | 1974 (£M) | (%) | 1980 (£M) | (%) | 1983 (£M) | (%) | 1985* (£M) | (%) |
|---|---|---|---|---|---|---|---|---|---|---|---|---|
| Long-term institutional | 25 | 3 | 100 | 8 | 290 | 16 | 800 | 17 | 900 | 13 | 950 | 12 |
| Banks | 200 | 23 | 500 | 42 | 840 | 47 | 2950 | 61 | 4800 | 68 | 5400 | 68 |
| Others (including trade credit, private mortgages and loans and hire purchase) | 655 | 74 | 590 | 50 | 650 | 37 | 1050 | 22 | 1400 | 20 | 1500 | 19 |
| Total liabilities | 880 | 100 | 1190 | 100 | 1780 | 100 | 4800 | 100 | 7100 | 100 | 7900 | 100 |
| Total liabilities at constant 1975 prices | 2924 | | 2967 | | 2211 | | 2454 | | 2806 | | 2830 | |

NOTE  * Forecast.
SOURCE  Collected from several sources in Centre for Agricultural Strategy (1978) *Capital for Agriculture* CAS Report No. 3 (University of Reading); and MAFF (1986) *Farm Incomes in the United Kingdom* (London: HMSO).

**Table 8.6**
**Advances by Banks and Specialised Land Mortgage Organisations[1]**

|  | Bank advances to agriculture, forestry and fishing (Feb. quarter)[2] | A as a percentage of total advances to UK residents | Agricultural Mortgage Corporation loans outstanding | Scottish Agricultural Securities Corporation loans outstanding (end of March) | $\dfrac{B + C}{A} \times 100$ |
|---|---|---|---|---|---|
|  | (£m) (A) | (%) | (£m) (B) | (£m) (C) |  |
| 1960 | 325 | 10.0 | 36 | 3 | 12 |
| 1965 | 505 | 9.5 | 64 | 6 | 14 |
| 1970 | 523 | 6.5 | 154 | 9 | 31 |
| 1975 | 1020 | 3.2 | 274 | 13 | 28 |
| 1980 | 2656 | 4.6 | 398 | 17 | 15 |
| 1984 | 5246 | n.a. 4.0 (1983) | 512 | 14 | 10 |

NOTES   1. Excludes the very small amounts advanced by LIC.
2. 1960–5 GB only.
SOURCES   EC (1975) *Monthly Digest of Statistics* (Brussels); CSO, *Economic Trends* (London: HMSO).

major share of the additional credit taken. However the rapid recent rise in bank lending has come, it appears, from a lack of liquidity. Since 1979 farmers have been faced with the necessity to extend their borrowings to finance normal farming activities (fertiliser, feed, wages and so on) in a period where moderate inflation has been combined with declining incomes.

More important than the absolute size of agricultural debt is the amount of burden it imposes on farm incomes and the way that this burden is distributed among farms. Despite rising interest rates, the cost of interest payments on the debt formed a remarkably constant 13 per cent of national Net Farm Income between 1953 and 1974, although the method of calculation may under-estimate the size of the burden in later years[8]. Recently, however, things have changed; borrowing, excluding that for land purchase (a rather artificial distinction but one commonly drawn in the official accounts for agriculture) has recently absorbed a much greater proportion

**Table 8.7**
**The Interest Burden on UK Farming**

|  | Pre-interest farming income (£, at current prices) (A) | Interest[1] on borrowing (£, at current prices) (B) | B as % of A |
|---|---|---|---|
| 1970 | 624 | 49 | 8 |
| 1971 | 684 | 47 | 7 |
| 1972 | 731 | 54 | 7 |
| 1973 | 1035 | 88 | 9 |
| 1974 | 919 | 124 | 14 |
| 1975 | 1119 | 124 | 11 |
| 1976 | 1422 | 139 | 10 |
| 1977 | 1409 | 153 | 11 |
| 1978 | 1442 | 190 | 13 |
| 1979 | 1464 | 323 | 22 |
| 1980 | 1511 | 464 | 31 |
| 1981 | 1836 | 468 | 26 |
| 1982 | 2318 | 501 | 22 |
| 1983 | 2002 | 494 | 25 |
| 1984 | 2602 | 569 | 22 |
| 1985 (p) | 1850 | 696 | 38 |

NOTE    1. Interest charge *excludes* loans for land purchase.
SOURCE    Derived from MAFF, *Annual Review of Agriculture* (London: HMSO).

of income, although the peak seems now to be past. During most of the 1970s, interest on this so-called 'commercial' borrowing represented less than 12 per cent of pre-interest farming income. However, since 1979 the proportion taken by interest has risen to about one quarter, with peaks of 31 per cent in 1980 and 38 per cent in 1985. As well as incomes being unusually low in these years in real terms, interest payments have been at peaks (see Table 8.7). 1985 witnessed a more extreme repetition of the conditions of 1980; income was lower and interest payments higher. These payments are determined not only by the volume of borrowing, which farmers can influence, but also by the rates charged on loans; both 1980 and 1985 corresponded with periods of historically high real interest rates. These are, of course, beyond the control of the agricultural industry.

### 8.2.3 Borrowing at farm level

At the farm level the borrowing pattern is by no means uniform although information is now rather dated. In England in 1969 the indebtedness of farming averaged 10.7 per cent of total liabilities, but over half the farms (54.9 per cent) had no liabilities, other than the short-term deferments of payment for purchases until the end of the month, a practice widespread in commerce.[9] Only five per cent of farmers accounted for 65 per cent of all borrowings. The most heavily indebted farmers (those with liabilities more than 30 per cent of assets) tended to come from the 40.5–121.5 ha 'working' size group and from the 40–49-year-old 'working' age group. They also tended to be full-time proprietors and to be tenants; but above all they tended to be relatively recent entrants and, in terms of total borrowings, to be owner-occupiers. This is a reflection on the rising price of land and a lack of availability of farms to rent, making it increasingly difficult to enter farming without borrowing heavily. Since 1969 the overall level of borrowing has risen in real terms, and probably fewer farmers operate without using some credit, most likely a bank overdraft. But there is undoubtedly still a wide range in the extent to which this facility is used.

More recent data from the Farm Management Survey (FMS) confirm the diversity of levels of indebtedness, although the FMS is not based on a random sample of farms so the results do not necessarily apply directly to farming at large. In 1985 half the farms in the sample were little indebted, with liabilities of less than one quarter of their farming assets, excluding the value of their land. Only 8 per cent were heavily indebted, with liabilities greater than the value of their non-land assets, and very few of these in fact owned no land.[10] The FMS also found that tenants as a group are more heavily indebted relative to their assets than are owner-occupiers although in absolute terms their liabilities are less. In 1985 tenant farms in the FMS had a liabilities-to-assets ratio of 27 per cent, owner-occupied farms 10 per cent and mixed-tenure ones 15 per cent; the percentages for each group had increased between 1979 and 1984, reflecting the rising borrowing of agriculture as an industry noted above. The difference in liabilities-to-assets ratio appears not to reflect any fundamental difference in attitude towards taking credit, but the effect of rising land prices which raise the value of land owned by established farmers without directly affecting their liabilities; among owner-occupiers one of the main influences on the ratio appears to be the time period when land purchase was made with those who bought their land longer ago being the less indebted.

Most farmers, particularly owner-occupiers, seem to be in a strong borrowing position; their equity is high and their main asset which forms collateral has a history of appreciating in real terms, at least when viewed over a run of years. Yet the absolute level of short-term borrowing of the financially stronger owner-occupiers is currently little different from that of tenants and they do not appear to exercise their borrowing power to finance higher levels of working capital or machinery stocks. The low levels of indebtedness of the general run of farmers can only be explained inadequately; the risk-aversion of farmers in the face of the considerable fluctuations in farm incomes which can occur is seen as a partial explanations for low borrowings.[11] Borrowing levels are assessed principally not on the relationship between assets and liabilities but more on the ability of the business to service the loans (the payment of interest plus any mandatory repayments of capital) not just in normal years but in those of unusual difficulties, resulting not only from poor weather or disappointing market prices but also from increasingly volatile rates of interest on loans. Although the land asset gives owners a degree of cushion against misfortune which tenants do not have, many owner-occupiers are reluctant to use their land as collateral more than is necessary since this implies a potential loss of control of it. Another element could be the lack of suitable on-farm investment opportunities with yields commensurate with the cost of borrowing. Another, linked to the bias towards the elderly in the population of farmers, might be the inertia towards change in the scale of activity which increases with age, known to apply to farming. The association between greater age (and stage of farming career) with lower indebtedness and reduced willingness to change the scale of farming activity is just one more manifestation of the integration of personal and business life in agriculture.

## 8.3   The Costs of Borrowing

Because of the effects of inflation and the way that farmers, as business operators, are allowed to treat interest payments on borrowed funds, there is frequently a very real difference between the nominal interest rates charged for loans and the effective burden they impose on the farm business. This difference can, in part, explain why farmers have shown surprising willingness to buy land on the rare occasions when it has come available in a location convenient to them, even though at first sight the investment appears unattractive in terms of the cost of the land, interest rates on loans from banks or mortgage corporations and the income-generating ability of the land.

For tax purposes, interest charges on business loans are deductable from income, thereby reducing the cost of borrowing. The higher the marginal rate of tax paid by the borrower the greater the effective reduction in the cost of borrowing. In Table 8.8 the standard rate of tax is used as an example for calculating the effective rate paid by farmers, but there must have been many individuals for whom the rate of tax on their higher slices of income would have been greater. When inflation occurs, the effective

**Table 8.8**
**The Costs of Borrowing**

| | Average[1] AMC mortgage interest rate[2] | | Standard rate of income tax (SRT) (%) | AMC interest rate x (1-SRT) (A) | % change in Retail Price Index (B) | Real rate of interest on AMC loans (A)–(B) |
|---|---|---|---|---|---|---|
| 1950–4 | 4.8 | | 36 | 3.1 | 5.3 | −2.2 |
| 1955–9 | 5.6 | | 32 | 3.8 | 3.3 | +0.5 |
| 1960–4 | 6.6 | | 30 | 4.6 | 2.6 | +2.0 |
| 1965–9 | 8.4 | | 32 | 5.7 | 4.3 | +1.4 |
| 1970 | 9.9 | | 32 | 6.1 | 6.4 | −0.3 |
| 1971 | 10.1 | | 29 | 7.0 | 9.4 | −2.4 |
| 1972 | 9.0 | | 29 | 6.4 | 7.1 | −0.7 |
| 1973 | 11.1 | | 30 | 7.8 | 9.2 | −1.4 |
| 1974 | 14.7 | | 33 | 9.8 | 16.1 | −6.3 |
| 1975 | 14.9 | | 35 | 9.7 | 24.2 | −14.5 |
| 1976 | 14.8 | (13.6) | 35 | 9.6 | 16.8 | −7.2 |
| 1977 | 14.3 | (11.7) | 34 | 9.3 | 15.9 | −6.6 |
| 1978 | 13.7 | (11.6) | 33 | 9.2 | 8.3 | +0.9 |
| 1979 | 14.5 | (16.1) | 30 | 10.2 | 13.3 | −3.1 |
| 1980 | 16.4 | (18.7) | 30 | 11.5 | 18.1 | −6.6 |
| 1981 | 16.1 | (15.6) | 30 | 11.3 | 11.9 | −0.6 |
| 1982 | 15.5 | (14.3) | 30 | 10.9 | 8.7 | +2.2 |
| 1983 | 14.4 | (12.2) | 30 | 10.1 | 4.6 | +5.5 |
| 1984 | 13.7 | (12.2) | 30 | 9.6 | 5.0 | +4.6 |
| 1985[3] | 14.1 | (14.7) | 30 | 9.9 | 6.0[4] (Jan–Oct) | +3.9 |

NOTES   1. Weighted. Fixed rate loans.
2. Figures in brackets relate to rates payable on short-term bank loans. Although the real rate of interest has been calculated based on AMC work rates, a closely similar exercise can easily be carried out using bank interest rates; the end results show a similar pattern.
3. Provisional.
4. January to October.

SOURCES   Harrison, A. in EC (1981) *Factors influencing the Ownership, Tenancy, Mobility and Use of Farmland in the United Kingdom* (Brussels). EC, *The Agricultural Situation in the Community* (various years) (Brussels). MAFF (1986) *Departmental Net Income Calculation* (London: MAFF).

interest rate is also reduced by the rate at which money loses its value. Table 8.8 shows that the effective rate paid by borrowers financing assets, whose value in nominal money terms kept pace with the rate of inflation, has been below 2 per cent for most years since 1950. During the 1970s and early 1980s the effective rate has been *negative*, except for 1978 when the rate of inflation was contained to 8.3 per cent, and again in the most recent years for the same reason. Land purchase will have accounted for a major share of the additional credit taken during the 1970s, when land prices rose at a rate generally faster than the Retail Price Index, making borrowing for this purpose even more attractive. Under such circumstances the burden of credit takes the form of initial stress on the farm business when service charges are large in comparison with farm income; with time, inflation erodes the relative size of servicing payments as incomes rise, and the liabilities-to-assets ratio of the business will be enhanced. The general low indebtedness of the industry has meant that there has been no shortage of potential land purchasers able and willing to accept the high short-term charges involved in land acquisition.

# Notes

1. Griffin, T. J. (1975) 'Revised estimates of the consumption and stocks of fixed capital' *Economic Trends* **264**; 126–9, and Griffin, T. J. (1976) 'The stock of fixed assets in the United Kingdom: How to make best use of the statistics' *Economic Trends,* **276**: 130–43.
2. Hill, Berkeley and Kempson, R. E. (1977) *Farm Building Capital in England and Wales* (Ashford: Wye College).
3. Gasson, R. and Hill, Berkeley (1984) *Farm Tenure and Performance* (Ashford: Wye College).
4. Centre for Agricultural Strategy (1978) *Capital for Agriculture* CAS Report No. 3 (University of Reading).
5. Black, C. J. (1965, 1966) 'Capital developement on farms in theory and practice' 1 and 2 *Farm Economist* **10**: 475–84, and **11**: 10–23.
6. Black, C. J. (1967) *Investment policy and farm buildings* Economics Division, School of Agricultural Science, University of Leeds.
7. For a discussion of the size and nature of the market in land, see Chapter 6.
8. Centre for Agricultural Strategy (1978) op. cit.
9. Harrison, A. (1975) *Farmers and farmbusinesses in England* Miscellaneous Study No. 62, Department of Agricultural Economics, University of Reading.
10. MAFF (1986) *Farm incomes in the United Kingdom* (London: HMSO).
11. Harrison, A. (1975) op. cit.

# Farming Structure 9

## 9.1 The Meaning of the Term 'Structure'

'Structure' is used as a term referring to the inner composition of something. We talk about the structure of a building, implying that it can be thought of in terms of subordinate parts which together make the whole. Again, the structure of an atom can be described in terms of electrons, protons and neutrons. The term need not only apply to material things; we can describe the structure of, say, a cricket club committee as consisting of a chairman, secretary, treasurer and other elected members. The purpose of knowing about the structure of anything goes beyond simple curiosity; such information gives insight into how the atom, committee or building functions or 'works' and, even more important, allows us to suggest explanations for why certain things have happened in the past and enables future responses to be predicted.

Agriculture is in a constant state of flux or adjustment. The structure of its output changes as certain products become more profitable – the rise in cereal production in the 1970s and early 1980s is a good example. The structure of its inputs has changed – notably, the reduction in the manpower it engages and the rise in its use of machinery. And the number of small farms has fallen sharply over the last forty years, changing the production structure of the industry. These changes all carry with them implications for rural communities through changing employment possibilities, changing the structure of society, and for the natural environment. Frequently these concomittant adjustments will be of greater national significance than the narrow agricultural change itself. The term 'agricultural adjustment' is given to the response which agriculture makes to new technology or altered price levels, and adjustments results in a change in the structure of agriculture.

**Table 9.1**
**Bases Frequently Used for Describing the Structure of UK Agriculture**

| Starting point | Initial subdivision | Criterion |
| --- | --- | --- |
| A. Farms | by size groups | i. land area<br>ii. capital value of farm<br>iii. labour used<br>iv. value of output<br>v. theoretical size measures European Size Units (based on standard gross margin) or standard man-days |
| | typology (e.g., dairy, cereal, mixed) | composition of value of output (actual or estimated) or of labour devoted to each type of product |
| | tenure (e.g., owner-occupied, tenanted, mixed, other) | balance between amount of land in each tenure |
| | economic status (e.g., high or low performers, part-time or full-time farms) | i. level of profit generated<br>ii. return on capital achieved<br>iii. level of efficiency<br>iv. time spent on farm by farmer |

225

B. Production (e.g., national milk output, cereal output, etc.)
  by size of producing unit — i. amount produced / ii. total size of farm
  by type of farm
  by channel of marketing

C. Resources
  i. Land
    by quality of land — i. soil characteristics / ii. rent levels/prices
    by size group of farm
    by type of farming
    by form of tenure
    by region

  ii. Labour
    by size of labour force — i. farm size / ii. farm type / iii. farm tenure
    by age
    by sex
    by employment status
    by farm characteristic

  iii. Capital
    by amount of capital — i. landlord type or tenant type / ii. owned or borrowed
    by composition
    by farm characteristic — i. farm size / ii. farm type / iii. farm tenure

### 9.1.1   Ways of describing agriculture's structure

The most common way by which to describe structure is in terms of farms – their numbers, sizes in area or capital worth, farming type (dairy, cereals, etc.) tenure and so on. Any number of subdivisions and cross-tabulations are possible – for example, numbers of farms could first be broken down by farm type, then subdivided into area size groups, then into tenure and then into high or low efficiency units. Much will depend on what the analysis is to be used for. But another basis of description is possible. For example, the nation's milk supply could be the starting point, with a description of how much comes from each region, how much from large producers and from small ones etc. Similarly, agriculture's labour force could be described by age, by sex, by self-employment or employer status, by size of workforce, etc. Some of the possible starting points are given in Table 9.1. It should be evident that Chapters 5 to 8 have already included descriptions of agricultural structure from the points of view of land, labour and capital and in terms of the industry's output composition. These will not be repeated. Here we are concerned with the units which organise the factors of production and transform them into saleable commodities – farm businesses. The structure we describe is thus the business structure of UK agriculture.

## 9.2   Farm Businesses

At first consideration a farm business might seem to be a fairly self-evident and easy unit to describe – each farm would correspond with clearly identifiable patches of land, worked by a labour force who could if necessary be named, and comprising the crops, animals, buildings and machinery found on the farm's land. As with any other businesses, farms have output which is sold (or in part consumed by the farm family) and costs of production. But in practice there is often a real problem in defining a farm business in relationship to a given area of land. If a farmer buys an adjacent farm, do the two become one larger merged farm? Does it alter the answer if they are managed as separate units or farmed as one? And if a farmer also owns a butchery business which uses the land to keep cattle bought at markets until they are needed for slaughter, where does the farm business end and where does the butchery business begin? It is possible to separate them? Such problems are met increasingly as farms amalgamate to form larger units and as an increasing proportion become 'part-time' in the sense that their owners also have some other business interests outside farming.

No one definition of a farm or farm business will be universally appropriate. However, perhaps the most useful is that given by Harrison:[1] a farm is defined 'so as to embrace such farming activities as fall within the compass of a given fund of capital. To count as a single business unit, there must be participation in a regular and at least annual assessment of the capital position with all sectors contributing to and competing for resources.' Hence, single ownership of several units of production probably geographically separate would not by itself be a sufficient condition to make them parts of one 'farm', but where one farmer buys a second farm business (say, on the retirement of its elderly operator) in the locality, runs them with a single stock of machinery and takes management decisions based on the amalgamated area, the formerly separate units have clearly become one. While this definition will run into complications at times, notably when dealing with part-time farmers and those splitting up farms in order to establish their children in businesses, it is generally workable.

Structure studies in the UK are made possible largely through the published results from the annual (June) official census conducted by MAFF. The basic unit in these official statistics is the 'holding'; this refers to the block of land which is operated as a single technical unit by the farmer.[2] While frequently interpreted as being synonymous with 'farm', non-official surveys of agricultural businesses repeatedly expose the discrepancies between what is returned in official censuses as a separate holding and what constitutes a farm, even allowing for reasonable latitude in the definition of the term 'farm'. This arises largely because, when farms amalgamate, farmers tend still to complete and return separate census forms for the combined units.

Although attempts have been made to encourage farmers to complete one census form for all the land occupied by them, it is by no means certain that this process has significantly improved on, let alone eradicated, the overstatement of small farms and understatement of large ones that a description of holdings produces. In England in 1969, the number of farms was estimated as 12 per cent less than the official number of holdings in spite of an understatement of the number of units of 300 acres (121 ha) and above. The bigger the farm, the more their numbers were understated.[3] A more recent survey in England[4] revealed that 6 per cent of holdings were parts of farms more than twice their size. In Scotland a similar situation exists.

Despite the problems of using a land-based definition of a farm and disparities between official statistics based on holdings and the numbers of independent farms, the size structure of UK agriculture is usually given in terms of holdings. Land area is intuitively simpler to grasp than blocks of capital expressed in money values, and statistics based on holdings are regularly and meticulously collected. Assuming that the farm/holding relationship is reasonably stable, holding data can be an acceptable indicator of changes in farming structure.

# Table 9.2
## Numbers of Holdings[1] by Area Size Group[2]

| | Under 2 hectares | 2–4.9 hectares | 5–9.9 hectares | 10–19.9 hectares | 20–29.9 hectares | 30–39.9 hectares | 40–49.9 hectares | 50–99.9 hectares | 100–199.9 hectares | 200–299.9 hectares | 300–499.9 hectares | 500–699.9 hectares | 700 and over hectares | Total |
|---|---|---|---|---|---|---|---|---|---|---|---|---|---|---|
| **United Kingdom** | | | | | | | | | | | | | | |
| 1979 | 16 640 | 29 385 | 37 361 | 45 436 | 31 201 | 21 772 | 17 085 | 43 384 | 24 433 | 6 803 | | 7 566 | | 281 056 |
| 1983 | 14 810 | 21 127 | 33 528 | 43 706 | | 67 055 | | 42 595 | 24 481 | | 14 646 | | | 261 948 |
| 1983 (%) | 5.7 | 8.1 | 12.8 | 16.7 | | 25.6 | | 16.3 | 9.3 | | 5.6 | | | 100.0 |
| **England** | | | | | | | | | | | | | | |
| 1979 | 13 288 | 18 307 | 19 239 | 21 766 | 16 602 | 12 515 | 10 500 | 28 144 | 16 664 | 4 718 | 2 865 | 744 | 629 | 165 981 |
| 1983 | 11 875 | 14 718 | 17 552 | 20 868 | 15 621 | 12 024 | 10 034 | 27 411 | 16 716 | 4 861 | 2 905 | 812 | 639 | 156 036 |
| 1983 (%) | 7.6 | 9.4 | 11.2 | 13.4 | 10.0 | 7.7 | 6.4 | 17.6 | 10.7 | 3.1 | 1.9 | 0.5 | 0.4 | 100.0 |
| **Wales** | | | | | | | | | | | | | | |
| 1979 | 651 | 2 670 | 4 026 | 5 074 | 4 055 | 3 174 | 2 539 | 6 032 | 2 420 | 456 | 261 | 74 | 66 | 31 498 |
| 1983 | 782 | 2 012 | 3 693 | 4 895 | 3 766 | 3 023 | 2 445 | 5 971 | 2 477 | 493 | | 400 | | 29 957 |
| 1983 (%) | 2.6 | 6.7 | 12.3 | 16.3 | 12.6 | 10.1 | 8.2 | 19.9 | 8.3 | 1.6 | | 1.3 | | 100.0 |
| **Scotland** | | | | | | | | | | | | | | |
| 1979 | 1 405 | 2 489 | 2 334 | 2 900 | 2 423 | 2 067 | 1 947 | 6 378 | 4 734 | 1 534 | 1 148 | 450 | 1 241 | 31 050 |
| 1983 | 1 505 | 2 695 | 2 443 | 2 910 | | 6 146 | | 6 216 | 4 669 | | 2 682 | | 1 670 | 30 936 |
| 1983 (%) | 4.9 | 8.7 | 7.9 | 9.4 | | 19.9 | | 20.1 | 15.1 | | 8.7 | | 5.4 | 100.0 |
| **Northern Ireland** | | | | | | | | | | | | | | |
| 1979 | 1 296 | 5 919 | 11 762 | 15 696 | 8 121 | 4 016 | 2 099 | 2 830 | 615 | 95 | | 78 | | 52 527 |
| 1983 | 648 | 1 702 | 9 840 | 15 033 | 7 889 | 3 973 | 2 134 | 2 997 | 619 | 109 | | 78 | | 45 019 |
| 1983 (%) | 1.4 | 3.8 | 21.9 | 33.4 | 17.5 | 8.8 | 4.7 | 6.7 | 1.4 | 0.2 | | 0.2 | | 100.0 |

NOTES 1. Excludes minor holdings data, i.e. those which fail to satisfy tests that determine whether they can be treated as commercial operations.

2. Total area includes crops and grass, sole rights rough grazing (except rough grazing on land owned by the Northern Ireland Forest Service), woodland and other land.

SOURCE MAFF, *Agricultural Statistics: United Kingdom* (London: HMSO).

**FIGURE 9.1**
**Distribution of Numbers of Holdings and Total Area by Size of Holding, UK 1982**

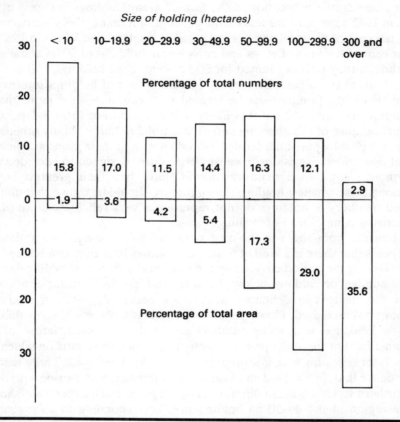

SOURCE   MAFF Agricultural Statistics (UK)

## 9.3   The Size Structure of UK Farming Based on Areas of Holdings

Table 9.2 shows the number of holdings in the UK and its constituent countries for two recent years. Figure 9.1 puts the numbers and areas occupied by holdings in a diagrammatic form. These illustrate both the wide variation in the sizes found, and also what changes are currently under way. Bearing in mind the above reservations, several points emerge.

First, Northern Ireland has a very different size structure from the rest of the United Kingdom, with proportionally far fewer large holdings but a relative concentration between 2 and 20 ha, and particularly between 12 and 20 ha. Second, although they are relatively few, large holdings account for a substantial proportion of the farmed area. Holdings in excess of 300 ha in 1982 accounted for less than 3 per cent of total numbers but occupied 36 per cent of the total agricultural area. In England and Wales in 1983, 77 per cent of the area of crops and grass was in holdings of 50 ha and above, although they only accounted for 30 per cent of all holdings.

Third, at the other extreme, the large numbers of holdings returned in the small size groups must be treated with caution; while more than a quarter are below 10 ha, it is likely that many do not form independent units because of the phenomenon of 'multiple' holdings. Many others will be occupied by part-time farmers who do not treat their farms as commercial operations. Revisions to census procedure, designed to exclude those without significant agricultural output, have had their greatest impact among these smallest holdings; the inference should be drawn that unqualified numbers of holdings do not represent a very reliable picture of the changing composition of farming.

Fourth, from the way holding numbers have changed over time, it appears that there is a kind of 'watershed' somewhere between 100 and 200 ha. During the period covered by the table and over the preceding decade, the number of holdings below this size band has declined (after allowing for the changes in definition which have occurred), while the number above has increased. However, a more detailed table would show that the early 1980s has seen rising numbers among the very small farms, of less than 5 ha, but these are predominantly not commercial units on which the operator relies for a significant proportion of his livelihood. There is some evidence that, in England and Wales, the watershed size is rising with time. Northern Ireland is again different in that it has a much lower watershed in the region of the 40–50 ha holding size, corresponding to its prevailing pattern of smaller farms.

The figures produced by the annual census are snapshots; they show cross sections at discrete points in time and give indications of overall changes. However, they miss an additional large number of small adjustments in land between farmers which tend to cancel each other. These go on as some farmers expand because of ambition, opportunity, high profits or family situation while others contract because they are becoming older, or their farming objectives change, or they are forced to sell all or part of their land to meet business crises. These changes could go on without involving any alteration in the overall numbers and sizes of holdings. There is evidence that such changes at the individual farm level may be substantial.[5] On a national level, the dynamics of the situation could only be revealed by a longitudinal analysis of changes over a period of time on individual holdings; MAFF has not yet published such an analysis.

### 9.3.1  Size distribution in other EC member states based on area

Similarities are to be found in the structures of other member states of the EC. In each the smaller farms are of great numerical importance but occupy little area, a position reversed among larger farms. Also each has a watershed in terms of changing numbers. In most countries holdings of less than 20 ha have been declining whereas those of over 50 ha has been rising, the UK being an exception with a higher watershed size. This difference reflects the markedly different size distribution of farms in the UK from that found in most other EC member states, a factor leading to difficulties in designing an agricultural policy which applies commonly across the whole Community. Measures appropriate to a farming system dominated by very small farms, such as product price support for commodities already in surplus, may seem perverse when adopted by the UK. Table 9.3 shows the size distribution of holdings in the countries of the EC, both in terms of numbers and the areas each size class occupies. One major feature is the much greater numerical importance of holdings of less than 5 ha in the other countries; in the whole EC their proportion of total numbers (47 per cent) is almost four times that in the UK. Only Denmark has a similar low level of incidence of these very small holdings, the proportions in the major agricultural countries of West Germany, France and Italy, which together occupy 60 per cent of the Community's farmed area, being much greater (35, 21 and 68 per cent respectively).

The UK is unique in the importance of its large farms, in this context implying those of 50 ha and over. These take up 82 per cent of the agricultural land area in the UK, double the proportion for the entire Community and much higher than in West Germany (20 per cent) and Italy (31 per cent). No other country even remotely approaches the UK figure, the nearest being France with 43 per cent. One-quarter of all the farms above 50 ha in the Community are situated in the UK, although less than a twentieth of all holdings are found here.

## 9.4  Alternative Measures of Size

### 9.4.1  Standard man-days

As is noted above, the physical area of a farm is not often the most appropriate measure of business size; because of factors such as land

**Table 9.3**
**Farm Size Structure in the European Community, 1980**

| Size of holding (hectares) | UK No. of holdings | UK Area | Germany No. of holdings | Germany Area | France No. of holdings | France Area | Italy (1972) No. of holdings | Italy (1972) Area | Netherlands No. of holdings | Netherlands Area | Belgium No. of holdings | Belgium Area |
|---|---|---|---|---|---|---|---|---|---|---|---|---|
| 1–5 | 11.8 | 0.5 | 34.6 | 5.4 | 20.6 | 2.1 | 68.5 | 21.6 | 24.0 | 9.1 | 28.4 | 4.7 |
| 5–10 | 12.5 | 1.3 | 18.7 | 8.9 | 14.5 | 4.2 | 17.2 | 15.8 | 20.2 | 9.5 | 19.8 | 9.4 |
| 10–20 | 16.0 | 3.4 | 22.7 | 21.6 | 21.1 | 12.3 | 8.4 | 15.3 | 28.9 | 26.7 | 26.6 | 24.9 |
| 20–50 | 27.1 | 13.0 | 22.3 | 43.9 | 30.4 | 38.0 | 4.2 | 16.8 | 23.9 | 44.8 | 20.9 | 40.2 |
| >50 | 32.6 | 81.8 | 3.9 | 20.2 | 13.3 | 43.4 | 1.7 | 30.5 | 3.0 | 14.9 | 4.2 | 20.8 |
| Total | 100.0 | 100.0 | 100.0 | 100.0 | 100.0 | 100.0 | 100.0 | 100.0 | 100.0 | 100.0 | 100.0 | 100.0 |
| No. holdings[1] | 249 242 | | 797 378 | | 1 135 000 | | 2 191 972 | | 128 960 | | 91 181 | |
| Average holding size (ha)[1] | 68.7 | | 15.3 | | 25.4 | | 7.4 | | 15.6 | | 15.4 | |
| % of total Community farmed area[2] | 18.5 | | 11.9 | | 31.1 | | 17.3 | | 2.0 | | 1.4 | |
| % of total Community persons employed in agriculture[3] | 7.4 | | 16.6 | | 21.3 | | 33.9 | | 2.8 | | 1.3 | |

| Size of holding (hectares) | Luxembourg | | Ireland | | Denmark | | Greece (1977) | | Eur (10) | |
|---|---|---|---|---|---|---|---|---|---|---|
| | No. of holdings | Area | No. of holdings | Area | No. of holdings | Area | No. of holdings | Area | No. of holdings | Area |
| 1– 5 | 19.4 | 1.8 | 15.2 | 1.9 | 11.1 | 1.3 | 70.9 | 39.0 | 46.6 | 7.2 |
| 5–10 | 10.9 | 2.9 | 15.9 | 5.2 | 17.6 | 5.2 | 20.6 | 30.0 | 17.2 | 7.6 |
| 10–20 | 14.5 | 7.8 | 30.3 | 19.4 | 26.5 | 15.4 | 6.6 | 18.3 | 15.0 | 13.6 |
| 20–50 | 38.5 | 47.6 | 29.8 | 40.4 | 34.7 | 43.0 | 1.7 | 9.7 | 15.1 | 29.6 |
| > 50 | 16.8 | 39.9 | 8.8 | 33.1 | 10.1 | 35.1 | 0.2 | 3.0 | 6.1 | 41.9 |
| Total | 100.0 | 100.0 | 100.0 | 100.0 | 100.0 | 100.0 | 100.0 | 100.0 | 100.0 | 100.0 |
| No. holdings[1] | 4 697 | | 223 300 | | 116 342 | | 731 710 | | 5 670 000 | |
| Average holding size (ha)[1] | 27.6 | | 22.6 | | 25.0 | | 4.3 | | 15.7 | |
| % of total Community farmed area | 0.1 | | 5.6 | | 2.8 | | 9.1 | | 100.0 | |
| % of total persons employed in agriculture | 0.1 | | 2.5 | | 2.3 | | 11.8 | | 100.0 | |

NOTES 1. Of 1 hectare and over.
2. Utilised agricultural area.
3. In the sector Agriculture, hunting and forestry & fishing.

SOURCE EC (1986) *The Agricultural Situation in the Community* (Brussels).

quality, location and personal circumstances of the occupier, different parcels of land of the same area can represent widely varying levels of agricultural activity. Perhaps the most obvious measure of the size of farm businesses would be the total sale value of their outputs, although because of practical difficulties in collecting data this measure is not used in the UK (although it is in the USA). The alternative most widely used in Britain over the 1960s and 1970s was Standard Man-Days (SMDs), which measured holding size by the estimated labour requirement, using standards derived from surveys of farm practices. In the SMD classification, holdings of under 250 SMD (275 SMD in an early series) were estimated as not being capable of providing full employment for one man, although in reality many of these holdings formed the occupier's sole occupation and implying that he was under utilised or suffering 'hidden' unemployment.

Again there is a problem of determining what is the threshold of significance as an agricultural producer – in effect, what is the difference between a domestic garden and a very small farm – and changes in the criterion used for statistical significance have altered over the years, with a consequent alteration in the numbers of holdings shown in official statistics. With this in mind, it is nevertheless quite clear that, when holdings were measured in SMDs, the distribution of holding numbers by size group was in sharp contrast to the contribution each size group made to total agricultural production, as implied by total estimated SMDs. In 1985, holdings below the estimated one-man size (250 SMD) formed 53 per cent of holding numbers but contributed less than 10 per cent of output (see Table 9.4). By contrast, in 1975 (the latest year available with a detailed breakdown and using the 1968 standards), the 40 000 holdings of 1200 SMD and over (approximating to those occupying 4 men and over) formed only 15 per cent by numbers, but accounted for over half agricultural production (56 per cent) in the UK. The relative importance of this size group was similar in England and in Scotland (61 per cent and 58 per cent of total output in 1975 respectively), but in Wales and Northern Ireland these large farms accounted for a much lower share of total output (28 per cent and 22 per cent respectively). In 1975 the largest holdings (4200 smd and over) formed only 2 per cent of holding numbers, but provided 21 per cent of total output; they dominated the country's production in six of the main enterprises – broilers (69 per cent), laying fowls (43 per cent), sugar beet (30 per cent), potatoes (29 per cent), wheat (25 per cent) and breeding pigs (21 per cent).

Table 9.4 also shows what has been happening over time. Numbers of farms which are just large enough to fully occupy one or two men (the 250–499 smd group) have been falling faster than have larger farms; a watershed farm size is not evident simply because the largest category in the table is set at too low a level for it to appear. As with the area breakdown, there is evidence that the very small 'part-time' holdings are increasing in number.

**Table 9.4**
**Distribution of Holdings by SMD size group, 1976 and 1985**

| SMD size | 1976 ('000s) | (%) | (% SMD) | 1985 ('000s) Prov. | (%) | (% SMD) | Change 1976–85 (%) |
|---|---|---|---|---|---|---|---|
| 1000 + | 32.0 | 12.6 | | 29.8 | 12.4 | | −7 |
| 500–999 | 46.6 | 18.3 | 90.8 | 40.5 | 16.9 | 90.5 | −13 |
| 250–499 | 54.0 | 21.2 | | 42.0 | 17.5 | | −22 |
| <250 | 122.0 | 47.9 | 9.2 | 127.9 | 53.2 | 9.5 | +5 |
| All sizes | 254.8 | | | 240.3 | | | |

SOURCE  MAFF (1982, 1986) *Annual Review of Agriculture* (London: HMSO).

*9.4.2  European Size Units (ESUs) and British Size Units (BSUs)*

The first of these measures has risen to prominence since the UK became a member of the European Community; ESUs are used widely for purposes of the Common Agricultural Policy. The unit is based on standard gross margins and approaches the value-of-output measure of farm business size, although it uses not actual gross margins (the difference between the value of output and the variable costs of production such as seed, fertiliser, feed, etc., but before deducting fixed costs such as rent or regular labour) but 'standards'. These are based on average actual gross margins found by surveys (based on the period 1972–3 to 1974–5) and worked out as a per hectare (for crops) or per head (for livestock) standard. The size of a farm in ESUs is found by multiplying its actual crops and livestock by these standards (expressed in European Units of Account), summing them, and dividing by 1000 (i.e., 1 ESU = 1000 European Units of Account of SGM at average 1972–4 values).

The fact that the standards are based on an actual period means that, with the passage of time, certain standards become outdated because of technical change or relative price movements. Inflation can be simply taken cared of, but problems arise when different enterprises change to varying extents. For example, the standards for sheep are now too low, because they were calculated before the CAP sheep regime was introduced, and these have greatly improved the output values (and hence real gross margins) of this enterprise. Of course the same sort of problem arose with SMDs; as the capital intensity of farming increased labour requirements became

Table 9.5
Farm size in ESU and BSU: Class Average Physical Characteristics ESU, 1982/3. BSU, 1984/5, England

| Size | Type of Farm | | | |
|---|---|---|---|---|
| | Specialist dairying | | Specialist cereals | General cropping |
| | Cows (no.) | Area (ha) | Area (ha) | Area (ha) |
| ESU[1] | | | | |
| 4–7.9 }'small' | 22 | 30 | – | – |
| 8–15.9 | 42 | 33 | 46 | 51 |
| 16–23.9 }'medium' | 65 | 52 | 77 | 66 |
| 24–39.9 | 97 | 72 | 112 | 95 |
| 40–99.9 }'large' | 162 | 127 | 204 | 176 |
| 100–249.9 | – | – | 462 | 386 |

| | Dairying | Cereals | General Cropping |
|---|---|---|---|
| BSU[2] | | | |
| 4–15.9 'small' | 29 | 30 | 40 | 50 |
| 16–39.9 'medium' | 67 | 59 | 90 | 78 |
| 40 and over 'large' | 156 | 146 | 263 | 262 |

NOTES   1. 1 European Size Unit is equal to 1000 European Units of Account of standard gross margin at average 1972–4 values.
   2. 1 British Size Unit is equal to 2000 European Currency Units of standard gross margin at average 1978–80 values.
SOURCE   MAFF (1984) *Farm Incomes in England 1982/3* (London: HMSO) and MAFF (1986) *Farm Incomes in the United Kingdom* (London: HMSO).

overestimated. To get over this problem the standards have been recalculated for the UK based on 1978–80 values and expressed in European Currency Units (or ecu), with one BSU equalling 2000 ecu of gross margin. This results in less confusion than might be apparent, since

**Table 9.6**
**Distribution of Farm Businesses by ESU Size Group, UK 1982**

| Size of business (ESU)[1] | Holdings ('000s) | (%) | Agricultural activity (% of total SGM) |
|---|---|---|---|
| < 4 | 79.0 | 34.5 | 2.8 |
| 4– 7.9 | 32.9 | 14.4 | 4.4 |
| 8–15.9 | 42.1 | 18.4 | 11.2 |
| 16–23.9 | 24.9 | 10.9 | 11.2 |
| 24–39.9 | 23.8 | 10.4 | 16.7 |
| 40–99.9 | 20.2 | 8.8 | 27.5 |
| 100 and over | 6.0 | 2.6 | 26.2 |
| | 228.8 | 100.0 | 100.0 |

NOTE    1. 1 European Size Unit is equal to 1000 European Units of Account of standard gross margin.

SOURCE    MAFF, DAFS and DANI given in Furness, G. W. (1983) 'The importance, distribution and net incomes of small farm businesses in the UK' in *Strategies for Family-worked Farms in the UK*, CAS Paper No. 15 (University of Reading: Centre for Agricultural Strategy).

inflation has also to be taken into account. In practice, 1 BSU approximates to 1 ESU across most farm types. BSUs will increasingly be used as a measure of size in the UK, although for European Community purposes the ESU remains current. The figure in Table 9.5 give some feel to the ESU and BSU by showing their approximate UK equivalents (class averages) in cow numbers or area.

Table 9.6 shows for 1982 the breakdown of holdings in the UK by ESU size group. Often the very small farms of below 4 ESUs are dismissed from significance; they are almost certainly too small to be able to generate enough income for a full-time farmer. Although they represent a third of all holdings by number, they generate less than 3 per cent of all agricultural activity, as measured by the country's total standard gross margin. Those in the 4–16 ESU band are often labelled 'Small Farms' – probably operated by one or two men – and these account for 50 per cent of all holdings above the 4 ESU threshold, occupy 21 per cent of the area of crops and grass and generate 16 per cent of the total business activity. Above 16 ESU the farm is almost certainly capable of supporting a full-time farmer. The 16–24 ESU and 24–40 ESU groups are often labelled the 'lower medium' and 'upper medium' size farms respectively, with those of 40 ESU and over called 'large' farms. As was found with area measures and SMDs, the bigger farms dominate the overall output; in 1982 the upper medium and large farms (24 ESU and over) although forming only one-third of total holding numbers (even ignoring the very small group), occupied 59 per cent of the crops and grass area and accounted for 70 per cent of the aggregate agricultural activity.

There are some notable regional differences. Small farms (4–16 ESU) are numerically relatively more important in Wales (71 per cent of holdings above 4 ESU) and Northern Ireland (78 per cent) than in the UK as a whole (50 per cent). On the other hand, large (40 ESU plus) farms are disproportionately important in Eastern England, where they contribute four-fifths of the total agricultural activity and in Western and Northern England (two-thirds). By contrast in Wales only one-third comes from this group and in Northern Ireland only one-quarter (the UK overall figure is 54 per cent). As a corollary, in Wales and Northern Ireland the medium-sized farm (14–40 ESU) is relatively more important (59 and 58 per cent respectively).

## 9.5  Types of Farm

Tracing the changing structure of agriculture in terms of types of farm in the UK is made difficult because of changes in the criteria used to classify farms since the 1960s. In any event, typology is bound to be somewhat arbitrary. However, several features and trends are apparent which are illustrated in Table 9.7. The first is the numerical importance of dairy farms in this country's agricultural industry, with approaching one-third of farms falling into either the specialist or the mainly dairy group. The second feature is the relative stability of the overall composition despite falling numbers of holdings, although there has been an advance by specialist cereal farms in the latter period, and of livestock farms in the earlier one. Third, there is a trend towards specialisation. This is noticeable in the switch in balance towards specialist dairy farms away from the mainly dairy group, the growth of specialist cereal farms, and the demise of the 'mixed' farming type.

There is an important link between sizes of farm and the type of farming engaged in. Very small farms (less than 4 ESU) are predominantly live-stock rearing; some two-thirds of the 79 000 holdings of this size in 1982 were of this type. Livestock rearing is particularly suitable for part-time farmers who dominate this size group. The small farms (4–16 ESU) are also mainly livestock rearers, except in Northern Ireland where they tend to be dairy farms. In the medium farms of 16–24 ESU dairying is the most

**Table 9.7**
**Distribution of Holdings Between Types of Farming**

| | England and Wales (based on SMD) | | England (based on SGM) | | |
|---|---|---|---|---|---|
| | 1964 | 1977 | 1977 | 1982 | 1984 |
| Specialist dairy | 19 | 28 | 19 | 20 ⎱ | 29 |
| Mainly dairy | 21 | 10 | 12 | 9 ⎰ | |
| Livestock | 15 | 22 | 19 | 17 | 22 |
| Pigs and poultry | 6 | 7 | 8 | 8 | 6 |
| Cropping: most cereals | 4 | 5 | 10 | 16 ⎱ | 37 |
| General cropping | 12 | 11 | 20 | 20 ⎰ | |
| Horticulture | 10 | 12 | 11 | 10 | 6 |
| Mixed | 13 | 5 | | | |
| | 100 | 100 | 100 | 100 | 100 |

NOTE    There are changes in definition between years. For the latest classification system see MAFF (1980) *Farm Incomes in the United Kingdom* (London: HMSO).

SOURCES    MAFF (various years) *Farm Classification in England and Wales* (London: HMSO); MAFF (various years) *Farm Incomes in England* (London: HMSO); MAFF (1986) *Farm Incomes in the United Kingdom* (London: HMSO).

important type, except in Scotland and Eastern England, where it is the cropping farms which tend to be of this size. Among the large farms (40 ESU and over) cropping predominates in the East and North of England and in Scotland, whereas in Western England both cropping and dairying are important.

Approaching the type/size relationship from the other direction, different types are associated with different farm sizes. Livestock production tends to be associated with the small and lower medium family farm of 4–24 ESU – just over half the UK total output comes from this size range. In dairying about 40 per cent comes from these family farms, but this figure is shrinking as the number of herds declines and the average herd size increases. At the other extreme, among cropping farms most of the output (84 per cent in 1982) comes from the upper medium and large farms, with a similar picture emerging for horticulture (80 per cent) and pig and poultry holdings (83 per cent).

**Table 9.8**
**Distribution of Dairy Cows Between Types of Farming**

|  | Percentage distribution of dairy cows | | | | |
|---|---|---|---|---|---|
|  | 1963 | 1968 | 1971 | 1974 | 1977 |
| Specialist dairy farms (where 75 per cent or more of the SMDs are attributed to dairying) | 35.4 | 44.0 | 51.1 | 56.5 | 65.7 |
| Mainly dairy farms (50–75 per cent) | 37.7 | 32.5 | 29.1 | 27.8 | 21.1 |
| Other holdings (<50 per cent) | 26.9 | 23.5 | 19.8 | 15.7 | 13.2 |

SOURCE   Derived from MAFF, *Farm Classification in England and Wales* (London: HMSO).

## 9.6   Numbers and Sizes of Enterprises

To complete the picture of the size structure and typology of UK agriculture, especially when tracing changes, one must look at the number of enterprises per farm and the changing size of individual enterprises. The process of increasing specialisation, noted above, has also been observed as a reduction of number of types of crops and/or livestock kept on the individual farm. While what constitutes an 'enterprise' is again arbitrary (is a single cow on a cropping farm kept solely to supply the farm house to be treated as a commercial enterprise?), it has been shown that the number of enterprises per farm fell from 3.18 in 1968 to 2.85 in 1974.[6] Surprisingly, the greatest proportional reduction was noted among the smaller farms, largely the result of the elimination of pig and poultry enterprises as technological advance and falling prices took these lines into large-scale production outside the scope of small farms.

The trend towards specialisation is well illustrated by dairy cows. An increasing proportion of the national herd (England and Wales) is found on specialist dairy farms – this rose from about one-third of the total number of cows in 1963 to two-thirds in 1977 (see Table 9.8). On the basis of the existing trend, by 1983 about four cows in five would have been found on 'specialist' farms. Another feature of the changing distribution of cows is that it is the larger holdings which have increased their stocking densities the most (number of cows per hectare), with a continuous relationship existing between the largest and smallest farms. The introduction of milk quotas in 1984 is bound to have had some impact on the changing pattern; with the initial rigidities of quota allocations being increasingly loosened by the introduction of schemes to allow the transfer of quotas between farmers, the trends are likely to re-establish themselves.

The annual June census of agriculture shows a remorseless increase in the average size of the enterprises, coupled with the almost equally persistent reduction in the number of herds, flocks or crop growers (see Table 9.9). The most dramatic change has been in egg production, where the number of flocks of laying hens in the UK has fallen from 188 000 to 46 500 over the period 1967–85 and the average flock size has almost quadrupled. Now two birds in three are in flocks of 20 000 hens or over. In dairying, the average size of herd has steadily increased by about two cows per year, at least up until 1984 when quotas effectively constrained further growth. This pattern is even seen with cereal growing.

Changes in enterprise size found in the UK are mirrored in all the other member states of the EC, albeit at lower absolute levels. For example, in the period 1973–84 when the average number of cattle per livestock rearing unit rose from 69 animals to 78 in the UK, in West Germany the equivalent rise was from 20 to 31, in France from 26 to 38 and in Italy from 9 to 15. (The Community average was 33 animals in 1984.) Rather as a watershed size of farm could be distinguished, marking the size below which numbers were declining from larger categories where rises were occurring, watershed sizes of enterprise can be seen in EC statistics. For the Community as a whole these are currently about 60 cattle for livestock-rearing enterprises, 200 pigs and 30 dairy cows. Each individual country will have its own watershed for each enterprise, reflecting among other things the existing size structures. In the UK, for example, the breakeven size of dairy herd in the decade to 1984 was about 70 cows, more than double the figure for the EC overall. Clearly this Community figure reflects the predominance of small herds in other member states; for the UK the average was 57 cows but the two countries with the most cows (France and Germany) had average herd sizes of only 17 and 14 cows in 1984, the average across the whole Community being 16 cows.

## 9.7 An Explanation for Changing Farm and Enterprise Sizes

No single simple factor will suffice to explain the evolving pattern of farm and enterprise sizes in UK agriculture. Rather, a number of elements come

# 242

**Table 9.9**
**Size of Enterprises, UK, 1967–85**

| | Laying fowl | | | Breeding pigs | | | Breeding sheep | | |
|---|---|---|---|---|---|---|---|---|---|
| | No. of flocks ('000s) | Average flock size | % of total in flocks of over 5000 (20 000 1977 onwards) | No. of herds ('000s) | Average herd size | % of total in herds of over 50 | No. of flocks ('000s) | Average flock size | % of total in flocks of over 500 |
| 1967 | 188 | 275 | 40 | 78 | 10 | 31 | 110 | 123 | 29 |
| 1972 | 114 | 471 | 67 | 56 | 17 | 51 | 86 | 145 | 34 |
| 1977 | 71 | 692 | 55 | 31 | 27 | 69 | 76 | 173 | 39 |
| 1982 | 54 | 824 | 62 | 22 | 39 | 78 | 80 | 188 | 43 |
| 1985 | 46 | 848 | 66 | 18 | 47 | 83 | 84 | 191 | 44 |

|  | Dairy cows | | | Cereals | | |
|---|---|---|---|---|---|---|
|  | No. of herds ('000s) | Average cows per herd | % of total cows in herds of over 50 (60 1977 onwards) | No. of growers ('000s) | Average per grower (ha) | % of total cereals area on holdings with over 40 ha cereals (50 ha 1977 onwards) |
| 1967 | 132 | 24 | 37 | 172 | 22 | 65 |
| 1972 | 99 | 34 | 54 | 132 | 29 | 71 |
| 1977 | 73 | 45 | 58 | 113 | 33 | 68 |
| 1982 | 59 | 55 | 68 | 100 | 40 | 73 |
| 1985 | 54 | 58 | 70 | 95 | 42 | 74 |

SOURCE  MAFF (various years) *Annual Review of Agriculture* (London: HMSO).

together and the relative importance of each will depend upon the time, location and type of farming in question.

Perhaps the least recognised motives for farmers wishing to expand is the most simple – larger farms can generally generate larger incomes. The more successful and ambitious farmers will naturally wish to grow, and the less successful will see expansion as a way of overcoming their problems. Expansion can be by intensification – stocking an existing farm more heavily or fertilising for higher yields – although such methods soon run into the problem of diminishing marginal returns; higher-performance farmers could be expected to already be at or near the point where they are using optimal quantities of variable inputs, and to go further would *reduce* their overall profitability. The alternative option is that of enlarging the area farmed (growth by extension). There are reasons for thinking that this would be preferred by farmers. By simply operating on a larger scale the risks associated with change are minimised. Similar intensities of land use can be employed, with little or no change in the type of management required. There may be a possibility of spreading fixed costs (especially of underutilised machinery and regular labour) over a greater volume of output, thereby reaping attractive profits. Farmers with spare machinery and other resources can afford to pay more than other farmers when land comes onto the market, and in practice the bulk of land with vacant possession is bought by local expanding farmers.

A second reason why some farmers wish to expand is that land is an asset which is attractive for investment purposes, enhancing expansion by purchase as a favoured path especially for the high-income farmer. As was shown in Chapter 8, there is a history of capital gain from land-holding; it is treated advantageously for tax purposes, with interest on loans for land purchase a deductable expense for income tax calculation. Capital gains are taxed at a lower rate or with deferable payment which again reduces the effective rate. Because of the advantageous treatment with respect to Capital Transfer Tax, land is a tax effective way of passing wealth to the next generation in a family. In short, over most of the postwar period a high income farmer would have found himself almost bound for purely tax reasons to buy land if it came available. This helps explain why when small farms come available for sale they are usually bought by large or medium-sized farmers, not by other small farmers. Only in the purely rented sector would amalgamation of similar-sized small farms be common, with landlords attempting to create a more viably-sized unit. As was noted in Chapter 6, the attractiveness of land as an investment also helps explain the trend towards owner-occupation in the UK.

A third reason why the size structure of farming is changing is technological change. As will be seen in Chapter 11, many of the new forms of machinery and buildings are only effective if they can be used over a sufficiently large volume of output. A rotary parlour only reduces milking costs per gallon if used with a large dairy herd; similarly, a large-capacity combine requires a large throughput to realise its potential cost saving per tonne of grain. Inevitably this implies the enlargement of enterprises through a combination of intensification, specialisation within farms, and increasing farm size. The attractiveness of the last option is enhanced by the other attributes of land given previously. Because these new techniques are also usually output-increasing, the effect of their adoption is to push down farm product prices, resulting in a cost-price squeeze. Operators using old techniques are forced to follow suit or to quit production altogether – either switching to other enterprises or leaving farming completely. Hence the rachet of technology ensures that the farms which are too small to use the new ways soon disappear as full-time viable units. In the absence of technological advance on some structural adjustment might still occur, but at a reduced rate; technological change is really the engine behind structural change.

Spreading fixed costs is one source of economies of size. Perhaps it is surprising that little attention has yet been drawn to economies of size in our explanation for the changing structure of farming. There is plenty of evidence that average efficiency improves with size up to the 2–3 man level, beyond which no further economies are evident, nor do diseconomies seem to appear.[7,8] However the notion of efficiency improving with size of farm is of little concern to the farmer; what he is interested in is the profits which can be earned, and he will wish to expand if by doing so his total rewards are increased. The efficiency of a farm is usually expressed in terms of the ratio between the value of its output and the value of the resources it uses. The income to the farmer is, however, the *difference* between these two values. It would be quite possible for a farm to be highly efficient in terms of its use of resources but too small to generate an adequate living for its occupier. On the other hand, farmers already operating beyond the point at which no further improvements in efficiency are likely to occur will still wish to expand if they see income advantages; overall efficiency might even fall a little. Efficiency is incidental to structural change, not its driving force.

## 9.8  Farms as Businesses: The Fusion of Business and Family

The size structure of the farming industry and, particularly, changes in the structure, cannot be satisfactorily understood without recognising a fundamental characteristic of UK agriculture – the fact that farm businesses are essentially family concerns. This, together with the sizes of business and the legal forms which farms take which are themselves the reflection of family farming, has great implications for the rate of change of farm size, the ability to raise finance for buying land and capital equipment, the types of enterprise found on farms, the amount of profit made and, indeed, almost all facets of farming.

A word is necessary here on the legal forms which businesses can take. The simplest is the sole proprietorship, where a single person is responsible for the affairs of the business and owns its assets (after having allowed for any borrowing). He (or she) retains any profits as his personal income and, if a loss is made, he is personally responsible for any debts. Profits will be subject to income tax. If necessary, personal assets (jewellery, televisions, stocks and shares) can be legally seized (through court action) and sold to defray any debts incurred by farming. In short, the personal and business affairs are inseparable. A partnership is normally just an extension of the sole proprietorship, with two or more partners being responsible; a partnership agreement will show how profits and assets are to be split between partners. Again, partners are personally responsible for debts; furthermore, if one partner takes a decision which has serious consequences for the firm the personal resources of other partners can be drawn upon to meet them.[9]

In the case of companies, however, the business itself has a legal entity. The profits accrue to the business, out of which the farmer-directors can be paid a salary in recompense for the management they do and/or a share of the profits which they take in their role as owners of the business. Profits will be taxed by corporation tax, although the directors will also have to pay income tax on the amounts they take as personal income. An important point is that, if the business runs up debts, the owners of the business cannot have their *personal* assets seized and sold to help meet the liabilities – their liability is limited to the business assets they own (i.e., the machinery, stock, etc.). A different set of rules covers small private companies and large public companies; one important distinction is that the latter can borrow directly from the public by issuing securities on the capital market while the private company cannot, having to raise credit from banks and other lending institutions in much the same way as partnerships or sole proprietorships. Changes in tax regulations can make it advantageous to change a partnership into a private company or vice versa, and this behaviour has been seen on a number of occasions in British farming.

British farms tend to employ only the simplest of business forms. This is because farms, even quite large ones, are by comparison with most other industries small firms in terms of their labour force, output or working capital (although not in total capital when the value of land is taken into account, a situation which makes farmers specially sensitive to capital taxation). The forms of business for English farms in 1969 are shown in Table 9.10, when 94 per cent of farms were found to be sole-proprietorships or partnerships (that is, without limited liability for their

**Table 9.10**
**Business Forms in Farming, England, 1969**

| Business form | 2–20 (%) | 20–40 (%) | 40–121 (%) | over 121 (%) | All sizes (%) |
|---|---|---|---|---|---|
| | | Size of farm (hectares) | | | |
| Proprietorships | 81 | 71 | 60 | 39 | 67 |
| Partnerships | 15 | 27 | 36 | 42 | 27 |
| Private companies | 3 | – | 4 | 17 | 4 |
| Public companies and Institutions | 1 | 2 | 1 | 1 | 1 |
| All forms | 100 | 100 | 100 | 100 | 100 |

SOURCE   After Harrison, A. (1975) *Farmers and farm businesses in England*, Miscellaneous Studies 62 (University of Reading: Department of Agricultural Economics and Management).

operators). These findings are supported by the 1975 EEC Structural Survey which discovered that proprietorships, partnerships and private (characteristically family) companies accounted for 94 per cent of holdings. A link between business form and farm size was evident in the 1969 study, with proprietorships dominating among small farms, partnerships tending to be employed for the larger businesses and company form for the largest of all. Private companies had been formed almost entirely for taxation reasons and were hardly ever instrumental in recruiting management or raising capital that could not have been equally well brought into the business if arranged as a partnership.

Overall, more than 19 farms out of 20 in England in 1969 proved to be genuinely family businesses in the sense that all the principals (where there was more than one, and taking partnerships and private companies together) were closely related by blood or marriage. Other countries in the UK appear to follow a similar pattern of family domination as found in England, and there is no reason to think that the picture has altered substantially by the mid-1980s. Frequently no distinction can be drawn between the personal wealth of the farmer and the business assets he uses. As has been shown in Chapter 7, farms are also largely dependent on families to run them, more so than in the past; about three-quarters of farms employ no hired workers at all (that is non-family workers). All this means that there is an almost inextricable fusion between the personal affairs of the farmer and his family and those of his business. Farmers can be found who are investing in buildings or land not primarily because of a desire to increase profits but because they have a son who is likely to succeed them, and conversely farmers can be found who are reducing the scale of their firm's activities because they are getting older and wish to take things more easily. Many farmers, especially owner-occupiers and those on medium-sized and larger farms, will be in a good financial position, able to change their farming pattern to suit their family circumstances.

However, changes in family circumstances such as a death of a farming principal, can also have serious repercussions for the viability of the farm firm. Frequently the stress to management which occurs when a father dies and a son assumes control is aggravated by the additional problem of taxation of the personal assets of the deceased. However, judicious tax planning and the use of family trusts (see Chapter 8) has enabled the farming community to be remarkably resilient to attempts at wealth redistribution in the UK. According to survey results, about three-quarters of farmers in the 1970s were the sons of farmers, with the majority of those owning land having inherited at least some of it.[10] As a consequence of the intermixing of business and family affairs, any legislation which aimed to restrict the freedom of farmers to pass their land to the next generation could be expected to have a major impact on farming practice.

The question is often raised as to why public companies have not come into farming (as opposed to land-owning, where a number are active, especially the financial institutions – see Chapter 6). Alternatively, why have farm firms not grown and turned themselves into public companies, as seems to be the pattern in other industries? There seem to be two main reasons:

1. most farm businesses are too small to bear the cost of floating shares. The reasons why farms remain small include the absence of marked economies of size, the difficulty of obtaining land to expand, and government policies of income support which have reduced the pressure for enlargement;
2. individual investors and stockbrokers could not readily obtain the necessary information about the business prospects of individual farmers, and the cost of collecting such information would be high in relation to the size of the farm business.

Of course, there are exceptions to the generality. There are a few large public farming companies, whose activities include farm management services for land bought by financial institutions. In specialist areas of agriculture (e.g., egg and poultry production) large-scale production provides the circumstances favourable to public company development (size and information availability). And public companies are involved in other minor ways, for example, fertiliser and chemical firms which run research farms. Other public bodies also run farms – Home Office remand centres and prisons, religious institutions and so on – but in terms of total farm numbers or agricultural area these are insignificant in the national picture. Recently farming partnerships between financial institutions and individual farmers, formed to gain taxation advantages and to circumvent tenure legislation, have attracted attention. However, the dominant form is undoubtedly the family-owned farm business, even among the larger units where it is sometimes disguised by legal arrangements designed to mini-mise taxation.

## 9.9 Farms Run in Parallel with Other Businesses

In Chapter 7 it was noted that about one-third of all farming couples (or single farmers) in England and Wales have some other job ('gainful occupation') outside farming. In Britain about three-quarters of part-time farmers gain their non-farm income as proprietors of second businesses and almost half of these have no obvious connection with agriculture; this situation contrasts with, say, Germany, where off-farm hired employment predominates. Self-employment in both farm and non-farm occupations could be expected to permit considerable flexibility in allocating time according to each business's need, a facility which might not be available to most manual or lower white-collar employees. Parallel proprietorship of a farm and non-farm business also means that there must be competition for capital funds, and that bank loans ostensibly intended to support activities in one sector are likely to have some impact in the other business under the farmer's control, although the extent to which loans to farmers have been used to develop other activities, and vice versa, is difficult to gauge. Land undoubtedly forms a useful collateral for borrowing, irrespective of the credit's use.

More farmers seem to think that 'part-time' farming is to the benefit of the farm business than to its detriment.[11] However, when two businesses share the same management and capital base, one being a farm and the other in some other industry which may or may not be related to agricul-ture, it becomes difficult, even meaningless, to treat the farm business in isolation. Furthermore, the continued existence of many of the UK's small farms, perhaps half, seems dependent on the farm family having some other form of income. One could speculate that many small farms would continue even if farming profitability suffered a major deterioration; indeed, their existence could be more sensitive to economic conditions in the other industries since it is on earnings from these that the farmer mainly depends.

## 9.10　Farm Size Structure and Policy

Some critics of government policy suggest that the fall in numbers of small farms has been accelerated by the various grants, tax concessions and forms of product price support which have favoured disproportionately the large-output, large-area, high-income farm. Such assistance has resulted in fewer people being employed in agriculture, but more capital. Technological changes have been taken up faster than might otherwise have been, with a consequential greater squeeze on the small farm. It is alleged that the disappearance of the small farm is much to the disadvantage of the national economy, the environment and social fabric. Small farms seem less likely to result in large areas of monoculture, objectionable from both aesthetic and conservation standpoints, appear to offer a greater number of jobs, an important feature in times when the spectre of unemployment haunts politicians, as now in Britain and many other industrialised countries, and through providing the first rungs on the farming ladder for the aspiring new entrant, give a higher proportion the opportunity to exercise the initiative and originality which flow from self-employment. Advocates of the small farm point out that the overall loss in national agricultural efficiency incurred by their continued existance is tiny,[12] and is more than compensated by the many wider attributes they possess.

Some of the advantages of a vigorous small-farm sector have been long recognised by governments. Before the turn of the nineteenth century there was concern over the disappearance of the small acreage holding and the loss of opportunity for farm workers to set themselves up as independent farmers. For much of the postwar period various UK (latterly Community) schemes have existed primarily to assist small farms. (These are covered in detail in Chapter 14.) The main thrust has been to make small farms viable by assisting them to become more productive through encouraging investment, e.g., in new equipment, or by covering part of the costs of farm amalgamation. The UK has not taken the much more positive steps to arrest structural change which some other member states of the EC have employed. Similar trends towards fewer farms are found in all EC countries except Ireland, where the reduction has been only modest. Denmark limits the amount of land individuals can buy and precludes companies and institutions completely. France controls the area which may be farmed but not ownership. Ireland, Italy and West Germany interfere in the land market with structural objectives favouring family farming in mind.[13]

We have argued earlier that structural change is inevitable, driven by the engine of technological advance. Attempts to protect small farms go against the long-term trend and are likely to prove increasingly costly, either in direct payments or in resource opportunity costs, as the structure of farming becomes increasingly separated from what could be sustained in a freely competitive market situation. The advantages which small farms bring will need to be reassessed constantly as the rising output from larger farms equipped with new technology has its inexorable downward effect on farm product prices, tightening the cost-price squeeze on farm incomes.

The fundamentals of change seem to be accepted in the UK where, despite a general sympathy for the small farmer, policy on agricultural structure has been limited to the assistance to investment mentioned above and to facilitating the process of exit by farmers on unviable holdings[14]. For those too old to benefit from such injections of capital into their farm businesses, retirement payments have been available in the form of lump sums and/or pensions. However, these latter measures have been largely ineffective as lubricators of structural change because they have failed to take sufficient account of the store which elderly farmers attach to the independence of their way of life, their enjoyment of the intrinsic processes of farming, and their attachment to their land, particularly in the light of the growing wealth it has represented and the tax advantages associated with it.

Radical moves to change the structure of farming into a state in which it was substantially more compatible with commercial reality at prices nearer internationally competitive levels have proved politically unacceptable, especially in the EC. The most notable of these was the proposal in 1968 by the EC Agricultural Commissioner, Sicco Mansholt, to retire significant numbers of farmers, to raise farm sizes and to withdraw some land from agricultural use, payments for these adjustments being accompanied by price cuts for commodities in surplus. Only in a very watered-down form were these proposals allowed to be put into practice (see the 1972 EC Directives outlined in the Appendix to Chapter 14).

In the UK the future of the small farm as an independent unit is seen more as one of part-time operation, with the farmer either working at some off-farm job, or carrying on some non-farming activity on his holding (holiday accommodation, farm shops, etc.). Under the present system of product price support and subsidising of capital inputs and the current tax regime the pressure for further structural change will continue. Further protection of the small farmer against the commercial forces pushing towards larger but fewer farms would require a greater value to be attached to the supposed virtues of smallness than policymakers currently think appropriate, at least in lowland areas. In the hills and uplands the case for maintaining the present or similar structure of farming is stronger, but the argument is in these areas even more one based on social and environmental reasons rather than agricultural ones.

# Notes

1. Harrison, A. (1965) 'Some features of farm business structures', *J. Ag. Econ* **16**:3, 330–47.
2. In the United Nations (1983) *European Handbook of Economic Accounts for Agriculture* (New York: UN), the holding is defined in the Programme for the 1980 World Census of Agriculture as 'all the land which is used wholly or partly for agricultural production and is operated as one technical unit by one person alone or with others; establishments or other units not including any agricultural land but producing livestock or livestock products are also to be considered as holdings.'
3. Harrison, A. (1975) *Farmers and Farm Businesses in England* Miscellaneous Studies 62 (University of Reading: Department of Agricultural Economics and Management).
4. Hill, Berkeley and Kempson, R. (1977) *Farm Building Capital in England and Wales* (Ashford: Wye College).
5. Edwards, C. J. W. (1980) 'Changing farm size and land occupancy in central Somerset', *J. Ag. Econ.* **31**:2, 249–51.
6. In the 275–599 SMD size band of farm the average number of enterprises per farm fell from 2.88 in 1968 to 2.49 in 1974, an average 13.5 per cent decline. By contrast, among holdings over 4200 SMD the fall was from 3.52 to 3.40, a 3.4 per cent decline. Britton, D. K. (1977) 'Some explorations in the analysis of long-term changes in the structure of agriculture' *J. Agric. Econ.* **33**:3, 197–209.
7. Britton, D. K. and Hill, Berkeley (1975) *Size and Efficiency in Farming* (Farnborough: Saxon House).
8. Power, A. P. and Watson, J. M. (1983) 'Total factor productivity and alternative measures of size', in R. B. Tranter (ed.) *Strategies for Family-Worked Farms in the UK*, CAS Paper No. 15 (University of Reading: Centre for Agricultural Strategy).
9. An exception exists in the case of 'limited' partnerships, but as this form is rare they are not considered here. Laws on partnerships are different in Scotland from those in England and Wales which are implied in the text.
10. Newby, H., Bell, C., Rose, D. and Saunders, P. (1978) *Property, Paternalism and Power: Class and Control in Rural England* (London: Hutchinson).
11. Gasson, R. (1983) *Gainful Occupations of Farm Families* (Ashford: Wye College).
12. Raising the performance of farms in the small SMD size groups to that of larger farms was estimated to raise total output by only about 2 per cent – see note 7 above.
13. Harrison, A. (1983) 'Family farm policies in the European Community: are they appropriate for the UK?' in R. B. Tranter (ed.) op. cit.
14. Revell, B. J. (1985) *EC Structures Policy and UK Agriculture* Centre for Agricultural Strategy, Study No. 2 (University of Reading).

# The Supply of Agricultural Products 10

## 10.1 Introduction

Farm businesses, described in Chapter 9, exist to bring together factors of production (Chapters 6–8) to generate a supply of agricultural products. Here we examine what determines the level of that supply in the short run, and in the next chapter turn to the technical advances which in the longer term contribute to farming's history of rising output. There is little in this section which will be new to readers familiar with elementary production economics, but for others a presentation of the determinants of supply will be necessary as a complement for the equivalent treatment of demand in Chapter 2.

Supply generally interests governments more than demand because, although the characteristics of both are at the root of the problems faced by the agricultural industry, in the implementation of policy supply is easier to manipulate than is demand. It is also less stable, primarily because of weather changes outside the control of farmers. This obviously effects some crops more than others, and crops more than livestock, although feed conditions will have some impact on grazing animals. Average yields in the UK for a run of recent years are given in Table 10.1, together with an indication of the general rise in yields over a 15-year period. In the USSR until the mid-1970s poor grain harvests resulted in the slaughtering of cattle, increasing beef supply in the year in question but curtailing it in subsequent years. In poorer countries weather-induced output fluctuations can result in unpredictable total food supplies for the population, reflected in starvation and in the mortality rate. In the UK this is no longer the case. The impact of variations in farm output is seen as price fluctuations for farm products, to a lesser extent in changes in food prices for consumers, and most especially in erratic farm incomes. There will be calls on govern-

**Table 10.1**
**Estimated Average Yields of Crops and Livestock Products, UK Calendar Years**

| | Unit | Average of 1967–69 | 1981 | 1982 | 1983 | 1984 | 1985 (forecast) | Average of 1983–85 |
|---|---|---|---|---|---|---|---|---|
| *Crops* | | | | | | | | |
| Wheat | tonnes/hectare | 3.91 | 5.84 | 6.20 | 6.37 | 7.71 | 6.29 | 6.79 |
| Barley | " | 3.61 | 4.39 | 4.93 | 4.65 | 5.59 | 4.91 | 5.05 |
| Oats | " | 3.33 | 4.30 | 4.43 | 4.32 | 4.89 | 4.24 | 4.48 |
| Potatoes | " | 24.90 | 32.31 | 35.83 | 29.87 | 37.03 | 35.58 | 34.16 |
| Sugar beet | " | – | 35.72 | 49.81 | 38.30 | 45.90 | 40.00 | 41.40 |
| Oilseed rape | " | 1.8 (e) | 2.70 | 3.30 | 2.53 | 3.43 | 3.01 | 2.99 |

| | | | | | | | |
|---|---|---|---|---|---|---|---|
| **Apples:** | | | | | | | |
| Dessert (a) | 9.59 | 8.90 | 12.73 | 11.59 | 11.78 | 10.79 | 11.39 |
| Culinary (a) | 8.57 | 7.60 | 13.64 | 12.23 | 16.77 | 13.28 | 14.09 |
| Pears (a) | 8.66 | 10.27 | 9.43 | 12.52 | 11.34 | 12.25 | 12.04 |
| Tomatoes (a) | 96.60 | 146.06 | 151.86 | 155.40 | 170.88 | 162.06 | 162.78 |
| Cauliflowers (a) | 18.40 | 18.74 | 16.29 | 16.72 | 20.15 | 19.41 | 18.76 |
| Hops | 1.46 | 1.61 | 1.75 | 1.51 | 1.55 | 1.49 | 1.52 |
| | | | | | | | |
| **Livestock products** | litres/ | | | | | | |
| Milk (b) | cow | 3673 | 4749 | 4934 | 4967 | 4749(c) | 4896 | 4871 |
| Eggs (d) | no./bird | 210.5 | 249.5 | 251.0 | 258.0 | 256.5 | 260.0 | 258.2 |

(a) Marketable output yields from cropped area, except for 1967–69 averages which are gross yields.
(b) Yield per dairy-type cow per annum. From 1977 based on an average population which includes estimates for dairy-type cows on minor holdings (previously called statistically insignificant holdings) in England and Wales.
(c) 366 days.
(d) Eggs per laying bird, including breeding flock.
(e) 1968–70.
SOURCE  MAFF (1979 and 1986) *Annual Review of Agriculture* (London: HMSO).

ments to limit the extent of such fluctuations by regulating the supply which reaches the market, which it can do by the mechanisms described in Chapter 13.

In practice it is difficult to separate stabilisation policies from others; the government will find that it can pursue related aims, such as supporting farm incomes, by controlling supply and thereby forcing up the prices payed by consumers, to the benefit of farmers. If the government chooses to support incomes by manipulating demand, such as by buying farm produce at guaranteed prices higher than would be found in a free market, it will need to be aware of the way that farmers are likely to respond to these enhanced prices; the cost to public funds could be great if supply expands rapidly.

In this chapter we begin by considering those factors which influence the supply of a commodity from farms to the market, given the existing state of technical knowledge. In the following chapter we relax this constraint and look at the impact of technical change.

## 10.2   The Elementary Theory of Supply

The supply of a commodity is the amount that farmers are willing to put on the market in a given time period.[1] A distinction should be drawn between this quantity and that which actually reaches the market in a given year. For a variety of reasons the amount that farmers intend to or wish to produce may not coincide with what eventually leaves their farms. Weather is the most obvious cause of this disparity in the agricultural context. There is always a time lag between a farmer deciding to supply a certain quantity to the market and that produce actually being available for sale and during this lapse events may have changed; a farmer may find himself producing more or less than he would prefer in the changed market conditions. The government may decide to intervene, directly or by giving legal backing to agencies acting on behalf of farmers such as marketing boards, and prevent some of the intended production from being put on the market, especially if this will drive down the price. It may constrain farmers in the amount they wish to supply by imposing quotas (as in the case of milk in the UK) or by permitting only a certain area to be cropped, as with sugar beet and potatoes. The buying of cereals and beef into storage by the Common Agricultural Policy's Intervention Board is one way of ensuring that intended supply does not reach the open market, although this policy can only be pursued for as long as storage facilities or some alternative method of disposal are available. Conversely, in times of poor agricultural production it is possible to increase supply for a time by running down stores. Consequently farmers' intended supply may differ from actual supply to the market. Nevertheless a review of those factors influencing supply intentions will help explain much of the real-world changes in farm output.[2]

Supply can be described either at the individual farm level or at various degrees of aggregation. For an industry such as agriculture, made up of a large number of independent businesses none of whom is large enough to influence the overall price of grain or milk or whatever, a description of those factors which influence supply at the farm level is a good guide to the supply behaviour of the whole farming industry.

The factors affecting a **single farm's supply** of a commodity, such as beef, can be listed as follows:

A. the price of beef, both at present and that expected to hold in the future.
B. the prices of other products that the farmer could produce with his resources, such as sheep meat or milk.
C. the costs of production associated with beef, such as the costs of concentrate feeds, medicines, labour and so on.
D. the objectives of farmers, such as their profit motivation, willingness to accept risk, and fondness for certain farm enterprises.
E. the state of technology, such as the genetic capability of the beef animals to respond to feed, the available husbandry methods and the known medicines. Technical change generally occurs at a rate much slower than the do the other factors.

At the **industry level** we may also have to take into consideration additional supply determinants, such as a change in the number of farm businesses or the way in which they are distributed between farms of different sizes. However, when considering the response of the agricultural industry in the short term, these can be set aside.

We will now briefly consider the way in which farmers respond to changes in each of the first four factors given above. In real life it is quite likely that all these are changing simultaneously, but for the sake of simplicity we will treat each in turn, that is, we will see how the supply from farms varies as each factor varies, all others assumed to stay constant.

## 10.3  Supply and the Price of the Commodity Itself

The most obvious factor influencing the quantity of a commodity which producers are willing to offer for sale is the price of the commodity itself. Generally speaking, the higher the price the greater the quantity they will want to produce. The relationship between price and quantity supplied is expressed in the typical supply curve shown in Figure 10.1. The reason why

## FIGURE 10.1
## A Typical Supply Curve

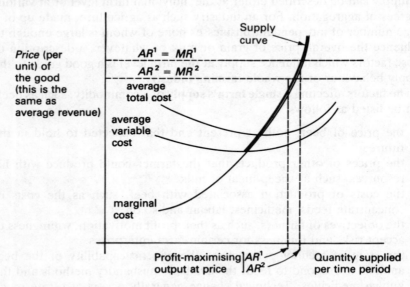

NOTE    1. Short-term decisions are based on variable costs. Hence the supply curve is shown as extending down to the minimum average variable cost. In the longer term producers will not supply unless price corresponds at least to the lowest level of average total costs.
       2. Average revenue is   $\dfrac{\text{Total revenue}}{\text{No. units of output}}$
       3. Under perfect competition average revenue also corresponds with marginal revenue.

the curve is of this shape is because of the way in which costs of production vary at different levels of output. A farmer, out to maximise his profit from, say, a cereal crop will find that he can achieve higher yields by using more fertilisers and other variable inputs. This is a relationship governed by the biological response of his plant material to its physical environment, some of which the farmer can alter (like the amount of nutrient given) and some which are beyond his control, at least in the short term, like soil quality and drainage. The farmer will know that high outputs can be achieved if large amounts of fertiliser are used, but that fertiliser applications are subject to diminishing marginal returns, that is, beyond a certain usage extra units of fertiliser produce a smaller and smaller response in the crop. There will be some high application level at which yield may even decline as problems of lodging and disease succeptability emerge.

In deciding the most profitable level of output to aim for the farmer will take into account the value of extra grain which could be produced by using more fertiliser and other variable inputs and compare it with the extra costs which are incurred by using these inputs. Because of diminishing returns the costs of every additional ton of crop produced from a given field rises; the higher the present yield the more costly is it to produce one more ton. The additions to total costs associated with producing this extra (marginal) ton is called its **marginal cost (MC) of production**. Figure 10.1 shows a typical rising marginal cost curve.

In formal terms the marginal cost of the $n$th unit of production can be defined as follows:

$MC$ '$n$'th unit *equals* Total cost of producing '$n$' units *minus* Total cost of producing '$n - 1$' units     (1)

Similarly, the extra revenue resulting from selling the additional unit is termed **marginal revenue (MR)**. If the farmer sees that the MR of an additional unit of output exceeds the MC of producing it, so that some extra profit will result, then he will go ahead and produce this marginal unit. He will continue expanding his output up to that point at which MC and MR are equal; to go further would incur losses on marginal units, lowering overall profit.

In an industry such as agriculture, if an individual farmer makes a greater volume of production available for sale he will not have any significant effect on the overall market price. The price he receives for each and every unit he sells will correspond to the average market price. Consequently, when an individual farmer is deciding how much grain (or whatever commodity he has in mind) to supply to the market, he will expand output up to the point at which his marginal cost of production has risen to equal average price. If the market price rises the farmer will respond by expanding the amount he is willing to produce, that is, he will increase his supply. Conversely, if market prices fall he will respond by reducing his production. The MC curve thus also shows the various levels of output which the farmer supplies to the market at a range of prices; in other words, the MC curve (more strictly as we shall see below, just part of it) is also the supply curve for the commodity under consideration.

The qualification to the above statement is necessary because a farmer will not be willing to supply any at all if price falls below a certain level. As well as marginal costs farmers will be aware of average costs of production, that is total costs divided by output. Costs also fall into two main groups; those that vary with the level of production, such as fertiliser and sprays in the case of cereals (**variable costs**), and those which in the short term must be faced whatever the level of output, such as rent (**fixed costs**). The price which a farmer receives must at least cover his average variable costs if he is to continue in production. Staying with the cereals example, grain production will stop if the price does not cover the lowest possible average cost of fertilisers and other variable inputs. In the longer term price will have to cover at least the lowest **average total costs**. Therefore the supply curve of a commodity corresponds with that part of the MC curve which lies above the lowest point of the average cost curve (see Figure 10.1).

The supply curve for the agricultural industry as a whole for a particular product is formed from the combined responses of all of the individual producers in the industry; in graphical terms this means the horizontal summation of many individual supply curves. While the individual curves will not be identical, as farmers face slightly differing costs situations, the combined curve will generally slope upwards from left to right. From this point onwards we will refer to the industry supply curve as if it were belonging to the 'typical' farmer.

### 10.3.1  Elasticity of supply

The industry supply curve shows that farmers are willing to supply a greater quantity at higher prices and less at lower prices. The extent to which they respond to price changes is known as **price elasticity of supply**, or more usually as simply **supply elasticity**. It is measured in a similar way as was demand elasticity (Chapter 2) and assumes that all other influences on supply remain constant. The coefficient is calculated as follows:

$$\text{Elasticity of supply } equals \frac{\text{percentage change in quantity supplied}}{\text{percentage change in price}}$$

A coefficient of less than 1, implying that a price change of $X$ per cent results in a less than $X$ per cent change in quantity supplied to the market, is termed inelastic. The slope of the corresponding supply curve will be steep (its gradient more than 1). Conversely, if a price change causes a proportionately larger change in intended output, the coefficient is large and the supply called elastic. The supply curve will have a shallow slope.

### 10.3.2  Time and supply elasticity

The willingness of farmers to respond to price changes will depend on a number of factors, described below, and whether a price fall or a price rise is being considered. On any one farm the response will differ from enterprise to enterprise. One important influence is the length of time period under consideration. Attempts to increase supply of any agricultural commodity runs up against the unhurriable characteristics of biological development. Over a period of two years, wheat production in the UK could be expanded considerably, say by 20 per cent, but it would take much longer to increase the level of beef output by the same proportion, even longer for top fruit. Because of these biological limitations the response to price rises measured over a longer period is usually greater than over a short one; in addition, the longer a price change lasts the more opportunity there is for farmers to respond to it.

Because farming is incapable of reacting immediately to changes in price by altering the level of output, the response of farmers is thought to consist of two stages, each with its separate time lag. The first is the expectations time lag, and describes the time it takes for farmers to recognise a price change and to be convinced that it is sufficiently permanent to be worth responding to. The second relates to the delay in adjusting to the now perceived and assessed price shift. Strictly what we are measuring when relating output changes to price movements is an estimate of actual response rather than of the amounts that farmers wish to supply under the new price relations. Nevertheless, it is the actual response which is of importance as far as governments are concerned when they are trying to manipulate agricultural markets with policy intentions in mind.

Table 10.2 gives, as an example, supply elasticities for cereals taken from empirical estimates. As expected, the coefficients are all positive, that is higher prices are associated with greater output. The long-term coefficients are larger than the short-term ones.

## 10.3.3 Cost structure and supply elasticity

Another factor determining the rate at which farmers respond to price changes is the cost structure of production. The distinction has already been made between variable and fixed costs of production. However it should be pointed out that the distinction is not absolute and will depend on the time span under consideration. While within a short period, say a month, the rent which a farmer has to pay, the interest he faces on a mortgage and his wage bill for regular labour, would be considered as fixed costs, the longer the time period in which costs are viewed, the more that become variable. The farm labour force can be adjusted, the size of farm changed and so on.

The implication of cost structure, in the form of the balance between fixed and variable costs, applies to the greatest extent in the response to falling prices. The higher the proportion of fixed costs, the smaller will be the tendency to cut output when output prices fall, that is, supply will be less elastic. As an industry, agriculture seems to have a comparatively large portion of its total costs in a fixed form, and many of the changes seen in the twentieth century have consisted of the use of greater quantities of fixed factors, sometimes net additions and sometimes replacing inputs whose levels of usage were moderately variable with those of a more fixed nature. The prime example is the increasing amount of buildings and machinery, reflected in the rising capital stock, and substituting capital for

**Table 10.2**
**Supply Elasticity Estimates for the UK**

| Estimates | Spring wheat | Spring barley | oats | mixed | all grains |
|---|---|---|---|---|---|
| 1. (1920s–50s) | | | | | |
| short | +0.33 | +0.63 | – | – | +0.12 |
| long | +0.46 | +1.75 | – | – | +0.52 |
| 2. (1950s–60s) | | | | | |
| short | +0.29 | +0.56 | +0.71 | – | +0.02 |
| long | – | +1.02 | – | – | +0.07 |
| 3. (1950s–70s) | | | | | |
| short | +0.55 | +0.18 | +0.84 | – | – |

SOURCE   Caspari, C., MacLaren, D. and Hobhouse, G. (1980) *Supply and Demand Elasticities* (Washington: USDA) (January).

labour; both result in a rising ratio of capital to labour. If variable costs are low, then it follows that prices will need to fall to a very low level before supply is stopped altogether; in the short term it will continue as long as prices cover average variable costs, even though all costs are not being met. Farmers will then be in a loss-minimising situation.

In Chapter 7 it was shown that within the labour force in agriculture the family is accounting for an increasing proportion of the total as hired labour is shed. Again this represents a rise in the share of costs which are fixed, in that it is easier to displace hired workers on a family farm in response to falling farm profits coming from lower prices than to shed family members. Consequently there will be a tendency to continue production and absorb falling profitability by taking a smaller income and tightening the proverbial belt. In supply terms this means a lower elasticity.

### 10.3.4  Transfer earnings of factors and supply elasticity

Next, there is the question of the opportunities for alternative uses of resources currently engaged on farms when product prices fall, in other words their transfer earnings. In Chapter 7 it was shown that farmers, especially those found on smaller farms, tend to be relatively old and not suitably educated for employment in other occupations. Because their transfer earnings are low they will tend to stay as producers when prices and incomes decline, giving agricultural production a low supply elasticity.

**FIGURE 10.2**
**An Asymmetric Supply Curve**

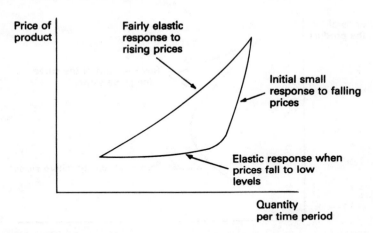

**FIGURE 10.3**
**A Ratchet Supply Curve**

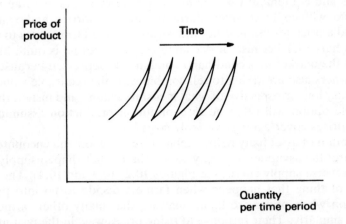

When the general economy is in a depressed state, with few suitable jobs available outside agriculture, the grin-and-bear-it syndrome will be even more marked.

With capital assets, a similar picture emerges; here the problem is

**FIGURE 10.4**
**A Reverse Supply Curve**

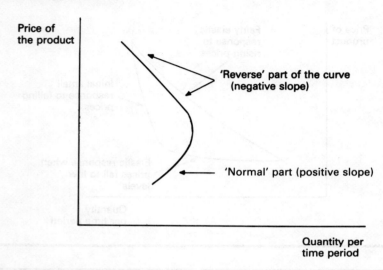

described as **asset fixity**. Once acquired and installed on farms, much of the buildings and equipment used by farmers has low resale or scrap values. The same will apply to much farm machinery; lorries and vans may command a price for use in other industries, but if a farmer tries to sell his combine harvester because of declining cereal prices he is quite likely to find that the market in second-hand combines is depressed because many other farmers also are trying to sell combines. If the realisable value of an asset is very low, irrespective of what it cost to acquire and install, then the preferable option will often be to continue in production assuming that product prices cover average variable costs.[3]

The notion of asset fixity helps explain three phenomena encountered in agriculture; the asymmetric supply curve, the ratchet-shaped supply curve and the reverse supply curve (see Figures 10.2, 10.3 and 10.4). The first is the sort of thing that happens when farmers decide to go into pigmeat production when prices are high, unaware that many other farmers are deciding similarly. Their response to rising prices was, in the past at least, quite elastic, with a ready investment in specialist pig housing. When prices fell as a result of the extra output from farmers as a whole eventually coming onto the market, farmers were reluctant to cut production. The scrap value of their new piggeries was low, and any margin above variable costs (mainly feed costs) went towards reducing the size of the loss by covering part of the fixed costs. Eventually pig prices fell to such a low level that even variable costs were not covered, when there was a rapid and widespread abandoning of pig production, giving the very elastic section of the response to falling prices. Naturally this curtailing of supply caused prices to rise again, whereupon farmers again took uncoordinated decisions to go into pig production, and the cycle restarted. (Price cycles are dealt with in more detail in Chapter 13.)

The ratchet supply curve (Figure 10.3) is more likely to be encountered in reality than the asymmetric curve; the latter is probably more a model than something clearly identifiable empirically in the UK. In the ratchet's case there is still the notion of different responses to rising and falling prices brought about by asset fixity, but here production does not contract completely to its original level, because prices start to rise again before all farmers have reduced their resources committed to this product back to the initial level. The long-run effect is a gradual expansion of output over a series of price cycles. On the ground this may take the form of farmers who have pulled out of pig production when prices fell restocking their buildings; they will be able to expand at lower cost than those with no existing equipment.

The reverse supply curve (Figure 10.4) is what would emerge if farmers responded to falls in prices by producing more. This sort of behaviour could arise in the case of a heavily-indebted farmer who has an intensive livestock enterprise; if prices fall he may try to protect his income and ability to service loans by stocking more intensely, thereby pushing through a larger number of animals in a year, each with a smaller margin. It is also the response in the short run that has been predicted, if cereal prices are reduced under the CAP, by those farmers in some cereal areas who have modern and partly under-used cereal tackle and storage facilities. Rather than easing the problem of cereal over-supply that the CAP faces, price reductions could in the short term make the situation worse. However in this case an expanded supply it is unlikely to persist; farmers in the longer term will not replace their grain equipment and will turn to other ways of employing their capital, thereby lowering the level of their production.

### 10.3.5  Institutional factors and supply elasticity

Finally, it should be noted that a response to higher prices may be constrained by imposed institutional factors. Dairy farmers are restricted by quotas from increasing their aggregate output, however high the price of milk were to go. At the individual farm level there may be some flexibility achieved by the trading of quotas. Even if quotas were obtainable by purchase from other farmers the response to milk prices would be muted by the cost of the quota. Similar restrictions have applied to potatoes, sugar beet and hops. Responses to falling prices are not usually constrained in the same way; quantitative restrictions are generally measures to deal with problems of over-supply. With other commodities, such as blackcurrants and peas for freezing, expansion may be dependent on securing a sales contract with a processor; growing the crop without such a contract may incur too much risk. But a contract may also limit the ability to reduce output if prices fall.

## 10.4 Supply and the Prices of Alternative Products

An important point arising out of Table 10.1 is that the supply elasticity coefficients for individual cereals are greater than for grain in general; this is explained by the switching that farmers make between wheat, barley, oats and so on according to their relative profitabilities, whereas they will be less willing or able to switch from cereal growing to other land uses. Discussion up to this point has been concerned with movements from one position on a supply curve to another in response to price changes of the product in question. However, with cereal enterprises that compete with each other for land within a farm, we see shifts in the entire supply curve taking place; a rise in the price of wheat will cause the supply curve of barley to shift to the left as farmers are willing to supply less at each barley price, while a fall in the wheat price will move the barley supply curve to the right as producers switch into growing more of it. This can be related in the **cross elasticity of supply** coefficient, defined as follows:

$$\text{Cross elasticity of supply} \quad equals \quad \frac{\text{Percentage change in quantity of } X \text{ supplied}}{\text{Percentage change in price of } Y}$$

Most UK farms are mixed, in the sense that they usually have a number of products, in comparison with stretches of monoculture in parts of North America. The various enterprises are competitors for the farm resources. The supply of one product will be influenced by what is happening to the prices of others. The coefficient for competitive products will be negative, the supply of, say, potatoes falling when the price of sugar beet rises (because of constraints such as acreage quotas the desire to switch between enterprises may not be reflected in what can actually be produced, but this complication will be ignored). The effect on the supply curve of potatoes in this instance is to shift it to the left; farmers are now willing to supply a lower quantity at each price than they were before the price of sugar beet rose.

Some agricultural products are inevitably produced together, such as calves and milk, or mutton and wool. With these types the supply of one is affected by the price of the other. For example, a rise in the returns to sheepmeat will cause more of that commodity to be produced and hence more wool also. The supply curve of wool will be shifted to the right. The coefficient of cross supply will be positive.

Enterprises on farms frequently form part of a system which takes advantage of complementary relationships between them. For example, a break crop on a cereal farm enables a higher output of grain. Under these circumstances the response to price changes of each will be more muted than if the enterprises were completely independent of each other. For example, it is conceivable that the break crop would still prove worth growing even if the product could not be sold. No farmer would grow the break crop alone: it only becomes attractive as part of a system.[4]

The interactions between products described above are the result of changes brought about by the price system, but there are other circumstances in which farmers are caused to alter their patterns of production which have similar repercussions. For example, when milk quotas were applied to the UK in 1984 there was an implication for the supply of other agricultural products, notably beef, as farmers, constrained by their quotas from further expansion of milk production and possibly even required to contract, looked for other ways of employing their resources. The supply curve of beef was shifted to the right. If the limits imposed by contracts in the growing of sugar beet in the UK were suddenly removed, the supply curve of other crops, especially potatoes, would be expected to shift to the left as growers switched to beet growing.

The existence of the cross-supply relationship is thus of obvious significance to policy-makers. Lower prices for cereals are often advocated on the grounds that the cost of supporting this commodity is too great; there are the costs of storing the excess production and of subsidising its sale on the world market. However, given that in the short term there is going to be no significant reduction in the amount of land in agricultural use and only a slow contraction in the number of farmers, there will an immediate search for alternative ways of gaining a livelihood. Farmers may switch to beef or oilseeds, commodities which may already be in surplus and whose disposal costs may be greater than cereals. In solving one problem policy-makers may create a bigger different one. The basic problem, of course, is one of general over-capacity in the farming industry at the sort of price levels that the market would result in if left to balance supply with demand, or even at those prices above this level that can be sustained using various forms of public support; this theme will be explored in Chapter 13.

We have seen (Chapter 9) that many small farms are run by their operators in parallel with some other economic activity, most often in the form of another business or profession. The existence of the other activity will have implications for the supply of farm products, although the relationship is not a simple one. If the non-farm business is prospering with rising prices for its products, the part-time farmer is likely to transfer more of his resources off the farm, shifting the supply curve of farm products to the left without any change occurring on the price of the farm products. Conversely, falling prices for the off-farm activity may result in a rising agricultural supply. The possibility of using resources on or off the farm would be expected to make the supply of farm products more sensitive to the level of price. On the other hand, empirical evidence points to part-time farms as acting primarily not as income generators but as places to live, long-term investments and so on. Under these circumstances the level of farm prices are of little importance to what is produced; the supply elasticity is therefore likely to be muted. So far the net implications for the UK have not been quantified, but with a rising proportion of farms of a part-time nature this will be increasingly necessary for policy-makers.

## 10.5   Supply and the Costs of Factors of Production

If the costs of producing a farm commodity rise, all else remaining unchanged including the price of the commodity in question, the profit margin remaining to the farmer to compensate him for the risks he takes in production is squeezed. There is some minimum level of profit which will be acceptable, termed 'normal profit', and if the actual profit falls below this farmers will withdraw from production. In other words, supply will shrink and the supply curve is shifted to the left. If for some reason actual profit rises to more than the normal level, new producers will be attracted into the particular line of production and existing producers will expand. The supply curve will be shifted to the right. To take a very simple example, a rise in the price of poultry feed will shift the supply curve of chicken meat to the left, while a fall in the cost of chemical fertilisers will shift the supply curve of cereals to the right.

How rising input costs affect the supply of commodities will in turn reflect the ease with which it is possible to substitute alternative inputs, and the relative cost of the particular input to total costs. A 10 per cent rise in chicken feed costs would have much more impact on poultry production than a similar rise in the cost of one brand of diesel fuel on cereal output. Also there is the question of how easy it is to make adjustments within farms. To a specialist poultry producer there may be no alternative other than a complete shutdown if feed prices rise significantly, whereas a mixed farmer with a cereal enterprise may easily respond to higher fuel charges by switching to rival diesel brands or, if all brand prices rise together, by making marginal changes in his land use pattern.

Some inputs are common to a wide range of enterprises, so that the effect on supply of each product may be broadly similar and muted. For example, it is unlikely that a rise in farm rents would cause a dramatic contraction in supply since all enterprises use land to some extent, although one might expect some switching after a time towards those types of production which required relatively small quantities, that is, the land-intensive enterprises. A much more marked response would be found in which there was an input specific to one line of production. If for example the cost of disease-control chemicals in potatoes were to suddenly rise to high levels, and it were not possible to grow the crop without use of these sprays, the supply curve of potatoes might shift considerably.

In the UK an increasing proportion of the inputs of farming come from off-farm sources. This is seen in the replacement of man and animal power by tractors and machinery using fossil fuels, the higher proportions of manufactured fertilisers and disease control using purchased chemicals rather than rotation and cultivation methods. Consequently a rising share of the revenues which farmers get from selling their products is passed on to other parts of the economy. In the accounts for the agricultural industry presented to Parliament in the Annual Review of Agriculture White Paper, about half the value of the industry's final output is taken up by purchased inputs (not counting the costs of hired labour, rent or interest charges). Deducting these (but not family labour) resulted in 84 per cent of final output in 1985 leaving the farm family, only 16 per cent remaining as income. In 1984, a year of particularly good yields, only 22 per cent remained as income for the farmer or family. It follows that a rise in input costs will make a greater impact on supply now than in the times when farming generated more of its inputs within the agricultural sector. However this may be countered by the lower elasticity resulting from greater asset fixity.

Within the EC there is an association between the level of agricultural development and the share of output going to reward off-farm suppliers of inputs. For example, in 1982 the two most prosperous countries in terms of **family farm income** per **family work unit** (that is, the income remaining after paying all costs and allowing for depreciation, divided by the number of working family members in full-time equivalents) were the Netherlands and the UK; the proportions of total output staying as family income were 23 and 19 per cent respectively. At the other end of the prosperity scale, Italy and Greece with incomes a third or less of these kept 51 and 53 per cent of their outputs as income.[5] This factor would tend to make supply elasticity low in the poorer countries, a situation made worse by the lack of suitable alternative agricultural enterprises.

## 10.6  Supply and the Goals of Farmers

A simple view of supply assumes farmers to be motivated by profit maximisation. In Chapter 7 it was shown that there are other motives and goals, with the objectives of producers a complex mix, many of them social or personal. Even within the range of financial goals there are targets other than simple short-term profit. For example, producers often take steps to avoid risk, and if higher levels of output correspond with more uncertain returns then increases in price are not going to call forth greater supply. On the other hand, if the government introduces a price-stabilising scheme or undertakes to place a floor in the market, the degree of risk is thereby removed. Producers are likely to generate a larger output on the strength of increased security without any change in the average price level, shifting their supply curves to the right.

As an extension of this notion, if farmers encounter an event which alters in some way their hierarchy of goals, this alone could cause a shift in the supply curve. Better communications in rural areas might engender a more market-seeking profit-orientated attitude and an increased level of supply without any change in product prices or costs. A law which restricted the ability of a farmer to pass his business to a family successor would be likely to alter the goals of the farming community and hence its output and supply to the market.

In the short term we can be reasonably confident that farmers display a stable set of goals, so that changes in supply can be attributed to relative cost and price movements. But taking a longer period, say a decade, the influence of this factor should not be ignored.

One other determinant of the level of supply in the longer term is the state of technology. Because of its importance as a root cause of many major changes which agriculture is undergoing, which in turn lie behind the need for government policy towards agriculture, technical change is treated as the subject of a separate chapter.

## Notes

1. A working definition of supply is given as 'The amount of a commodity offered for sale in a particular market during a specific time interval at the prevailing values of prices and any other relevant conditioning variables,' Colman, D. (1983) 'A review of the arts of supply response analysis', *Review of Marketing and Agricultural Economics* **51**:3.
2. For a more expanded treatment of supply see Hill, B. (1980) *An Introduction to Economics for Students of Agriculture* (Oxford: Pergamon).
3. A fuller discussion of asset fixity is contained in Hill, B. E. and Ingersent, K. A. (1982) *An Economic Analysis of Agriculture* (London: Heinemann).
4. This point is made in diagrammatic form in (2) above.
5. Figures taken from the EC Farm Accountancy Data Network.

# Change in the Supply of Agricultural Products: Technological Advance 11

The output from agriculture in all industrialised countries seems to be on an inexorable upward trend. While in the EC high prices for farm products, heavily supported by Community aid, are often blamed for overproduction, in real terms prices of many crops and livestock have declined. At first sight this seems at odds with what has been described in the preceding chapter in which supply in the shorter term was seen to be directly related to prices of farm products; we might have expected that lower prices would cut supply. Clearly there must be some other explanation why expansion continues, in the UK at about 2–3 per cent per year. Furthermore, there are suggestions that greater expansion occurs at times when farmers' profits are under most pressure from depressed market prices or higher costs. Much of the explanation lies in the area of technological advance.

## 11.1  Technological Advance

The term 'technology', interpreted literally, means the sum of knowledge of the means and methods of producing goods and services;[1] 'technological advance' implies a growth in this knowledge. Knowledge transcends the boundaries created for descriptive convenience between industries, and

271

many of the most significant influences on agriculture have come from discoveries or developments in the engineering and chemical industries. However, it is not simply the advance in knowledge which is of significance, but the fact that some of the new 'means and methods' are incorporated into actual production, carrying implications both for farming and for the country's wider economic and social fabric. Some advances in knowledge do not prove practical, at least in the short term; the automatic, driverless tractor is a case in point, although future developments may make it attractive to commercial farmers. Others are seized on and applied rapidly, such as new heavier-yielding varieties of cereals. The factors influencing the rate at which 'advances' are taken up by farmers will be considered later.

The essential nature of an adopted technological advance is that it improves the relationship between inputs and outputs so that, at the farm level, the margin between costs and revenues is greater than under the old system. This can be achieved in several ways, described below. Because farm firms are in business for a run of years, and because agriculture is by its nature subject to fluctuations, innovations which merely reduce these fluctuations, and the risks which go with them, may also be attractive to farmers. However, the net result at the agricultural industry level is that the aggregate supply curve is pushed permanently to the right, manifest by greater output and a downward pressure on farm product prices (see Figure 11.1). New techniques, or 'innovations', are therefore watched closely by policy-makers because in large part they dictate the future structure of the agricultural industry and the shape of governmental agricultural policy.

## 11.2 Types of Innovation

A useful classification of innovations is according to their impact on inputs and output at the farm level. Each will carry different implications for structure and policy. Table 11.1 shows that advances can either increase output or, at least initially, leave it unchanged, while they can also be input-saving, input-requiring, or leave inputs unaltered.

The simplest types of advance to understand are those which primarily either affect output *or* inputs but not both, at least in the initial stages. For example, a heavier-yielding variety of cereal seed will increase output but not the amount of other resources used (land, labour, machinery, fertiliser). A better way of organising office procedures (paying bills, contacting merchants to sell produce, etc.) will not have a direct impact on the farm's physical output but might mean that the farm secretary may be employed for fewer hours, saving on inputs and costs. In practice, of course, there may be indirect effects. The new variety of seed may be able to respond

## FIGURE 11.1
## The Effect of Technological Advance on the Supply Curve of an Agricultural Product

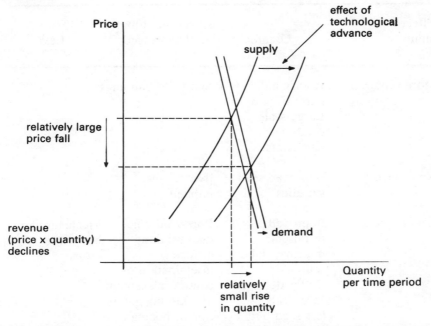

NOTE The effect of technological advance being taken up by farmers occur over a period of years. During this time Demand is also likely to have increased a little because of rising incomes and population growth.

profitably to higher levels of fertiliser, thereby resulting in the use of more inputs as well; and the wages not paid to the farm secretary can be used to finance higher stocking rates on the farm. The examples given in Table 11.1 only take into account the initial impacts.

A reduction in one form of input may be partly compensated by more of another. This may happen, for example, in the hoeing of field-scale root or vegetable crops if a small army of labourers working with hand tools is replaced by a tractor-mounted multiple hoe; the amount of capital involved may rise but a great deal of labour is displaced so that the total cost of hoeing will be less. A similar situation has occurred with the mechanised harvesting of blackcurrants; with mechanisation it is quite conceivable that output might fall (by less effective picking or through bush damage lowering next year's crop) but the technology would be used if costs were sufficiently reduced. If risk reduction is also considered, a farmer might

**Table 11.1**
**Classification of Technological Advances, with Approximate Examples**

| Effect on inputs | Greater | Effect on output Unchanged | Less |
|---|---|---|---|
| More required | Rotary milking parlour. Large-scale combine harvesters. | Short-life buildings | |
| No change | Improved seed varieties. | [No advance implied] | |
| Less required | Improved fertilisers; improved labour training. Better designed seed drills use less seed but improve yields. | Improved office organisation; artificial insemination of cows; replacement of hand hoeing by tractor hoeing | Mechanised harvesting of crops |

NOTE   The empty cells could be filled with examples of risk-reducing innovations, but these might cloud the issue. For example, risk-averse farmers might adopt a lower-yielding cereal variety which required more fertiliser if it offered great resistance to unforeseen droughts or diseases. Again, mechanisation of crop harvesting might require more inputs and produce lower output but farmers might be released from the uncertainties associated with gang-labour.

prefer to dispense with the uncertainties of hired gangs of labour to harvest crops of vegetables or fruit if dependable machine-harvesters become available, even if this means using more inputs and/or getting lower yields.

Other types of innovation from the start imply changes both in inputs and outputs. Without an increase in both the innovation is not attractive. Examples are the rotary milking parlour and the large-scale combine harvester. The initial costs of these items is greater than the types they replace, but their attractiveness lies in the large capacities they can handle

**FIGURE 11.2**

**A Technical Advance with Higher Fixed Costs**

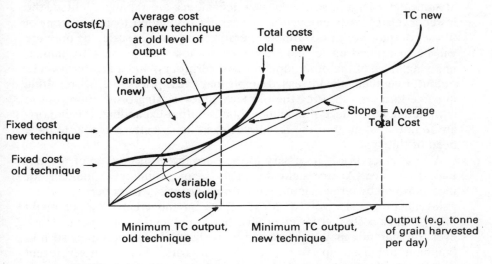

1. The TC lines represent total costs of the two techniques (old type machine and new larger-capacity machine) at different levels of output. These are comprised of two components: fixed costs (related to initial acquisition, such as depreciation, interest charge, which do not vary with output) and variable costs (related to the volume of output, such as fuel costs).
2. The fixed costs of the new higher-capacity machine are typically higher than the old, low capacity one.
3. Average (total) costs of operation are given by the slope of a line drawn from the origin to the TC curve. The lowest-possible ATC of each technique is where this line is tangential to the TC curve.
4. The new technique can produce a lower ATC than the old as long as a greater output is achieved. At the output at which the old machine was operating at its lowest cost, the new machine has *higher* average costs. Only if output is increased will lower average costs result. Farmers will tend to expand to reap this benefit. If the price of the product (grain) falls as a result, the implication is that the old-type machines will be made unviable and fall out of use whereas the new types will survive.
5. The slope of the TC curve indicates the marginal cost of harvesting, that is, the extra cost incurred by harvesting one more ton. At the old-type machine's least-cost level of operation, the new type clearly has a lower MC than the old. MC of the new type at *its* least-cost output is also lower than the old type's MC at its least-cost output.
6. For either technique, average fixed cost at each level of output would be given by the slope of a line drawn to the fixed cost line, and average variable cost by the slope of a line drawn from the fixed-cost point on the cost axis to the TC line.

– potential number of cows milked per day or tons harvested. If this capacity is utilised the average milking or harvesting costs are below alternative methods (see Figure 11.2). For those farmers already operating at sufficiently large a scale, the new techniques are rapidly adopted; the lower levels of costs enjoyed are generally responded to in the form of further output. For those farmers operating on a smaller scale, the pressure will be to expand up to that size necessary to enjoy the new technique, implying more of the other inputs (cows, fertilisers) and a greater level of output. Eventually the increased supply may cause prices to drop, resulting in some farmers still using the old techniques ceasing production, but at least initially the effect of the innovation is output-expanding. It is important to note that most of the recent advances in agriculture appear to have been of this type.

A brief reference is necessary to the pure factor-using type of technological advance. At first sight this might seem an unlikely candidate for incorporation by farmers into their farming systems, except perhaps where risk-reduction was involved. More inputs but no greater output might imply higher costs and lower profits, hence preventing its uptake. However, the paradox may be explained by realising that some inputs such as buildings are used up over a number of years.[2] For example, recent developments of low-cost housing for dairy cattle pose an attractive alternative to higher-cost concrete and asbestos structures, but the cheaper alternative is less durable; for illustration, let us assume a life of 20 years as opposed to 60 years for the concrete and asbestos building. If, say, the new-technology housing costs only half that of the traditional structure and the farmer's planning horizon is 20 years, corresponding to the life of the new-style housing, the new technology may be chosen as being the preferable investment. However, over 60 years this type of housing will need replacing three times whereas the old-technology building would last this long without renewal. Hence, looking back and assuming that costs indicate the quantity of resources, the technological advance has, over 60 years, resulted in *more* resources being used to provide the same amount of housing. This example begs the question of the incorporation of further advances in technology over the period and, indeed, the whole process of valuing resources. In agriculture such examples are rare, although in food packaging it is by no means an irrelevant issue, with new materials of the throwaway variety replacing re-usable glass bottles or metal containers.

While some sorts of innovation do not in themselves need an increased output to be realised, in practice this almost always happens in some form. A better method of manufacturing fertilisers will lower their price, more will be used and cereal output will expand. Even improved farm office organisation can release the farmer for more attention to field operations, with an increase in production at the whole-farm level.

## 11.3   Technological Change and Input Substitution

We have already seen (in Chapter 10) that the combination of labour, capital, land and other inputs which farmers use is a reflection of each input's marginal productivity and its price. More expensive labour will cause farmers to attempt to substitute capital goods for labour without any technological advance being required, simply to bring the ratios of the costs of labour and capital back into line with their marginal productivities. This is shown in diagrammatic form in Figure 11.3, which also illustrates the effect of a technological advance.

While it is theoretically possible for a technological change to leave the balance between factors unchanged, in reality there is usually some factor substitution. The normal pattern has been that new machinery has enabled labour to do more, increasing its marginal productivity. This shifts the balance which farmers aim for in favour of a greater use of capital per man. Because the average and marginal costs of production will be lowered, farmers will wish to expand output, which will require more inputs than if output were limited to the pre-advance level. If costs are reduced to the extent that great expansion is called for, the overall impact at the farm level *may* be to use more labour (and a lot more capital). For example, the introduction of mechanised mushroom growing might so lower production costs and hence prices in the shops that extra labour could be required to cope with the expanded output demanded by the market. However, more typical for farming is the case where more output is produced using more capital but less labour, shown by diagram D in Figure 11.3. If individual farms could easily expand their areas then the labour force on those farms might not be reduced. In practice farms cannot change their areas readily, so labour tends to be displaced as technical advance proceeds, or at least not to be replaced when workers leave or retire. Also, at the aggregate level, because the total area of land is fixed, even if farms which grow in order to reap the benefits of new technology do not reduce their labour forces they will tend to displace labour on the land they take over.

## 11.4   Industry, Regional and National Consequences of Technical Advance

If a higher yielding variety of cereal becomes available, or a waste-preventing trough for feeding pigs or a labour-saving device for cleaning milking equipment, it is not too difficult, knowing the innovation's characteristics, to predict the outcome for the enterprise in question. However, the consequences for whole farms are less clear cut; labour saved from

**FIGURE 11.3**
**Technological Advance and Input Substitution**

**A**
*Change in relative input costs*
*no technological advance*

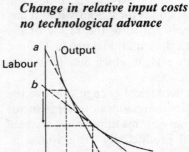

*aa* cheap labour to capital ratio
*bb* dear labour to capital ratio

**B**
*Technological advance, unchanged*
*ratio of labour to capital; same output*

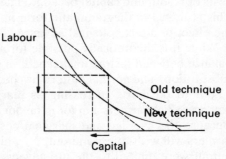

The new technique uses less of
both inputs to produce the same
output

**C**
*Technological advance; changed ratio*
*of inputs, same output*

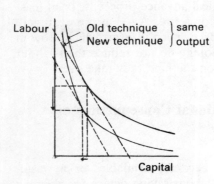

To produce the same output
the new technique uses much
less labour and a little less
capital

**D**
*Technological advance; changed*
*ratio of inputs; more output*

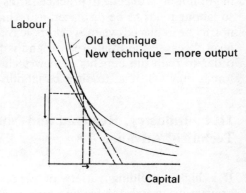

To produce a greater output,
less labour and more capital
is used

dairying may leave milk output unchanged but might be used to expand other enterprises or may simply be wasted if no other uses for it can be found.[3] At more aggregate levels the outcome is even harder to predict because of the multiplicity of factors to be considered. Too narrow an approach can be wildly misleading, as was shown in early-nineteenth-century England by the Luddites who tried to destroy new textile machinery which appeared to threaten their employment; in practice the machinery resulted in more, not less, employment although admittedly of a different sort.

### 11.4.1 *Aggregate input use and output*

At the industry level, technological advances which at the farm level may require more inputs and generate more output do not necessarily carry quite the same implications. For example, many advances require large-scale farming to reap their benefits, but the total UK area of farm land as an input is not expandable. Consequently these innovations have been taken up by the already large farmer, by farmers specialising so that land is released from other enterprises, and by some farmers taking over larger acreages from others who leave the industry or reduce their scale of operations. As was seen in Chapter 6, most of the land coming onto the market in recent years has been bought by local farmers expanding their areas, farmers who most frequently already occupy medium or large farms. These adjustments are reflected in the changing farm size structure of agriculture noted in Chapter 9. With the input capital, however, no such natural limitation on the amount available applies and it is clear that increasing quantities are used.

On the output side, again the expansion by some farmers will be in part offset by others quitting the enterprise. With those products facing very inelastic demand, and with very little demand growth arising from population changes, the result of a tendency to increase supply coming from technological advance will be a severe drop in price and the shedding of productive resources as they search for more profitable employment. This has been seen in potato growing where rising yields coming from improved varieties has been accompanied by a reduction in the area of land used for potatoes. Despite controls on the areas grown and support buying when the market is heavily depressed, the price system operates relatively unhindered for potatoes. In contrast, where there is regular and vigorous market intervention by Community agencies which endeavor to keep up prices by buying and storing farm produce – producing cereal, beef and butter mountains and wine lakes – resources are retained in lines of production to a greater extent than if prices (and hence profitability) were allowed to decline in response to expanding supply. Despite such market support it is evident that, in real terms, agricultural product prices are declining. The upshot is that *some* expansion of output and *some* shedding of resources normally accompany technological advance, with the balance varying between commodities. In the UK part of the rising supply has displaced imports, the degree of self-sufficiency in all major products having risen in the postwar period.

Taking a view even broader than agriculture, at the national level all technological improvements can be considered as output-increasing since, even if they do not result in a rise in the industry to which they most directly relate, they release resources for greater production elsewhere. This effect on national output assumes, of course, that the released resources do not remain idle. This process has already been encountered in Chapter 5 when the role of agriculture in development was touched upon. Labour shed by agriculture has proved an important resource for other industries, and in many countries this applies to capital also. These are only released, however, when improvements in farming mean that food production has expanded through technological improvements to the point that it becomes possible to withdraw the resources without endangering the national food supply.

Where agricultural productivity has failed to improve at the desired rate, as in the USSR, the release of resources has been hampered and food supply has presented problems. In the less-developed countries the take-up of technological advances in farming may be even more vital, since they have to contend not only with a desire for a growth of industries other than farming, requiring resources, but also with feeding a rapidly expanding population. Fortunately some of the necessary technical advances are already known in richer countries, so that the poorer ones do not need to rediscover irrigation techniques, pest control, storage environments, etc. However, a danger exists that they will copy the technology of industrialised countries where labour is relatively scarce and capital relatively plentiful, the opposite of what obtains in many low-income countries. In the third quarter of this century increasing care has been taken to ensure that the technologies which such countries try to employ is appropriate to their relative factor endowments (especially their plentiful labour supply) and the skills of their populations.

### 11.4.2 Factor substitution

In Britain since the mid-1950s the cost of machinery and buildings has been subsidised in various forms as part of government policy to improve productivity and farm incomes, making labour appear even more expensive and encouraging its reduction. It has been suggested that the development of labour-saving technology has been encouraged by such capital subsidies and by the type of research which has been undertaken by government agencies. Calls have been made to remove this capital-favouring distortion by scaling down grants and tax allowances, to which the UK government responded in part in 1985. The calls have been on the grounds not only of improving nationally the use of resources (by approaching more nearly the economically-desirable state where factor prices represent their relative scarcity to society), but also because more people engaged in agriculture means more people living and working in the countryside. In turn this implies that rural communities and the services on which they depend (schools, transport, shops, etc.) become more viable. Under the spectre of continued substitution of more capital for less labour these necessary attributes of a thriving rural society are threatened, where they have not already been lost. These arguments have been given further impetus by the greater rates of general unemployment seen in the 1980s (13 per cent in 1985 in contrast with 4 per cent in 1970), in which labour displaced from farming is less likely to be taken up by other industries.

Technological change which alters the balance between the inputs used has implications for the share of industry's total earnings that each takes. For example, a labour-saving machine would result in a decline in the share of net revenue going on wages; it might also reduce the absolute wage bill. This would depend on whether labour was shed, or retained and used to increase output, and also on what happens to wage levels when the supply of labour is increased by workers displaced from agriculture. In practice, labour-saving technological advance seems to proceed simultaneously with rising labour costs, so that substitution of capital for labour comes from both causes.

### 11.4.3 Incomes

In hazarding a prediction of the implication of a technological change on incomes in agriculture the following characteristics should be in the forefront of consideration:

1. the effect on output of the product and the resultant fall in its price
2. the effect on total costs of production
3. the nature of the short-run supply for individual factors of production
4. the effect on supply and demand of other products.

The income of farmers will correspond with the margin between the revenue coming from selling output and the sums paid out to other sectors of the economy for fertilisers and fuel (and where hired labour and tenanted land is involved, wages and rent). We know that most agricultural products face an inelastic demand, so that an increase in output of *x* per cent causes the market price to fall by *more than x* per cent, with the result that the returns coming back to farmers actually fall with an increase in output. Innovations which increase output volume without raising total costs must, on this basis, *lower* the income of the industry as a whole although, for the individual who first spots the innovation and applies it, his single income will benefit in the short term.

Innovations which reduce costs without any effect on output raise income at the industry level. With those which lower total costs but raise output the income of the industry could go either way, depending on the relative magnitudes of cost saving and revenue contraction due to the expanded output encountering inelastic demand.

In practice the full impact of rising output on revenue is frequently cushioned by governmental intervention. For example, the extra grain coming from heavier-yielding varieties in the EC has not reduced cereal prices by anything like the amount that might have been expected in a 'free market': a rough estimate for 1985 suggests that they might be some 20 or 25 per cent lower in the absence of the Common Agricultural Policy.[4]

Technical change can spread its impact through the price mechanism far beyond those farms which take up and apply the new technology. For example, improved transport and storage of tomatoes from Spain might seriously affect the incomes of British growers. A new strain of wheat, suitable for growing in East Anglia because of an improved toleration of dry conditions, by expanding cereal output could depress prices and hence have income-reducing consequences for cereal growers in the wetter West of England where there is generally no call for such a plant variety. The pleas of such growers and farmers for import bans or other product price support mechanisms can be easily understood – they feel that they should not be penalised for the innovative activities of others. But the market system operates in a way which ensures, in an industry such as agriculture which has many independent operators, that the benefits of lower-cost production methods get taken up, with ultimate benefit to society, although in the process it is inevitable that farmers wedded to old technology come under severe economic pressure.

## 11.5   The 'Treadmill' of Technological Advance

The notion that innovations spread remorselessly and inevitably through the agricultural industry has been encapsulated by the idea of a 'treadmill'. In the words of Cochrane, the American economist who is linked with the development of this concept, 'The average farmer is on a treadmill with respect to technological advance. In the quest for increased returns, or the minimisation of losses. . .he runs faster and faster on the treadmill. But by running faster he does not reach the goal of increased returns; the treadmill simply turns over faster.'[5]

## FIGURE 11.4
## The Treadmill of Technological Advance

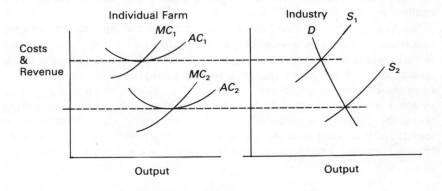

NOTES 1. AC and MC curves represent the average and marginal costs of production.
2. The new technology has a lower level of costs than the old ($AC_1$ and $MC_1$ are higher than $AC_2$ and $MC_2$).
3. In a perfectly-competitive industry, to which agriculture approximates, competition will ensure that farms operate at the lowest point of their $AC$ curves.
4. The industry's supply curve is composed of the $MC$ curves of individual farms. Anything which shifts the $MC$ curve to the right will also shift the industry $S$ curve.
5. Innovators adopt the technique as it means they can earn more than 'normal' profits ('normal' profits being those just adequate to compensate for the risks involved in production).
6. When significant numbers of farmers adopt the technique the industry $S$ curve will change, and prices will fall as output rises.
7. Falling prices will make the old techniques increasingly unviable and force all farmers to adopt the new technique eventually.
8. Innovators will have their surplus profit eroded, so they will search for new technological advances in this or other products.

The more progressive farmer will always be on the look-out for means of improving his profit, and this generally results in higher output. The average farmer will thus find himself faced with product prices that tend to decline in real terms, so that he has to adopt the new technology in order to guard his income position and ensure his survival. It is the *average* farmer who is on the treadmill; it is the innovator who is responsible for its direction of turning. The unaware or those unwilling to adapt find themselves squeezed out of the industry as their old-style technology proves unviable at the lower level of prices and their incomes too low to continue in farming (this argument is explained in greater detail in Figure 11.4). There will be a residue of these hanging on to farming because they have few or no opportunities for employment elsewhere and an unwillingness to invest their capital in activities other than farming. In many cases they will be the elderly, the small farmer and those in the remoter regions. It is towards this disadvantaged group that agricultural policy is frequently directed, with the intention of supporting their incomes or aiding their outflow from farming (see Chapter 14).

While technological change can result in a lower income to the industry as a whole, it would be wrong to conclude that this *inevitably* leads to lower farm incomes at the individual level. If innovations require farms to be larger, then a reduction in aggregate income could result in a maintained or even increased average income per farm if a sufficiently large number of people leave the farming industry. However, average figures tell us nothing about the distribution of incomes. In the process of adjustment it seems inevitable that there will always be some farms which are unviable, with incomes squeezed to the level that their occupiers wish to leave farming.

## 11.6   Rates of Diffusion of New Technology

Some innovations spread through farming faster than others, and some experience a period of dormancy – while they are known about, they have to await some event or change in prices to trigger them off. Some farmers are habitual innovators, always the most progressive and the first to buy new types of machines, while others cling on to old methods so that their farms come to resemble museums. This prompts the question of what factors influence the rate of take-up of innovations and which farmers adopt them first. This is of obvious importance if, for example, it is government policy to promote an efficient modern agriculture and thereby to aid the incomes of the farming community.

### 11.6.1   Information

The first step necessary for the take-up of any new technology is the dissemination of information about it. Commercial advertising literature, agricultural shows and the farming newspapers and broadcasts play an important part, and these avenues can be augmented by public information services (MAFF advisers and their literature) and public-sector research station bulletins. As will be seen below, different groups of farmers tend to use different channels of communication, with the more progressive tending to be most likely to use source material (scientific reports) whereas the least progressive will tend to rely on the farming press and talk from farming friends and relations. If the government wishes to affect, say, the small old-fashioned farmer, it will need to choose the appropriate means of communication or its efforts will be wasted. This should force the designers of agricultural policy into clarifying at whom they are aiming – the large progressive farmer who dominates total agricultural output or the back-woods man who may have an unacceptably low income.

### 11.6.2 Uncertainty

Any innovation by its very nature will involve some element of uncertainty. Even when extensive field trials of some new crop variety or machine have been undertaken, there may be surprises when it is put into the hands of commercial farmers. For this reason alone one might expect innovations to be taken up first by the bigger, diversified, more prosperous farmers since usually they are best able to take the risk. As more and more farmers take the product up more becomes known, the uncertainties are reduced and the more risk-averse farmers (that is, those less willing and able to take risks) grow willing to try the new method.

The economic climate can be expected to affect the rate of innovation; in times when incomes are subject to instability and uncertainty because of some influence such as changing government policy farmers might be expected to steer clear of adding to their instability by adopting innovations of an unproven nature. While simple cost-saving exercises might still be found attractive, major changes and those requiring heavy capital expenditure would be avoided until more certain times returned.

### 11.6.3 Capital aspects

It is commonly found that innovations such as new seed varieties which simply replace one form of working capital with another are taken up far faster than changes which require new capital assets and the abandonment of the old. One useful classification of technical changes is into those which:

A. result in capital saving, such as the development of the artificial insemination scheme for cattle in the UK, which meant that most smaller dairy farmers no longer needed to buy and keep a bull;
B. imply little or no capital change, such as an improved type of fertiliser or new seed variety;
C. require much additional capital, such as a new type of milking parlour.

While the first two can be rapidly taken up, the third spreads much more slowly. Not only must an expensive new building be financed, perhaps competing with other potential investments on the farm, but existing outmoded equipment may still be viable. For example, once a milking parlour has been installed its cost becomes of little more than historic interest; its scrap value is probably zero. In Chapter 10 it was shown that this asset fixity can help explain why farmers frequently keep on producing when the prices they receive for their products fall. It also has relevance to the rate of technology uptake. If a new type of parlour comes on the market, the farmer decides whether to continue using his existing equipment on the basis of its current and future operating costs (labour, feed, etc.) in relation to the price of milk. An investment appraisal could well favour the continued use of the outmoded equipment, although with its gradual wearing out and, most probably, a fall in the price of milk brought about by expansion by those farmers who have just acquired the new technology, there may come a time when abandonment in favour of investing in the new equipment becomes the preferable option. This could mean scrapping the old before it reaches the end of its physical life.

### 11.6.4 Management demands

Management ability, describable as the power to predict and foresee outcomes and plan accordingly, must be present before any innovation can be put into effective operation, but some new techniques or products clearly demand greater or different management skills than others. Some innovations require little additional expertise – a heavier yielding variety of oats would probably be within the capabilities of any current oat grower. A whole new crop, such as grain maize, lupins or evening primrose would require rather different skills, and an intensive, environmentally-regulated piggery yet others. Where the skills required are close to those already being exercised, the innovation is the more likely to be taken up rapidly.

It is interesting to note that farmers are increasingly called upon to use management skills which, to the layman, appear to be primarily non-agricultural. For example, tourism finds an increasing place in the income-generating activities of farms in the West of England and in the upland areas of Britain. Farm computers, electronic animal supervision, processing and packaging, and financial control all require talents in general business management rather than specifically agricultural skills, suggesting that increasing levels of farmer education are required to cope with the advance of technology. Consonant with this in the United States it seems that innovation proceeds most rapidly in areas where formal education is most available, where farm families have more time and money for travel, study and meeting fellow managers of high ability. In the UK it has already been noted that formal higher education is more commonly encountered among the occupiers of large farms, though the line of causality is not self-evident (Chapter 7). Probably the larger farmers, who are will placed to be innovators, are also in a favourable position to ensure that their children receive a formal education, which reinforces the rate of innovation on large farms.

**FIGURE 11.5**
**Technological Advance Requiring Changed Input Prices for Its Uptake**

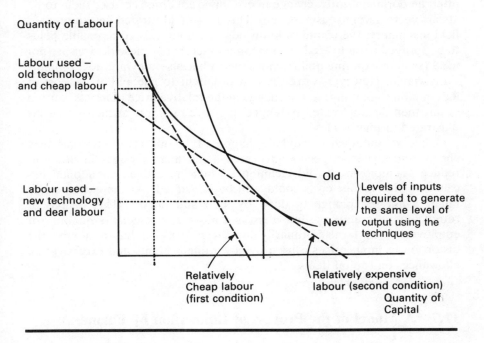

NOTES 1. The two solid curves (isoquants) represent the quantities of inputs (labour and capital) required to produce a given level of output under the old and new techniques.
2. The dashed lines (isocost lines) show the combinations of labour and capital which can be acquired for a given sum. The slope indicates the relative prices of the two inputs. Isocost lines further from the origin represent higher levels of costs.
3. When labour is relatively cheap and the old technology is in use, the availability of new technology will not affect the way production takes place. To use the new technology would incur *higher* costs, as can be seen above; the first isocost line would need to be moved further out from the origin before it could touch the new isoquant.
4. If labour prices go up the slope of the isocost line changes. Under this new regime the new technology is the lower cost way of producing the given level of output. Again, to use the old system would mean moving the isocost line further out.

## 11.6.5 *Factor and product pricing*

The point was made earlier that an advance in knowledge does not necessarily imply that changes will necessarily take place in the production techniques actually used by farmers. There are countless inventions which never prove practical – such as man-powered flying machines and Brunel's atmospheric railway – and agriculture has its fair share of these. Even if they can be made to perform satisfactorily in the technical sense, they may often lie dormant until a change in economic conditions causes them to be attractive to farmer-businessmen. This is well illustrated by automated field machinery; the technical knowledge has long existed to enable fields to be sprayed or cultivated by unmanned tractors but these devices are not used (save in experimental circumstances) because they are not commercially viable. However, were the cost of labour to rise and/or the cost of automation to fall, that is, if a change in the relative price of inputs were to occur, then the automated system could prove the more attractive. This is illustrated in Figure 11.5.

Of course, the reverse could also happen. If labour were to become very cheap, and some politicians advocated a fall in real wages in order to reduce unemployment, old techniques which have been abandoned because of high labour costs could well be reinstated. At the extreme one might again see ploughing done by men and horses. This degree of retrenchment is unlikely, but there is a move to protect the quantity of employment on farms, primarily for social reasons, but removing the distortion to input prices caused by government grants and excessive tax allowances on capital items.

## 11.7 A Model of the Process of Innovation by Farmers

The process by which technological advance is taken up is thought to be common to all innovations, and consists of four basic stages: first comes the awareness of an innovation; second, the formation of interest and the gathering of information; third, the 'mental acceptance' of the innovation, that it is something which should be taken up; and fourth, the actual adoption of the innovation.

These stages sweep in waves through the farming community, but not uniformly to all farmers. Some will acquire knowledge earlier than others and some will accept the desirability of a new technique faster than their neighbours.Some farmers will be in a position in which they can incorporate the innovation into practical farming more rapidly than others. The spread of knowledge is likely to be faster than the rate of physical uptake,and this will depend on the type of innovation – where it originates, whether it is commercially promoted, the sorts of management required, capital aspect and so on. The progression of each step through the farming community in generalised form is shown in the top part of Figure 11.6. It is commonly found that the awareness and adoption curves are both approximately sigmoid (in the form of the letter S).

**FIGURE 11.6**
**The Adoption Process**

### A  *The steps involved*

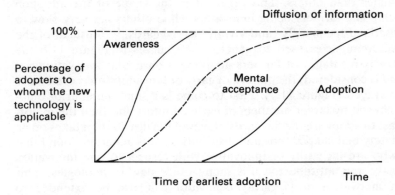

NOTE    The adoption and mental acceptance curve need not reach 100 per cent if some farmers reject the innovation and eventually leave production.

### B  *Innovations adopted at different rates*

NOTE    1. The adoption curve is roughly symmetrical.
        2. The speed of uptake by the first 10 per cent of farmers will vary between innovations.

We have already explained that the rate at which innovations are taken up by farmers will be affected by factors such as the amounts of capital they require and the necessary management skills. The lower diagram of Figure 11.6 illustrates this, with the first 50 per cent of farmers taking different times to adopt the innovation. The 'fast' curve might be illustrated by a new wheat variety and the 'slow' curve by rotary milking parlours.

### 11.7.1  Farmers as fast or slow adopters

Our earlier review of what factors could affect the rate of spread of an innovation led to the suggestion that some farmers would be more ready to adopt changes than others. This reflected in the shape of the adoption curve, with some farmers being innovators while others are very slow to adopt, described technically (and vividly) as 'laggards'.[6] If instead of the cumulative frequencies used to generate the curves in Figure 11.6 the number (or percentage) of farmers adopting a particular innovation *per time period* is considered, the pattern of early or late adoption is even more clearly portrayed (Figure 11.7). First to come is a small number of innovators, followed by larger numbers of early adopters and then the majority. The last to adopt are the 'laggards'. Overall the distribution takes on an approximately bell-shaped 'normal' distribution. It is generally found that farmers who are laggardly or innovative with respect to one innovation tend to have this attitude towards many other new technologies. This notion of innovative or laggardly behaviour can also be extended to communities, and those areas of the country or types of farmer which are fast or slow adopters (the two, of course, are linked) tend to hold their relative positions over the years.

There are some recurring personal characteristics of individual farmers who are persistent innovators, laggards or between the two extremes. These are set out in the lower part of Figure 11.7. Size of farm, income level of education and off-farm experience are immediately apparent as linked with innovative behaviour. Perhaps less obvious is the marked association with the channels through which communications are received. This is of great relevance to government policies of stimulating incomes through encouraging the uptake of productivity-improving innovations. If the intention is to target this policy on the small elderly farmer there is little point in publishing lots of scientific literature since the target farmers do not read it. Rather it will be necessary to approach such farmers through the channels they do use (neighbours, friends, relatives and the mass media), a more difficult task than that posed by the big farmer who is already aware of information sources, able to accept the risks associated with innovation and encouraged by the taxation systems and publicly financed grant aid to invest in them.

## 11.8 The Future Direction of Technological Change

Finally we must raise a question about attempts to steer the direction of technological advance and to stem its flow. It has been stated earlier that most advances in agriculture seem to be of the output-increasing and input-using type, and especially of the sort which uses more capital but less labour. Continued movement in this direction would seem strange when most agricultural products in the EC are in chronic surplus and when there is a sizeable pool of unemployment. Part of the reason is the insulation given to farmers from the market consequences of rising levels of production by various forms of support, described in Chapter 12 and 13, and the tax and grant environment which, by subsidising capital but not labour, creates a favourable climate for labour-saving capital-using innovations. Also, because agricultural scientists have tended to regard more output per acre or more milk per cow as an indicator of their success, the direction of technological advance has been towards this sort of development. The breeding of a lower-yielding variety of cereal (which raises profits through requiring disproportionately less fertiliser and other inputs) does not seem to have received much serious consideration.

In the 1980s such matters have received much public airing with the result that bias in technological advance towards expansion and labour-saving is increasingly looked for and criticised in the hope that it may perhaps be curtailed. However, the operation of the treadmill would seem to ensure that, as long as farmers are free to expand their individual levels of production, technical advances will be taken up and output will continue to rise unless prices fall sufficiently to cut back supply. Later it will be evident that governments have frequently tried to protect farmers from the price and income consequences of expanding supply through supporting product prices, a situation which has, in the context of the European Community, grown to be unacceptably expensive and has also resulted in embarrasing agricultural surplusses. Once this defensive policy has been embarked upon it is difficult to reverse. The political consequences of a drastic lowering of product prices and hence of farm incomes, and of the changes in the numbers of farms these are likely to produce, rule a speedy return to market-clearing price levels out of court at present.

The alternative is some form of direct control of growth in output, such as quotas; these not only limit the use of present technology but may for a time discourage the uptake of further innovations. While in the short term quotas may protect the interests of farmers, there is a cost to the rest of society in that consumers have to pay prices higher than are necessary and hence suffer a reduction in their real incomes. By not taking advantage of technical advances, resources are trapped in agriculture which may have better alternative uses elsewhere in the economy. In the longer term a domestic farming industry ossified with outdated technology will become increasingly uncompetitive in a world where innovations know no respect for national boundaries, needing major and therefore painful adjustments if prohibitive support costs are to be avoided. In short, the treadmill of technological advance has an inevitability about it which ultimately overcome attempts by individual policies to resist it.

# FIGURE 11.7
## Fast and Slow Adopters of Technological Change

### A  Numbers of Farmers Adopting an Innovation per Time Period

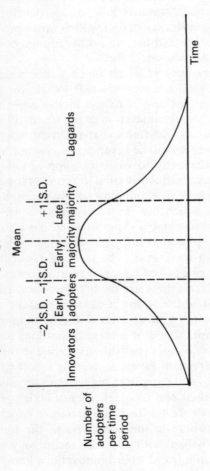

### B  Characteristics of Farmers in Five Categories of Adoption

| Adopter category | Personal characteristics | Salient values and social relationships | Communication behaviour |
|---|---|---|---|
| Innovators | Highest social status; largest and | 'Venturesome', willing to accept | Closest contact with scientific |

| | | | |
|---|---|---|---|
| | most specialized operations; wealthy; often young; well educated; often experience in non-farming environment. | risks; some opinion leadership; cosmopolite. | information sources; interaction with other innovators; relatively greatest use of impersonal channels of information. |
| Early adopters | High social status; often large and specialized operations. | 'Respected'; regarded by many others in the community as a model and influential; greatest opinion leadership of any adopter category in most communities. | Greatest contact with local change agents (including extension or advisory services, commercial technical advisers, etc.); competent users of mass media. |
| Early majority | Above-average social status; average-sized operations. | 'Deliberate'; willing to consider new ideas only after peers have adopted; some opinion leadership. | Considerable contact with change agents and early adopters; receive mass media. |
| Late majority | Below-average social status; small operations; little specialisation; relatively low income. | 'Sceptical'; overwhelming pressure from peers needed before adoption occurs; little opinion leadership. | Interaction with peers who are mainly early or late majority; less use of mass media. |
| Laggards | Little specialisation; lowest social status; smallest operations; lowest income; often oldest. | 'Tradition'; oriented towards the past; avoid risks; little if any opinion leadership; almost isolated socially. | Neighbours, friends, and relatives with similar values are main information source; suspicious of change agents. |

NOTE   This table is based on findings in several British studies, and confirmed by many studies in other countries.

SOURCE   Jones, G. (1972) 'Agricultural innovation and farmer decision-making', in *Decision-making in Britain: Agriculture* (Bletchley: Open University).

# Notes

1. The description appearing in Bannock, G., Baxter, R. E. and Rees, R. (1978) *The Penguin Dictionary of Economics* (2nd ed, Harmondsworth: Penguin).
2. Heady, E. O. (1952) *Economics of Agricultural Productions and Resource Use* (New York: Prentice-Hall) pp. 303–304.
3. In that case, the question must be posed as to whether the new cleaning equipment was a wise investment. One motive for making the investment in dairy cleaning equipment might have been to reduce the boredom and drudgery of the operation, improving worker relations, or perhaps it was assumed beforehand that some other ways of absorbing the released labour time would emerge.
4. The model developed by the Agricultural Economics Department of the University of Newcastle upon Tyne suggests that cereal prices in the absence of the CAP would be expected to fall by about 22 per cent (wheat by 15 per cent and barley by 22 per cent).
5. W. W. Cochrane (1958) *Farm Prices, Myth and Reality* (Minneapolis: University of Minnesota Press).
6. For a wide-ranging account of the diffusion of innovation process see Jones, G. E. (1967) 'The adoption and diffusion of agricultural practices', *World Agricultural Economics and Rural Sociology Abstracts* **9**:3, 1–34.

# Incomes and Agriculture 12

The incomes of farming families are central to the explanation of agriculture's response to changing product prices and costs, to structural adjustments such as increases in farm sizes, and to the very existence of governmental policy towards agriculture. If an 'income problem' did not exist, or at least was not thought to exist, then much of the case for a separate British ministry for this industry and a Common Agricultural Policy within the European Community would collapse. In this chapter we will consider the characteristics of incomes in farming and the measures currently employed in their assessment. The ground will then be prepared for the analysis of agricultural policy contained in Chapter 14.

## 12.1 The Agricultural Industry's Income

According to the UK's national accounts, agriculture as an industry generates a declining proportion of the nation's total Gross Domestic Product, a feature shared by virtually all industrialised countries. This point was made in the Chapter 5, and was seen to derive from the static demand for farm products in growing developed countries in contrast with rising demand in other sectors, combined with the tendency of improvements in farming technology inexorably to expand supply. It was also shown that the income of the industry as a whole might fail to increase, and could even decline, as prices of farm products are pushed down by this rising supply. Whether the incomes of individual people engaged in farming fall depends on how rapidly numbers in agriculture contract; even if total income was static or declining the average income per head could rise if the number of farmers was reduced sufficiently quickly.

The income of the whole UK farming industry is calculated and presented together with many other statistics in the Annual Review of Agriculture White Paper. For those directly or indirectly connected with the industry the White Paper's publication is an eagerly-awaited event; the income estimate is singled out for comment by the Minister of Agriculture and is seized on by the farmers' unions, journalists, academics and members of the other sectors which supply or buy from agriculture (bankers, chemical manufacturers, etc.), as an indicator of farming's economic health.

The figures used to calculate aggregate farming income in recent years are given in Table 12.1. The revenue which farmers receive for selling their crops and livestock is termed the **total output** of agriculture. A small adjustment is made to this to allow for items such as production grants and changes in the stocks held by farmers; the resulting estimate is termed the industry's **gross output**. On the input side, the equivalent figure in **gross input**, which is the sum of all the items purchased by farmers from other sectors of the economy, and which are used up within a farming year, such as fuel, fertilisers, processed feedstuffs and accountancy services, also adjusted for changes in the stocks held on farms at the end of the year. Deducting from gross output the gross input gives **gross product**. For capital items such as machinery and buildings, which normally yield up their services over a run of years, an estimate of their wearing out (capital consumption) is made and deducted from gross product to leave **net product**; this is the figure which represents agriculture's contribution to the national income, although gross product is often used for this purpose (see Chapters 5 and 15) as the calculation of capital consumption is open to dispute. Out of this net product farmers have to pay for the labour they employ, rent on tenanted land and for interest on borrowings. What is left, **farming income**, is the sum available for remunerating farmers (and their spouses) for the physical work they conribute to farming and for a return on the capital represented by the farm and its assets. A series of farming income estimates at current and constant prices starting in 1970 is given in Table 12.2. Note that a convention is adopted whereby interest charges on borrowing for land purchase are *excluded* (a division which must be somewhat arbitrary), and no account is taken of any increases in the value of land, which as a productive asset should arguably have changes in its value taken into the income calculation. This will be reconsidered later.

An essential point to recognise is that this income calculation is not based on actual farm incomes; rather, it is a national figure derived by deducting from estimates of the value of the industry's output the estimates of its costs. There is no single 'right' way of doing the calculation and on both theoretical and practical grounds some room for alternative approaches exist. Recent major changes in methodology mean that it is difficult to compare recent income figures with those before 1970.[1] However, these changes mean that now the accounts for agriculture shown in

**Table 12.1**
**The Output, Input and Income Account for UK Agriculture**

|  | 1976 | 1978 | 1980 | 1982 | 1984 | 1985 |
|---|---|---|---|---|---|---|
|  |  |  | £ million |  |  | (forecast) |
| *Output* |  |  |  |  |  |  |
| Farm crops | 1 464 | 1 525 | 2 120 | 3 120 | 3 624 | 3 235 |
| Horticulture | 629 | 750 | 913 | 1 013 | 1 252 | 1 246 |
| Livestock | 2 202 | 2 754 | 3 288 | 3 799 | 4 257 | 4 300 |
| Livestock products | 1 692 | 2 065 | 2 500 | 2 973 | 2 965 | 3 005 |
| *Total output* | 6 014 | 7 158 | 8 867 | 10 941 | 12 190 | 11 857 |
| Gross output (A) | 6 121 | 7 295 | 9 002 | 11 201 | 12 550 | 11 883 |
| *Inputs* |  |  |  |  |  |  |
| Feeding stuffs | 1 567 | 1 774 | 2 188 | 2 615 | 2 858 | 2 601 |
| Other livestock costs | 109 | 175 | 151 | 171 | 193 | 187 |
| Crop costs | 563 | 684 | 851 | 1 013 | 1 242 | 1 262 |
| Machinery | 380 | 493 | 668 | 833 | 940 | 1 091 |
| Other | 570 | 731 | 963 | 1 175 | 1 308 | 1 369 |
| *Total expenditure* | 3 189 | 3 857 | 4 820 | 5 806 | 6 541 | 6 510 |
| Gross input (B) | 3 179 | 3 879 | 4 844 | 5 815 | 6 553 | 6 465 |
| Gross product (A–B) | 2 942 | 3 416 | 4 158 | 5 386 | 6 000 | 5 418 |
| – Depreciation | 607 | 821 | 1 133 | 1 269 | 1 356 | 1 418 |
| Net product | 2 335 | 2 594 | 3 026 | 4 117 | 4 641 | 4 000 |
| – Labour | 882 | 1 102 | 1 445 | 1 695 | 1 897 | 1 997 |
| – Interest | 136 | 184 | 464 | 501 | 569 | 696 |
| – Net rent | 32 | 55 | 69 | 105 | 142 | 154 |
| Farming income | 1 284 | 1 255 | 1 047 | 1 817 | 2 033 | 1 154 |
| *In real terms* |  |  |  |  |  |  |
| (1980=100) |  |  |  |  |  |  |
| Gross output | 114 | 109 | 100 | 102 | 105 | 92 |
| Gross product | 119 | 110 | 100 | 106 | 108 | 91 |
| Net product | 129 | 115 | 100 | 112 | 115 | 93 |
| Farming Income | 206 | 161 | 100 | 143 | 146 | 78 |

SOURCE  MAFF (1986) *Farm incomes in the United Kingdom* (London: HMSO); and authors' calculations.

**Table 12.2**
**UK Farming Income**

|  | Current prices (£m) | In real terms (1980 = 100) |
|---|---|---|
| 1970 | 567 | 201 |
| 1971 | 640 | 202 |
| 1972 | 682 | 200 |
| 1973 | 952 | 256 |
| 1974 | 803 | 184 |
| 1975 | 994 | 186 |
| 1976 | 1283 | 206 |
| 1977 | 1263 | 175 |
| 1978 | 1255 | 161 |
| 1979 | 1147 | 129 |
| 1980 | 1047 | 100 |
| 1981 | 1368 | 117 |
| 1982 | 1817 | 143 |
| 1983 | 1508 | 113 |
| 1984 | 2033 | 146 |
| 1985[1] | 1154 | 78 |

NOTE    1. Forecast.
SOURCE    MAFF (various years) *Annual Review of Agriculture* (London: HMSO).

the Annual Review of Agriculture White Papers are on the same basis as those for other UK industries within the system of national accounting and give a fair representation of the contribution which farming as an economic activity makes to the national economy. In addition to farming income the Annual Review contains other indicators (**farm business income** and **cash flow** of farmer and spouse) which reflect different aspects of farming activity; these are described in the notes to Figure 12.1. The European Commission also publishes industry-level income estimates for UK agriculture based on net value added, which is equivalent to the national net product concept. For simplicity, here we concentrate on the most-quoted and long-established farming income series.

From Table 12.1 it can be seen that purchased inputs (mainly feeding stuff, fertilisers and machinery) currently absorb over half of the total revenue received by the industry from selling its output. When provision

## FIGURE 12.1
## UK Farming Income and Other Economic Indicators, 1972–85

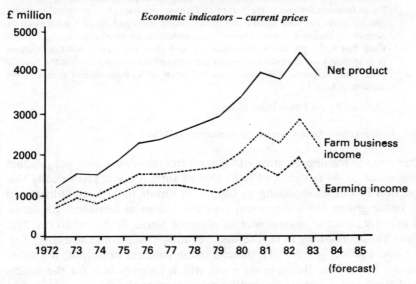

£ million      *Economic indicators – current prices*

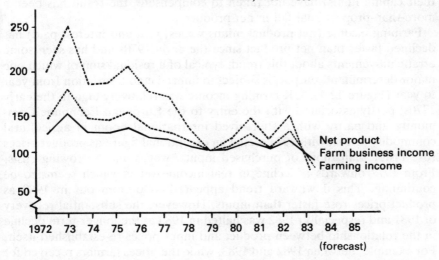

*Economic indicators – in real terms, 1980 = 100ᵃ*

NOTES    1. **Net product** is a measure of the value added by the agricultural industry to all the goods and services purchased from outside agriculture after provision has been made for depreciation.

*continued on p. 300*

*continuation of notes from p. 299*

2. **Farming income** is the return to farmers and spouses for their labour, management skills and own capital invested after providing for depreciation.
3. **Farm business income** is the return to farmers, spouses, non-principal partners and directors for their labour and management skills and on all capital (own or borrowed) invested in the industry, after providing for depreciation.
4. **Cash flow** is the pre-tax revenue accruing to farmer and spouse *less* cash outlays (i.e. spending on material inputs and services and on capital items) in the specific year. The definition has now been extended to include capital grants. (Not illustrated here.)

a. Deflated by the Retail Price Index.

SOURCE   MAFF (1986) *Annual Review of Agriculture* (HMSO).

has been made for depreciation of farm capital, the remaining net product is only some 36–40 per cent of the value of total output. Historically this proportion has been shrinking as bought-in inputs have been substituted from home-grown (or indigenous) inputs – chemical fertilisers for farmyard manure, tractor horsepower in place of horse flesh and so on. Net product has been falling in real terms over the period from the early 1970s to the mid-1980s (Figure 12.1); despite rising volumes of output, the value of output has fallen. Because the costs which farmers face for the inputs bought from other sectors (including an allowance for the wearing out of their capital items) have not fallen to compensate, the result has been a more-than-proportional fall in net product.

Farming income (net product minus wages, rent and interest paid) has declined faster than net product since the early 1970s and has seen some erratic movements about this trend, typical of a residual sum of which one major determinant, output, is subject to unpredictable variation from year to year (Figure 12.1). UK farming income was relatively high in the early 1970s, partly associated with the entry to the European Economic Community and partly with a short-lived international boom in agricultural commodity prices, with 1973 seeing an exceptional figure as product prices rose faster than those of purchased inputs, wages and borrowing costs. From 1977 onwards a decline in real income set in which seems to be continuing. This downward trend appeared to bottom out in 1980 as product prices rose faster than inputs. However, the substantial recovery of 1981 and (especially) 1982 was halted in 1983 as the longer-term decline in the relationship between product and input prices re-established itself. For example, between 1982 and 1983, while the prices farmers received for their outputs rose on average by 4.5 per cent, the prices of their inputs rose by 6.6 per cent. In 1984, although this cost-price relationship worsened, the farming income estimate rose because of an exceptionally good harvest, especially of wheat. In 1985 the unusually poor summer caused farming income to fall to a level even lower than 1980.

## 12.1.1   Interpreting aggregate income measures

The estimate of farming income relates in essence to the income generated from the activity called farming; that is, it is based on the industry irrespective of who receives the income or where production takes place. It measures this sector's contribution to the national income, and a good case could be made out that the calculation should be undertaken as part of the annual assessment of the economy's performance irrespective of any special interest which governments may have in the prosperity of farmers.

Nevertheless, at its inception in 1942, it seems that the White Paper's aggregate income figure was supposed to be a proxy for the income of farmers. The various price rises which had been given during the Second World War to encourage higher output were thought to be having a major but unknown effect on farmer incomes. Direct measurements at the individual farm level were impractical, so a calculation was made at the aggregate level, and this suggested that the industry's income had risen fourfold in four years. Subsequently this aggregate figure became interpreted as an important barometer of agricultural prosperity, with farming pressure groups (in the form of the NFU) attempting to achieve the recoupment of any rises in costs through more government support of prices or cost subsidies. But a moment's thought will reveal that the aggregate farming income figure is at best an imperfect indicator of the incomes of individual farmers. While it may be satisfactory at representing farming's overall contribution to the nation's income, little credence can be given to any average income per head derived simply by dividing the aggregate income by the total number of farmers. Additionally, there are many reasons why changes in the fortunes of individual farmers may differ widely from those indicated by the aggregate measure's movement.

First, all farmers are not solely dependent on farming for their livelihoods. Approaching one-third of all UK farmers or farming couples are part-time in the sense that they have another source of earned income, and there may be unearned income as well (pensions, investment dividends and interest, rent from property, etc.). Information on the personal incomes of farmers – that is, incomes they receive from agriculture and all other sources – in aggregate is not readily available, but Table 12.3, (which refers to the total income of full-time farmers and part-timers who are primarily dependent on their farming for their incomes), indicates that non-farming sources provide at least a third of total income. Some 17 per cent of the total came from investments, 5 per cent or so from pensions, and 15 per cent from earnings as hired employees.

Self-employment income earned by the farmer and his spouse formed only 63 per cent of the total, and not all of this would have been profits earned by farming. Left out altogether from this table are those farmers whose main source of profits is their other, non-farm, business. Hence the

Table 12.3
**Composition of Total Income of Couples and Individuals in the Trade Group Agriculture of the 1978/9 Survey of Personal Incomes**

|  | % total income |
|---|---|
| Self-employment income | |
| Husbands (and single persons) | 54 |
| Wives | 9 |
| Employment income | |
| Husbands | 8 |
| Wives | 7 |
| Other (including pensions) | 5 |
| Earned income | 83 |
| Rents | 2 |
| Building Society interest | 5 |
| Other interest and dividends | 11 |
| Investment income | 17 |
| All income | 100 |

NOTE   Figures may not sum due to rounding.
SOURCE   Inland Revenue, given in the Hill, Berkeley (1984) 'Information on farmers' incomes: data from Inland Revenue sources', *J. Ag. Econ,* **35**:1, 39–50.

aggregate farming income figure substantially underestimates the overall personal incomes of farmers and farming couples.

Second, farming is not a homogeneous industry. It contains a wide range of farm sizes and types, and the income from farming of a Welsh 50 ha livestock-rearing farmer will be very different from a 1000 ha cereal farmer in East Anglia. A fall in the aggregate figure may hide significant rises of income in certain sectors. This point is effectively made by Table 12.4 which illustrates the changes of profitability of different types of farming. Between 1977–8 and 1984–5 the incomes of dairy and lowland livestock farms fell by half, but incomes of cereal and cropping farms increased substantially. The fall in aggregate industry income was clearly no reliable indication that all types were experiencing decline. Even among farms of similar sizes and types there will be some who will be doing better than others, either by chance in particular years or consistently because of management ability, standards of equipment or local geography. Com-

**Table 12.4**
**Changes in Income by Sector of UK Farming**

(Index of average net farm income in real terms, 1982/3=100)

|  | Dairy | LFA Cattle and sheep | Lowland cattle and sheep | Cereals | Other cropping | Pigs and poultry |
|---|---|---|---|---|---|---|
| 1977–78 | 107 | 118 | 188 | 68 | 60 | 149 |
| 1978–79 | 114 | 128 | 201 | 86 | 114 | 200 |
| 1979–80 | 70 | 59 | 90 | 63 | 109 | 140 |
| 1980–81 | 70 | 69 | 107 | 66 | 66 | 126 |
| 1981–82 | 87 | 118 | 124 | 64 | 95 | 144 |
| 1982–83 | 100 | 100 | 100 | 100 | 100 | 100 |
| 1983–84 | 63 | 92 | 91 | 100 | 124 | 72 |
| 1984–85 | 59 | 96 | 58 | 118 | 98 | 171 |
| 1985–86[1] | 50 | 50 | 25 | 45 | 30 | 110 |

NOTE   1. Forecast.
SOURCE   MAFF (1986) *Annual Review of Agriculture* (London: HMSO).

ments are often made by those who visit farms regularly of the wide disparities of performance achieved on farms which may appear superficially similar.

Farms operate under a variety of tenure arrangements, some facing land charges and others not, and there are different levels of borrowing and different cost structures (e.g., farms vary in their balance between labour and machinery). These and other factors can cause substantial deviations in the patterns of income between the aggregate figures and those characterising substantial groups of farms.

In short, while the aggregate farming income figure may reasonably be used to indicate year-to-year changes in the overall income accruing to farmers from the activity of farming, care should be taken not to use it for other purposes for which it is not appropriate, although this is often done by politicians or pressure groups in support of their cases. It certainly should *not* be interpreted as being the sum of all the personal incomes of farmers and their spouses. The fact that the aggregate figure may be falling is not sufficient to justify government support of agriculture on the grounds that individual farmers are suffering unacceptably low incomes or even a major drop of their living standards. To discover whether this is so one must move from the aggregate to examine incomes at the individual farm level.

## 12.2    Incomes at the Farm Level

When a farmer talks of his income, or politicians try to promote policies directed at farmers' incomes, they are essentially concerned with *personal incomes*. Personal income can be described as the potential spending power a person has in a given time period. This will come not only in the form of earnings from 'gainful activities' such as employment or from running a business but also from owning assets such as property or pension rights or flowing from investments, rent from houses and land, pensions and capital gains.[2] It is convenient at times to partition this broad definition of income into 'earned' (from employment or self-employment), 'unearned' and 'capital gains'. Society tends to regard each in a different light, and taxation rates are usually different, frequently with a surcharge on unearned income and with lower rates applied to capital gains.

In looking back over a protracted period, perhaps a decade or a farming career, the income of a farming couple in terms of potential spending ability will thus have both current income and capital gain components. An outside observer might conclude that in retrospect a farmer may have done quite well in comparison with what he might have achieved in some other occupation. However such a broad view of income may well not be appropriate when assessing, for example, whether a farmer has enough cash with which to buy the necessities of life for his family; in this circumstance the outside observer might only consider the weekly cash income of the farmer. To enquire whether farming alone would be capable of generating an adequate income, off-farm earnings might be set on one side. In short, the appropriate concept of income to use would depend on the circumstances.

### 12.2.1    Characteristics of concern

Governments of industrialised countries the world over employ policies aimed at farm incomes. The reasons are similar in all cases. The main concerns over farmers' incomes at the individual level seem to be three-fold: a) that farmers' incomes are inherently *unstable* over time; b) that farmers on certain types or sizes of farm have incomes which put them in *poverty*; c) that incomes in agriculture are unacceptably low in comparison with those in other industries – the *comparability* concern. It will be

## FIGURE 12.2
**Output of Potatoes in Physical and Value Measures**[1]

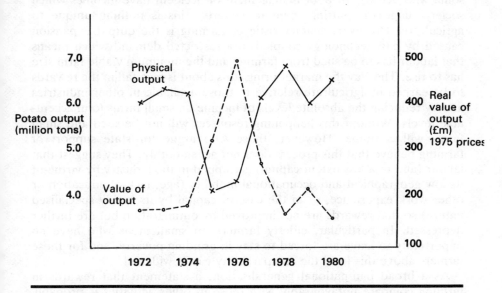

NOTE 1. Value of crop is assumed to equal revenue to farmers from its sale.
SOURCE Derived from *Annual Abstract of Statistics* (HMSO, various years).

necessary to choose an income measure appropriate to each strand of concern.

On *a priori* grounds one might expect instability, poverty and poor comparability to be characteristics of farm incomes in general on a world scale. The output of agriculture is subject to random variation from year to year because of weather differences, pest incidence and so on. Where a changing supply encounters an inelastic demand the result is widely fluctuating product prices and hence unstable farm revenues and farm incomes. This is easily shown using as an example the output and revenue from potatoes in the UK. As Figure 12.2 shows, the revenues coming in to farmers can be in an inverse relationship to the volume of output. In the UK, in 1975 and 1976 the output of potatoes fell as a result of drought conditions, but the price of potatoes rose greatly so that the total revenue (and hence income once costs had been deducted) was greater than in a year of normal yields. Those farmers who were less affected by drought were naturally even better off. On the other hand, for those farmers where the crop was completely destroyed, income was reduced too.

On the poverty and poor comparability aspects of income, what has already been stated on the structure of agriculture and its relative decline would be enough to suggest that both might be found among farmers. In an atomistic industry with many independent operators there are bound to be some who because of poor management or accident have incomes which society judges as putting them in poverty; this is nothing unique to agriculture. However, characteristic of farming is the output expansion caused by new technology coupled to a restricted demand which means that labour has to be shed from farming and the minimum viable farm size has to rise. The way the market brings this about is by a fall in the rewards to be earned in agriculture relative to those available in other industries and by reducing the absolute levels of income on small farms to unacceptable levels. Without this happening resources will not be shed and small farms will continue. However, those who argue for state support of farming believe that this process does not act smoothly. They suggest that labour (and to a less extent capital) is trapped in the industry by virtue of its low geographical and occupational mobility (age, lack of education or other work experience, or in the case of capital by its highly specialised nature) so that rewards are not improved by outmigration but are further depressed. In particular, elderly farmers on small areas who have no opportunity to leave are forced to stay in grinding poverty, and for those farmers above this level the comparability gap is widened.

As a broad pan-national generalisation, a statement that rewards in farming compare unfavourably with those in other activities is probably acceptable. But in the UK, as in most rich industrialised market economies, government involvement with agriculture has so influenced the markets for farm outputs and inputs that it would be foolish to assume that farmer rewards are either absolutely or relatively low. Rather, one has to re-examine the characteristics of farmer incomes without preconceptions to ascertain whether they exhibit instability, poverty or poor comparability.

## 12.3  Farm Incomes in the UK

Despite the central part played by incomes in the justification for government and Community agricultural policy (see Chapter 14 for the formal statements on which UK and EC policy is based), surprisingly little empirical evidence is available to answer such important questions as, how many farmers are in poverty and on what sizes and types of farm are they found? In which areas do they occur? Are they poor even after taking into account other sources of income? Are they owners of their farms, and therefore wealthy although having low current incomes? Are the rewards from farming low compared with alternatives when all the aspects of the farming business are taken into account, such as capital growth, taxation and risk? And are the fluctuations in profitability greater in farming than, say, the holiday trade which is also affected by weather?

The agricultural industry is unusual in that there is an annual survey of a sample of farm businesses, started in 1936 and paid for by government although carried out mainly by universities, which monitors economic developments at the farm level. In the UK the Farm Management Survey collects accounting information from some 3500 farms, many of which will have cooperated in the Survey for a number of years. From the costs and returns, which of themselves are valuable as indicators of short-terms changes, a net income figure is calculated (Net Farm Income, NFI) for each farm. For several reasons this NFI figure should not be taken as corresponding closely to what either the farmer or an outside observer might reasonably consider to be actual income. For example, to enable all the sample farms to be considered together, owner-occupiers are treated as if they are tenants and a rent figure imputed and treated as a cost (although of course no such payment is actually made). Similarly, interest on any borrowing is also ignored. No account is taken of any other income source than the farm. Despite the title given to the annual reports (for example, *Farm Incomes in the United Kingdom*) the FMS is concerned with much more than incomes; paradoxically the least satisfactory figures it generates are those on incomes. Nevertheless it is a frequently quoted data source of farm level information. Equivalent surveys take place in all EC member states, data gathered from each being passed to the European Commission's Farm Accountancy Data Network (FADN) for use in policy formation.

## 12.4 Instability of Farm Incomes

Of the main characteristics of income, *instability* is perhaps the most easy to obtain information for the UK. Despite the drawbacks given above, the net farm income estimates arising from the Farm Management Survey can be useful indicators of farming income instability, and information for three farming types is given in Figure 12.3; the lines refer to *average* incomes within groups and are in real terms, that is, after taking inflation into account.

It is obvious from Figure 12.3 that, while there was some income instability in the late 1950s and 1960s, since about 1970 this has been greatly amplified, caused primarily by changes in the value of output, not of costs. The great expansion of incomes initiating this less stable period corresponded with UK entry to the EC; Figure 12.4 shows the unstable and disparate fortunes of major farming types since the general decline in agricultural incomes set in towards the end of the 1970s. The lines in

## FIGURE 12.3
## Index of Real Income per Farm (England and Wales)

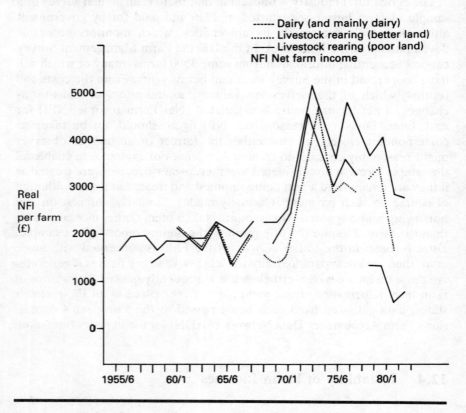

SOURCE  Simplified from Jenkins, T. N. (1982) *Financial Results from the Welsh Farm Management Survey, 1956/7 to 1980/1* (Aberystwyth: University College of Wales).

Figures 12.3 and 12.4 refer to average incomes of a wide range of farm sizes; disaggregating further suggests that instability is relatively more marked among small than large farms, but again group averages are usually employed to illustrate this. Unfortunately, little evidence exists by which the instability of incomes of individual farms over time can be judged; present systems of data handling do not facilitate a longitudinal time-series analysis. However the general consensus is that instability is a feature of farmers' incomes, and various schemes aimed at stabilising prices of farm outputs have been devised. Benefits are seen in income stabilisation *per se* than in more efficient farm planning and better resource utilisation.

**FIGURE 12.4**
**Trends in Net Farm Income in Real Terms by Type of Farm, UK, 1977/8 to 1985/6**

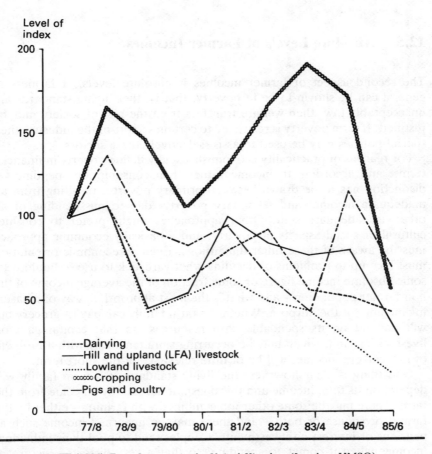

SOURCE   MAFF (1986) *Farm Incomes in the United Kingdom* (London: HMSO).

As we have seen already, a substantial minority of UK farmers have some non-farming income from a job or business; overall between a quarter and a third. This proportion is higher among the small farms, with at least two farms out of five having some earned income. While there is no evidence yet available for the UK, in the USA and Canada off-farm income has been shown to be not only growing in relation to farm income, but to be more stable than farm income, so that in the UK it is likely that fluctuations in farm incomes, particularly on small farms, are dampened by the moderating influence of off-farm activities. In addition, farmers may have unearned income from investments, rents or pensions. While some in the first category may be no more stable than farming, the latter groups may again improve overall stability; occupiers of small farms (of less than 4 ESU) are not infrequently found to be old-age pensioners.

## 12.5   Absolute Levels of Farmer Incomes

The second aspect of farmer incomes is absolute levels. If farmers in general can be shown to be in poverty, that is, their living standards are unacceptably low, then welfare transfers from the rest of society may be justified. If farm poverty is restricted to certain sectors of the industry, then special policies may be used to assist selectively these sectors.

For reasons of practicality economists usually define poverty in financial terms, and according to income rather than consumption spending. A distinction has to be drawn between primary poverty, resulting from an inadequate income, and secondary poverty, due to misspending of an otherwise adequate sum. Other approaches might prefer to consider cultural or social aspects of poverty, and a basically economic approach must be aware of these other dimensions. Even an economic orientation must face up to problems of deciding what yardstick to use – should it be some absolute income figure or some fraction of the average income of the non-farmer population – and what is the most appropriate way of measuring income for the purpose. Whether a farm family can pay its grocery bill will depend on its spendable cash resources, so that capital gain on livestock or land, which may be occurring simultaneously with a problem of low current income, will be of little relevance in the short term.

One thing is clear however; the living standard of a farm family will depend on its total income and this does not necessarily all come from the farm. There may be some other job combined with farming – either by the farmer or spouse or both – and there may be unearned income such as pensions or interest on investments. Couples tend to pool their individual incomes when it comes to spending, so that a very distorted picture of potential living standards might emerge if only the income of the farmer were considered. For welfare purposes the sensible unit is not the individual but the household; normally this will be a couple with or without dependent children, or single farmers. If there are other members who are earning and who pool their incomes they might also be included. Grown-up children of farmers frequently live with their parents even if they work off the farm, and elderly relations or brothers and sisters sometimes live together. This raises the problem of defining a household for this purpose. Members will have different degrees of financial integration and it will not always be appropriate to regard all the members who live together as contributing to or sharing a single income. In practice, earnings of relations are usually excluded from these studies.

Having decided that the appropriate unit of income in a poverty context is total household income (from farm and non-farm sources) and on an appropriate parameter of income (probably excluding capital gains) the next step would be to settle on some yardstick against which the farmer household income can be judged. In the US there are official poverty lines of income for farm and non-farm families, the farm level being 85 per cent of the non-farm level to reflect different living costs. There is dispute over the most acceptable method of calculation of these lines, [3] and the arbitrary decision on the level of the division between those who are poor and those who are not is primarily not one of economics but a political one. Nevertheless, the poverty line forms a useful administrative criterion for public assistance.

### 12.5.1    Sources of information

Although in the UK agricultural policy is fundamentally concerned with the incomes of farmers, in practice the main source of information on individual farms (the MAFF-sponsored Farm Management Survey) has restricted its income measure to a very narrow farm base. Its main measure, net farm income, relates essentially to that income generated by the farming business, and any non-farming income accruing to the farmer or spouse is ignored. Because of this and other conventions used in its calculation described earlier, the published NFI figures are not at all suitable as indicators of living standards. However, the NFI data is useful in indicating the folly of talking about 'farmer incomes' as if all farmers were equally prosperous (or poor); not only do the occupiers of large farms tend to have higher farming incomes, but there is a wide range of incomes within the same size group and even within the same type within that size group. From Table 12.5, it is clear that while the median[4] farm income rises with farm size, there is considerable overlap, with the most prosperous farms in the smaller size bands having incomes greater than the least prosperous, much larger farms. The variation within farming types can be demonstrated effectively with specialist dairy farms, the most numerous type in the FMS: within the 600–1199 SMD size band (the two-to-four-man-farms) the median income in 1976–7 was £7647, but one-fifth of the farms made less than £4665, and another fifth more than 11 682.[5]

Another source of information about farmers' incomes is the Inland Revenue's annual Survey of Personal Incomes (SPI), based on a sample of tax returns, and in which individuals and couples with some income from self-employment in agriculture can be identified. Its approach comes much nearer to using the household as the unit, as the incomes of couples are

**Table 12.5**
**Dispersal of Net Farm Incomes: Figures from the Farm Management Survey of England and Wales, 1976/7**

| (SMDs) | Median (£) | Size of farm (all types excl. hortic.) Bottom 20% below (£) | Top 20% above (£) |
|---|---|---|---|
| 275–599 | 4 861 | 2 382 | 7 595 |
| 600–1199 | 9 227 | 5 019 | 13 969 |
| 1200–1799 | 14 796 | 6 440 | 23 689 |
| 1800–2399 | 18 730 | 8 081 | 30 750 |
| 2400–4200 | 26 046 | 12 525 | 44 868 |
| Av. 275–4200 | 9 708 | 4 388 | 20 365 |
| very small farms < 275 | 2 621 | 445 | 3 786 |
| very large farms > 4200 | 48 287 | 21 847 | 86 637 |

SOURCE   MAFF (1978) *Farm Incomes in England and Wales* (London: HMSO).

added together. Also it embraces all sources of income, that from the farms, from other businesses, from employment, pensions, dividends, rents and so on; although capital gains on land are not counted (as in NFI). There are problems with sample coverage and with the income measure used, basically the income on which tax is assessed after allowances; the accelerated capital allowances used to calculate taxable income distort and depress the final income figures. Unlike the long-established FMS, the SPI data are not widely available, although major advances are to be expected in the availability and flexibility of analysis from this source. Compared with results from the FMS there are fewer cases of negative incomes (i.e., losses) in the SPI (1 per cent in the SPI as opposed to 8 per cent in the FMS), which might be suggestive of loss-making farmers also having other sources of income. Another very interesting feature of the SPI is that, when the incomes of all taxpayers (farmers and non-farmers) are arranged in bands, the percentage of incomes belonging to farmers rises with increasing incomes, from about 1 per cent of all incomes in the range £ 1000 to £ 8000, to 7 per cent among those of £ 20 000 and over (Table

**Table 12.6**
**Distribution of Incomes: All Incomes and Those in Agriculture and Horticulture**

| Range of total income (£) | All incomes 1977–8[1] Number of cases[2] ('000s) (B) | % | Agriculture and Horticulture 1978–9[1] Number of cases[2] ('000s) (A) | (%)[3] | (A) as percentage of (B) |
|---|---|---|---|---|---|
| 1 000–2 999 | 8 670 | 40 | 93 | 37 | 1 |
| 3 000–4 999 | 7 280 | 33 | 63 | 25 | 1 |
| 5 000–7 999 | 4 670 | 21 | 47 | 19 | 1 |
| 8 000–9 999 | 720 | 3 | 16 | 6 | 2 |
| 10 000–14 999 | 431 | 2 | 19 | 8 | 4 |
| 15 000–19 999 | 105 ⎫ | 1 | 7 | 3 | 6 |
| 20 000 and over | 86 ⎬ | | 6 | 3 | 7 |
| All 1000 and over | 22 000 | 100 | 250 | 100 | |

NOTES 1. The years chosen for comparison are nominally different but in reality approximate to the same period. Agricultural incomes are predominantly taxed under Schedule D, where the tax is normally assessable on the profit of the accounting year of the business ending in the previous tax year. However, the bulk of earned income (86 per cent) in the 'all incomes' column was assessed under Schedule E where tax is payable on the income of the income tax year for which it is charged.
2. Individuals and married couples.
3. Note that these percentages relate to only those incomes of £1000 or over.

SOURCE All incomes, Inland Revenue, (1980) *Survey of Personal Incomes 1977–78: Agriculture and Horticulture 1978–79*, special Inland Revenue analysis. Given in Hill, Berkeley (1984) 'Information on farmers' incomes, data from Inland Revenue sources', *J. Ag. Econ.* **35**:1, 39–50.

12.6). This does not suggest a sector of the population suffering particular poverty problems.

It must be concluded that neither the FMS nor the SPI in their present forms enable the incomes of farm families to be assessed satisfactorily for welfare purposes. All they can show is that some farmers have higher incomes than others, according to the criteria of income in use. We hear little about those farmers with high incomes (and there were 21 per cent in the FMS in 1982–3 over £ 20 000 and 7 per cent over £ 40 000) but much more about those farmers who suffer low incomes. Yet without a much more satisfactory approach to the measurement of incomes it is impossible to say how many have incomes which, according to some agreed criterion, fall below the threshold of poverty and justify help from the rest of society. This is an important and major gap in our knowledge about the UK agricultural industry.

## 12.6    Income, Wealth and Economic Status

Income so far in this chapter has been treated as the potential power a person (or household) has to spend or save over a given time period. Any wealth the person may have has been treated as being of no consequence other than through the income it may generate through interest on investments, rents from property or capital gains (or losses). The last could be set aside in certain circumstances. Essentially we have taken the view that when measuring people's incomes we should not include any spending they might make 'out of capital', that is by running down their savings or selling off assets. However, wealth *does* represent potential spending power and contributes to the economic status of its owner; two elderly people with equal but low incomes will be in very different circumstances if one owns land or jewellery and the other does not. The property owner has the option to sell and spend the proceeds, or perhaps in the case of house property, raise a loan against the house's deeds. The term **economic status** is used to indicate potential spending power taking both income and wealth into account.

In countries where low farm incomes are a matter of public concern, it is nevertheless found that farm families often hold large quantities of wealth. It is obvious that in attempting to assess the economic status of farmers it would be foolish only to have regard to their incomes and ignore their wealth. For the UK, it appeared that all the owner-occupied farms in the 1977–8 FMS, even those in the small farm (275–599 SMD) group, had net worths which put their occupiers among the richest 6 per cent of the population, and occupiers of most sizes and types of farms were in the top 2 per cent or less.[6] Both income and wealth are important determinants of economic status, and the question is how to incorporate them into some unified measure which has validity for public policy.

One method of doing this is to convert farmers' net worths (that is, assets less liabilities) into an income-flow equivalent. In broad terms this involves posing the question, 'If a farmer were to sell his farm and realise his wealth and he invested the proceeds with a pension fund, what annual income (annuity) could he get from the fund given his age and life expectancy?' This annuity is then added to his current income to give his potential spending power, his 'economic status'. This shows the amount which the farmer could spend on consumption if he chose to exercise the options available to him. Although his current income may be small, a farmer could not be considered to be poor if he commanded the means of living comfortably for the rest of his days.

In working out the size of the annuity there are problems of selecting appropriate asset values, rates of interest and returns to capital needed for the annuity formula. This assumes, as does any annuitising process, that no net worth will remain at the death of the person. However, some would take the view that farmers treat land as being held 'in trust' as an estate for their survivors or heirs so that they are not free to regard their net worth as a potential source of current income by annuitising. The notion of steward-ship is quite widely held among traditional English landowners, an attitude to succession which tends to lead its advocates to deny the existence or significance of land as a source of personal wealth, for the owner of land emphasises his 'inability' or at least his unwillingness to sell.[8] Succession by a family member has been found to be the intention of three farmers out of four, although there seems generally to be a failure to distinguish between the right to assign occupancy, which itself may have a market value, from the other rights which constitute ownership.[9]

If the 'stewardship' argument is accepted, then only that part (if any) of net worth which is not held in trust for successors should be annuitised; in the extreme, where no annuitisable net worth is available, the economic status of the farmer is solely his current income including any part formed by current returns on his capital. However, a bland acceptance of this case put forward by farmers would not be satisfactory; land is generally con-sidered to be part of their personal wealth.

Estimates of the economic status of farmers in north America clearly show the significance of including wealth.[10,11] In the US, the proportion of farm families below the $2500 poverty line in 1966 was 32 per cent, but more than half of these were raised above the line if annuitised net worth was incorporated. This also narrowed the farm/non-farm gap. Annuitising net worths appeared of particular importance to the elderly farmers who tended to have low farm incomes but high net worths and short life expectancy, and the findings prompted the exploration of possible mech-anisms by which farmers could realise the annuity value of their wealth, such as the provision of special banks or mortgage companies to cater for them. In Canada, annuitised net worth has formed a large and increasing proportion of the economic well-being of farmers.[12,13] In 1967, the average farm annuity added approximately 47 per cent to total economic well-being (farm and non-farm sources of income) and in 1977 the annuity added 57 per cent. A preliminary exercise for England and Wales, suggested that, of those FMS farms with a NFI of less than £ 2000, about half were lifted above this line if the annuitised value of their land was added, taken at tenant-land prices, and the all-farm level of income raised by one-third.[14]

Taking an economic status view of farming rather than just an income one has major complications for the case that farmers require assistance from governments. Not only do the numbers of cases who appear to justify support fall, but the mechanisms which could be used alter. Rather than product price support and income supplements, attention is diverted to ways of making more liquid the wealth farmers currently have. In an economic climate where the typical consumer and taxpayer is less wealthy than the typical farmer (although we are here playing a dangerous game of generalities) such measures are likely to be viewed more favourably than those which involve income transfers in the form of higher food prices and taxes on one hand and farmer grants and subsidies on the other.

## 12.7 Comparability of Rewards from Farming

Taking finally the **comparability** of rewards between farming and other occupations, one is immediately confronted with problems of choosing the yardstick by which to make the comparison and therefore of defining income. In a seminal book in this area, Bellerby (1956) points out that 'the great majority of farmers exercise the functions of labour, management risk bearing and property ownership, and their total revenue is therefore compounded of wages or salaries, profit, interest and, in many cases, rent'.[15] How many of these elements one includes in a definition of income will depend on the purpose for which the comparison with other sectors is required. For a manpower planning exercise, gauging the productivity of labour in various industries with a view to moving workers from one to another in order to raise the nation's total output, one might take into account only that income coming from gainful employment.

However, a farmer's decision to stay in agriculture or to move to another line of business will reflect the returns he gets and anticipates from *all* the resources he controls. At the farm level the question of comparability takes the form of 'What are the rewards open to me and my assets in alternative occupations?' If the alternatives are better than currently being earned, that is, there are high **transfer earnings**, one would expect the farmer to switch, although there may be impediments to overcome such as the costs of selling the farm. In making the comparison with rewards in other industries the farmer will look at his 'full income' which will take the form not only of money income from his business, but also of capital gain on his business assets such as land, and the value of fringe benefits such as living close to his job.

No official estimates of the comparable income position of UK farmers is published, although the comparability of returns of the typical owner-occupier has been explored independently using the concept of **occu-**

**pational income.**[16] The starting point to this is the current income of the farm, to which is added the average capital gain over the previous ten years. This gives an economic income estimate. From this is deducted a charge for capital (including land) based on the returns available on government securities – representing the most plausible alternative investment available to land-holding farmers which possesses similar degrees of financial risk as land ownership. The sum left over is the occupational income, representing the earnings accruing to the farmer after allowing a reasonable opportunity cost for capital.

The occupational income figures form a basis for comparability. First, they enable the statement to be made: 'If a farmer had given up his farm at the start of a given period and had invested his money in government securities, he would had to have earned this sum (the occupational income) to make him as well off as he has become by being a farmer.' And secondly the question may be posed: 'How does this occupational income figure compare with alternative job opportunities?'

Some indication of the relative position of farmers' incomes in the mid-1960s to mid-1970s is given in Table 12.7, in which the average occupational incomes from owner-occupied farming are compared with manual industrial earnings, non-manual industrial earnings and general managerial earnings in the rest of the economy for the same periods. A ratio of below one suggests that the average rewards from farming were greater than were being earned elsewhere. Only on the smallest farms were the ratios greater than one, suggesting that their occupiers would have been better off had they sold up and taken non-farm jobs. On the larger farms opportunity costs were covered, and by many times over. It is hardly surprising, then, that despite the low current rate of return that incomes represent on the total value of their farms (that is, ignoring capital gains), medium and large farmers were reluctant to consider leaving agriculture.

Much depends on what possible alternative non-farm occupations are selected for a 'fair' income comparison. For the period in question average farming occupational incomes were, for the two-to-four-man farm (600–1199 SMD), within the top 15 per cent of professional earnings, while for the larger farms (1200–4199 SMD) they were in the top 5 per cent of earned incomes. It must be highly doubtful whether the farmers, with their limited educational and non-farm work experience, would have been capable of generating such incomes outside agriculture. Even the relatively poor showing of the incomes on the smallest farms may be less unfavourable than first appears. In view of the high average age of farmers on small farms (see Chapter 7) it is debatable whether the comparison with the earnings of industrial manual labour is a fair one and may result in an overestimate of the true opportunity cost of farmers' labour. In practice even the occupiers of small farms may have covered their opportunity costs.

**Table 12.7**
**Ratio of Average Non-Farm Earnings to Occupational Earnings[1] on Full-Time Owner-occupied Farms – All Types (excluding Horticulture)**

| | Farm size (SMDs) | | | |
| --- | --- | --- | --- | --- |
| | 275–599 | 600–1199 | 1200–4199 | Average full-time 275–4199 |
| **1965–9** | | | | |
| Non-manual industrial earnings | 1.60 | 0.86 | 0.37 | 0.81 |
| Manual industrial earnings | 1.24 | 0.67 | 0.29 | 0.62 |
| General managerial earnings | – | – | – | – |
| **1970–4** | | | | |
| Non-manual industrial earnings | 1.83 | 0.70 | 0.33 | 0.71 |
| Manual industrial earnings | 1.54 | 0.59 | 0.28 | 0.60 |
| General managerial earnings | 2.35 | 0.90 | 0.43 | 0.91 |

NOTE   1. Based on an assumed capital return equivalent to medium dated British government securities. Accrued capital gains on farmland is included.
SOURCE   Hearn, S. (1977) *Farm Income and Capital Gains: Implications for Structural Change* (unpublished Ph.D. Thesis: Ashford, Wye College).

None of these comparisons take account of the advantages of being self-employed or the non-pecuniary rewards of farming. These have been shown to be particularly valued by smaller farmers. Regrettably no more recent estimate of the comparable income position of farmers seems to have been made.

### 12.7.1   Capital gains as a form of income in comparability estimates

The estimates of the 'economic income' of farmers, given above, embrace both current income and spending power coming from rises in the value of assets – capital gains. It is important to note that it is real capital gain (that is, after allowing for falls in the value of money) and not the nominal gain which makes a positive contribution to real income, since it is this which represents an increase in purchasing power.

Part of the long-term return to the business of farming in Britain since the Second World War has clearly come from changes in asset values, and to exclude rising farmland prices is to ignore a major source of explanation for farm business behaviour, borrowing, investment and personal spending. For some, it is the chief reason why they are in agriculture. Comparisons of rewards with the rest of the economy becomes a nonsense if a major part is excluded, yet in measuring incomes in the farming industry, capital gain is often ignored. While it may be in the interest of farmers, when presenting their case for higher product prices, to play down capital gains, their existence has not escaped the notice either of outside investors or that of the farming community itself. Between 1970 and 1981 the net worth of UK agriculture (that is, the value of its assets minus its liabilities) increased by 49 per cent in real terms, despite an increase in total liabilities of 10 per cent and a rise in bank lending of 57 per cent. The MAFF Assets and Liabilities Survey, an adjunct to the Farm Management Survey which traces the growth and development of net worth, suggests that, up to the 1980s, particularly large increases in net worth have occurred at times when net farming income was under pressure.[17] In Chapter 6 it was shown that land prices have been on an upward trend, but that a run of years with declining prices has occurred since 1979, linked to the poor prospects for farm incomes in the 1980s. Short-term trends are to be expected, but owners of land who have held their land since the early 1970s have still reaped some capital gain despite recent falls in price.

It should be noted that a fall in the value of debts can also increase net worth and give a capital gain; this frequently happens in times of rapid inflation when falling money values reduce the real size of borrowings in relation to the value of the assets (which rise in money terms).

There is some feeling among farmers that capital gains are just 'paper profits' which exist only in some unreal world and are of no benefit to the farmer. This must be challenged, and although capital gains differ from current income in terms of their degree of **certainty**, and their **liquidity**, they nevertheless constitute part of real personal income. Capital gains represent spending power the farmer potentially could exercise without diminishing the value of his wealth.[18]

The additional **uncertainty** arises because gains may disappear; a farmer who sells land when it reaches a peak will realise the capital gain, but if he hangs onto land longer a fall in land prices, something outside his control, may reduce or wipe out the gain.

Capital gains are less **liquid** than current income. Sale of an asset is perhaps the most obvious way to realise a gain – that is, turn it into spendable form – but this may be both a costly and a lengthy process. In the case of land, selling hectares could affect the future viability of the business, although many examples exist of farmers selling small outlying parcels of land (often for building houses) and making the necessary adjustments in their farming systems by altering their enterprise mix or intensity. Other ways of realisation are possible: some farmers have used

sale-and-leaseback arrangements to allow them to acquire funds for investment (frequently in more working capital for further expansion) without
losing occupation of the land. But for most farmers, capital gain on
farmland will be made liquid through the form of extra borrowing using the
land as security, although credit institutions can be expected generally to
favour spending on capital assets and to discourage borrowing for purely
personal consumption. No doubt part of the additional credit taken by
farming over the last few years to weather the decline in sector income has
taken the form of realising past gains on landholding.

In the United States and Canada the importance of capital gain in
contributing to well-being is widely recognised.[19] Until 1982, when falling
land prices were experienced, estimates of income showed that including
real capital gain greatly reduced or even reversed the farmer/non-farmer
income gap in the US.[20] A parallel situation has existed in Sweden. For the
UK the 'occupational income' exercise, quoted above, presents a similar
picture: excluding capital gains resulted generally in **negative** occupational
incomes (that is, farmers were not even covering the opportunity cost of
their capital) but including capital gains produced positive occupational
income figures which compared favourably with off-farm ways of using the
resources.

If capital gain can be argued to add to current income, then conversely
capital losses should be treated as a negative figure in any income calculation. This is undoubtedly true, although there may be disagreement about
how precisely they should be added or subtracted. The accepted view is
that, because land is in practice held as a long-term asset, then gains and
losses should be averaged over, say, a ten-year period. Most of the
assessments of the full or economic income of farmers have taken place
against the background of rising real land prices (farming's principal asset)
so that capital gain has increased current income from farming profits.
However if land prices start a downward trend, as seems possible in the
UK in view of the decline in farming profitability which is expected to
persist, and has already started in the USA, real incomes should reflect
these capital losses. If capital gain has resulted in many people staying in
farming who otherwise would have left, capital losses might unleash a
major change in farm structure and land ownership as these people revise
their decisions.

### 12.7.2 *Income comparability for farmers who do not own land*

So far the discussion of comparability has assumed that the farmer owns his
land, but in a substantial minority of cases, about 4 out of 10 in England
and Wales, some land is rented, and in 2 out of 10 all is rented. Capital gain
on tenanted land goes not to the farmer but to the landlord, and a major

source of economic income accruing to owner-occupiers is thus not available to tenants. However, the amount of capital which tenants have locked up in the farm is also less, so that the opportunity cost of their capital is also lower (in absolute amounts of interest foregone).

Two approaches to comparability have been widely used for tenant farmers, or for owner-occupiers, by treating them as tenants and imputing a rent: estimates of 'labour income' and 'return on tenant capital'. Both attempt to split the overall reward from farming into separate labour and capital components, the first by imputing a charge for the use of tenant capital (for example, 8 per cent on the value of livestock, crops and machinery) so that the residual is the income left for remunerating labour, the other by charging for the labour of the farmer so that the residual is the return on tenant capital.

Until recently (1984), the European Economic Community has published comparable income figures based on labour income per labour unit – the estimated labour income of holdings divided by the total by the number of agricultural labour units on them (both the farmer and hired labour, with part-time workers expressed in full-time equivalents). This was compared with incomes in other sectors as indicated by the average gross wage of non-agricultural workers. While this may be appropriate in countries where hired workers are rarely found and where farms are small, it is obviously a nonsense for many farms in the UK. On a farm with two workers paid £6000 a year and the farmer earning £12 000 (after interest charges on tenant capital), the average of £8000 represents the real position of neither part of the labour force. Also it would seem far more sensible to compare farmer incomes with those of other self-employed persons rather than with wage earnings, since the latter do not perform any entrepreneurial risk-bearing function. In practice the EC's comparable income figure has been used not for assessments of relative welfare positions but as part of the mechanism for approving grant aid under EC structural policy on the modernisation of farms.

The second approach to comparability is to base it on returns on capital. This is frequently done in the following way: from an individual farm's net farm income as calculated by the Farm Management Survey a sum is deducted to represent the physical labour input of the farmer and spouse, based on the cost of hired agricultural labour. The result is the reward to the farmer for his entrepreneurial activities of management and risk taking and a return on his capital (called **management and investment income**). Although still a hybrid of rewards, this figure when viewed against the value of farm capital (excluding land) is frequently interpreted as the appropriate return on capital by which the profitability of farming should be judged. However, it is difficult to give 'representative' figures achieved by UK farming. There have been great variabilities in the rates of return, particularly on small farms; for example small dairy farms suffered a severe drop from +19.7 per cent in 1972–3 to −9.3 per cent two years later. The rates are generally lower in the 1–2 man size (275–599 SMD) farm than on those of over 4 man-size (over 1200 SMD). Additionally, rates differ markedly between farming types. In the mid-1970s the overall rate of return was around 22 per cent, and 33 per cent on the most profitable half of farms. For 1985 an average of around 8 per cent seemed likely, with the 'premium' farms about 12 per cent and cash cropping farms on the very best soils achieving double the average figure.[21] An important feature, thus, is the wide variation between farms in the same size and type groups.

Figures of this order do not suggest unacceptably low overall returns in farming. However, the practice of imputing a labour charge for farmers at the level of hired manual labour probably overstates the true opportunity cost of elderly farmers (primarily on small farms), the nominal return thereby understating the real return on their capital, while at the other end of the size spectrum, management talent is probably undervalued, so that the real return is lower than it is shown.

## 12.8   A Retrospect on Farm Incomes

This chapter shows that the study of farm incomes is remarkably limited in its ability to provide answers to many of the fundamental questions about the living standards of the farming population. While there is a satisfactory annual estimate of the aggregate income generated by the agricultural industry, this cannot reveal information on the personal incomes of farmers at which much agricultural policy is directed. The official annual survey of farm businesses, which might be expected to yield the relevant information, restricts its coverage to income from farming only and adopts conventions which prevent the results being useful for indicating farmers' living standards. Other potential data sources are as yet inadequately developed. One can only hope that improvements will take place in collecting the information so necessary to the efficient implementation of public policy.

If there is one consolation, it is that the UK is in not much worse a situation over the knowledge about its farm incomes than are most of its fellow members of the European Community. All member states make industry-level estimates of the income of their agricultures using an agreed procedure and supply this information to the Statistical Office of the EC for use in policymaking, such as in preparing proposals for the annual round of price fixing for farm products.[22] However only in Denmark and West Germany are regular and reliable assessments of the total personal incomes of farmers and their spouses carried out. The official European Community farm income monitoring system (FADN), which helps policy-

makers of the Common Agricultural Policy by providing farm-level information for all member states on a harmonised basis, that is using the same
conventions as far as is practical, restricts its interest to the income from
farming, although it readily admits that a number of further factors
(non-farm income, taxation, housing costs, welfare payments and so on)
bear heavily on the standards of living of the agricultural population. This
attitude stems largely from the lack of data at national level in most major
states, although there has also been a reluctance to acknowledge that the
CAP is in large part a social policy directed at poverty; the shape of the
CAP would be very different if it were solely concerned with matters of
production and efficiency. Community policy-makers thus only have access
to incomplete information on the income position of the farmers they are
trying to assist.

Reform of the CAP to avoid surplus production and to contain its cost,
discussed in Chapter 14, would be relatively simple were it not necessary to
take into account the effects on farm incomes. Because of the lack of basic
information on the full income position of farm families it is not possible to
advance far along those lines generally agreed as desirable, namely of
targeting aid to those farmers with unacceptably low incomes and accompanying this with a reduction in the prices of farm products in order to
bring supply and demand back into better balance.

# Notes

1. The changes are described in Outlaw, J. E. and Croft, G. (1981) 'Recent
   developments in economic accounts for agriculture', *Economic Trends* **335**
   (September) 95–103.
2. More formally, personal income has been described as 'the sum of 1. the
   market value of rights exercised in consumption and 2. the change in the store
   of property rights between the beginning and end of the period.' Simons, H.
   (1938) *Personal Income Taxation* (University of Chicago Press).
3. Madden, J. P. (1975) 'Poverty measures as indicators of social welfare', in
   Wilber, G. L. (ed) *Poverty: New Perspectives* (University of Kentucky Press).
4. The median income is the income below which half the farms lie and above
   which half lie.
5. Including Breeding Livestock Value Appreciation.
6. Peters, G. H. (1980) 'Some thoughts on capital taxation', *J. Ag. Econ.* **31**:3,
   381–97.
7. A farmer who sold his land but retained occupancy and farmed as a tenant,
   exchanging the proceeds from land sales for an annuity, would have a smaller
   annuity than if he had sold up completely but would still derive an income from
   farming, although he would of course now have to pay a rent for his land. As
   for life expectancy, one approach is to treat the farmer and wife together as the
   consumer unit and base the annuity on the expected life of the partner
   anticipated to live the longer.

8. Newby, H., Bell, C., Rose, D. and Saunders, P. (1978) *Property, Paternalism and Power: Class and Control in Rural England* (London: Hutchinson).
9. Harrison, A. (1975) *Farmers and Farm Businesses in England* Miscellaneous Studies 62 (University of Reading: Department of Agricultural Economics and Management).
10. Carlin, T. A. and Reinsel, E. I. (1973) 'Combining income and wealth: an analysis of farm well-being', *Am. J. Ag. Econ.* **55**:38–44.
11. Chase, L. and Lerohl, M. L. (1981) 'On measuring farmers' well-being' *Can. J. Ag. Econ.* **29**:225–31.
12. Chase, L. (1980) 'Inflation, capital gains and farmer's economic well-being' *Can. J. Ag. Econ.*, Proceedings of Annual Meeting, 67–77.
13. See note 11 above.
14. Hill, Berkeley (1982) 'Concepts and measurement of the income, wealth and economic well-being of farmers' *J. Ag. Econ.* **33**:3, 311–24.
15. Hearn, S. (1977) *Farm Income and Capital Gains: Implications for Structural Change* Unpublished Ph. D. thesis, Wye College.
16. Ibid.
17. Capstick, C. W. (1983) 'Agricultural policy issues and economic analysis'. *J. Ag. Econ.* **33**:3, 263–78.
18. Hill, Berkeley (1982) op. cit.
19. See note 7 above.
20. See *Am. J. Ag. Econ.* vols 61–3, (1979–81), and *Can. J. Ag. Econ.* Proceedings of the 1980 Annual Meeting.
21. Nix, J. S. (1985) *Farm Management Pocketbook*, 16th edition (Ashford: Wye College).
22. See a number of chapters dealing with many aspects of data collection within the EC in Dubgaard, A., Grassmugg, R. and Munk, K. J. (1984) (eds) *Agricultural data and economic analysis* (Maastricht: European Institute of Public Administration).

# The Price System and Agriculture 13

## 13.1 Introduction

The economy of the UK is fundamentally market-based and capitalist in nature. This means that the forces of supply and demand dominate in shaping the pattern of what is produced in the UK and who can acquire the fruits of production. Factors of production are predominantly owned by private individuals. In the agricultural industry, while land and capital are overwhelmingly privately owned, there is considerable interference by the government and other public bodies in the market for many farm products, and this also extends into the markets for agricultural inputs. Most agricultural prices have long since ceased only to reflect economic forces and have become the result of political pressure at national and international level. This chapter seeks to explain the characteristics of the markets for farm products by bringing together the demand and supply factors explained in Chapters 2–11. Then we consider the influence of government on the market and the mechanisms by which the market is manipulated. The question of why government intervenes in agriculture and the objectives of agricultural policy are dealt with in Chapter 14.

## 13.2 Price in a Free Market

A market economy is well suited to handling the demand and supply of goods and services whose production and consumption have no spill-over effects on the rest of society. Manufacturers of, for instance, pork pies will adjust their level of supply according to the price they can obtain, and their response will depend on their marginal costs of production in the form of

325

wages, bought raw materials and so on (see Chapter 10). Consumers who purchase and eat the pies do so without affecting those people who do not chose to buy them. With such goods the price which the market system generates through the free interplay of demand and supply results in a use of national resources which reflects satisfactorily the desires of consumers, given the initial distribution of spending power.

However, there are situations where the uncontrolled interaction of supply and demand will not produce a pattern of production and consumption which is the best that society could devise. It might mean, for example, that poor people in society received a standard of diet which in the general view was inadequate. Some sort of government interference, possibly transfers of income from the rich to the poor or by food subsidies to those on low incomes, directly as with the use of food stamps as in the USA or less directly through a cheap food policy as was practiced in the UK in the 1950s and 1960s, would receive general assent. Government interference could also be justified if single firms grew to control a significant sector of the market. As we have seen in Chapter 9, the structure of farming does not suggest that any single farmer could ever gain or exploit monopoly power in the market for farm products, but there might well be need to prevent the accretion of market share in the industries supplying farmers or later in the marketing chain (see Chapter 3). Next, there could be inadequate information in the markets to allow reliable choices to be made. For example, in Chapter 11 a case was made that the smaller farmers are not aware of the latest technology, and that a government-sponsored advisory system could help them keep technically up-to-date (although this might not enable them to stay in farming if their resources were inadequate). Lastly, there is the question of externalities in production or consumption.

The production of many commodities generates costs to society which the supplier does not voluntarily take into account in his supply plans, and farm products seem to be increasingly falling into this group – the polluting effect of fertiliser and sprays in water courses and the loss of visual attractiveness resulting from hedgerow removal have already received comment (Chapter 6). These spill-over costs to society (called external costs, sometimes external diseconomies) will not be reflected in market prices since these will be based only on the private costs faced by farmers. There may also be spill-over benefits, such as where a farm's field drainage scheme benefits a neighbouring farmer as well without him paying anything. Spill-over effects also occur with consumption; visitors to the countryside often deposit litter which at the least is unsightly and at worst is dangerous to other visitors and livestock, imposing costs on the rest of society. Society may take action to ensure that externalities *are* taken into account by imposing controls (on the use of certain insecticides) and fines (for the pollution of rivers, noise from lorries), taxes or, in the case of external benefits, subsidies or grants (drainage grants may be an example here, although such aid has been criticised for encouraging schemes which also do severe damage to natural wildlife).

While the government frequently interferes with the market, by and large this is not to correct for the market imperfections given above, but because it is pursuing some more limited aim, such as raising the incomes of farmers or, as has happened in wartime in the UK, to expand production and ensure an adequate food supply.

## 13.3  Basic Demand and Supply Characteristics

We have seen from Chapter 2 that as people become more prosperous they spend a lower proportion of their incomes on food. Generally the volume of food purchased does not change but the types of foodstuff alter and the expenditure on foodstuffs rises. Spending increases because the preferred foods (in the past at least) have been fatty, sugary and sold with a high degree of processing and packaging. Thus, most if not all of increases in spending on food are accounted for by intermediaries in the food chain rather than by farmers (see Chapter 3).

Apart from families with more than three children and low incomes (constituting at most 20 per cent of the UK population), demand for food and the derived demand for agricultural products is largely unresponsive to income and price increases. As life-styles and dietary habits change, there is substitution between types of food and the food industry is energetic in introducing 'new' food products which will encourage consumers to switch brands and products. Even here, the impact at the raw material level is very limited, with sugars, starches and fats often simply being re-cast into different forms.

Demand for food from farms (at the 'farm gate') is a reflection of retail demand. Since only 40–60 per cent of consumers' spending is received by farmers, and the remaining costs (distribution, packaging, etc.) tend to not vary, the price and income elasticities of demand at the farm level are even lower than those in the shops.

The last point to note regarding demand relates to non-food products. The demand for wool (and timber, in so far as it can be regarded as a farm product) reflects the level of industrial demand. Prices tend to go up and down with the level of economic activity in the broader economy.

On the supply side we have seen from Chapters 10 and 11 that the output of farm products in the UK suffers from variability in an unpredictable way from the influence of weather. Apart from this, farmers are limited in their ability to expand output in the short run because of the restrictions imposed by the biological nature of agricultural production. However, in the longer term they are generally able to respond to economic forces.

There is a history of technological advances which spread through the industry with the inevitability of a treadmill, with the further repercussion of inexorable expansions of output.

Supply tends to increase with a ratchet-like progression. Given time, farmers respond to rising prices by expanding their supply. Price falls have less effect on output because farmers are already committed to skills, to investments and to markets for particular production processes, implying lower elasticity when price falls.

## 13.4 Price Characteristics of Agriculture

The consequences of the characteristics of demand and supply given above are that agriculture faces a particular set of price problems. These include annual price fluctuations, price cycles, and the long-term decline in farm product prices relative to the prices of the inputs that farmers use (the cost-price squeeze).

### 13.4.1 Year-to-year price variations

The prices of many farm products are subject to year-to-year fluctuations caused by the weather's influence on the supply which faces a highly inelastic demand (see Figure 13.1). Because of the inelastic nature of demand, years of high output result in a drop in prices which is greater than the rise in output; consequently the revenue coming in to farmers from selling their production will be less in years of good yields than in poor yield years. As costs are to a large extent unaffected by weather, variations in revenue will result in disproportionately larger variations in income (that is, revenue minus costs). This point was made in Chapter 11 when empirical evidence from potatoes in the UK was cited.

There is a tendency for farmers, faced with price changes which are beyond their power as individuals to affect and whose causes it is difficult to ascertain from the perspective of a single farm, to believe that this year's prices are a good guide to those that will exist next year. Consequently there will be a tendency for a year of high prices to be followed in the next year by a moderate expansion of supply and low prices to be marked by contraction. This induced instability, or over-sensitive response to transient influences, exacerbates the inherent supply variations.

**FIGURE 13.1**
**Year-to-year Variations in Price**

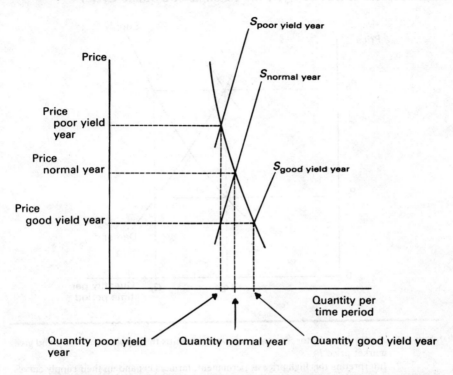

NOTE    The revenue to farmers is indicated by the size of the price x quantity rectangles. The revenue in the poor yield year is greater than in the good yield year.

### 13.4.2   *Price cycles*

Because of asset fixity and other related problems of resource immobility there is a likelihood of cyclical expansion and contraction of supply (see Chapter 10). Coupled with inelastic demand the result is cyclical price movements which cause farmers to enter and withdraw from production but which serve no economic function in the resource-allocating sense. Rather they result in a waste of resources as farmers continuously attempt to adapt to changing price signals. The most famous example of agricultural price cycles is with pigmeat, but blackcurrants afford another.

A price cycle is most likely to occur in an atomistic industry, such as agriculture, where there are a large number of operators who each take

## FIGURE 13.2
## The Progress of a Price Cycle (an Example of a Stable Cycle)

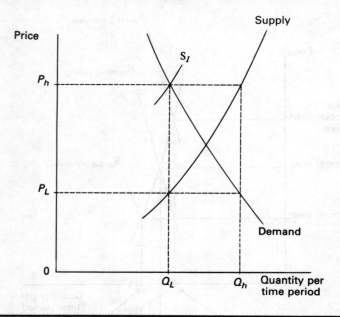

NOTES
1. A temporary factor, such as poor weather, shifts the supply curve to $S_I$, and gives market price $P_h$.
2. Interpreting this high price as permanent, farmers expand up their supply curves, and supply $Q_h$.
3. $Q_h$ will only be taken off the market when price falls to $P_L$.
4. Interpreting this low price as permanent, farmers contract supply to $Q_L$. This again forces price up to $P_h$.
5. Depending on the relative steepness of the $S$ and $D$ curves the cycle could be a contracting or expanding one. For simplicity we have have assumed the case in which the cycle continues to alternate between the same highs and lows.

decisions about production with no coordination with the decisions that others are taking simultaneously; where there is a lag between the decision to change the level of output and that output coming onto the market to effect price; where demand is inelastic; where there is a problem of asset fixity; and where there are random disturbances to supply or demand to set off cycles. In farming all these conditions exist, though for some commodities more than others. The passage through a typical price cycle is illustrated in Figure 13.2 and explained in the notes relating to it.

Because self-sustaining cycles are wasteful of resources, governments may intervene to try to dampen or eliminate them. This can be done by attacking their root causes listed above. By improving information reaching farmers, so that they are aware of the existence of cycles, action will be encouraged which will lead to the reduction of cycles; for example, farmers who are aware that there are regular and predictable periods of high prices and periods of low will time their production decisions to catch the former and avoid the latter. Rather than relying on the free market for selling their output, farmers can be encouraged to make longer-term contracts with buyers; in the UK these have developed notably in both the pig sector and blackcurrants. The government can also put a floor in the market by acting as a buyer if prices fall to levels at which cyclical-starting action by farmers is anticipated. If this government-bought commodity is stored it could re-enter the market when prices were tending to rise; in other words the government would be operating a stabilisation programme. Alternatively, by pursuing policies of general price support the government may as a by-product remove the events which trigger cyclical behaviour.

### 13.4.3 Agricultural terms of trade

In Chapter 11 it was shown that there is a tendency for technological advances to spread throughout the agricultural industry, and for most advances to be of an output-increasing type. The out-turn is that the supply curve for agricultural products moves to the right faster than does the demand curve, resulting in a long-term fall in the prices received by farmers (real prices, that is after allowing for inflation). After allowing for inflation, the real price of wheat in 1981 was only 66 per cent of its price in 1938, and milk a similar proportion.[1] Taking all farm products together, the real prices they command in the market, after taking into account any subsidies received, declined from an index in 1952 of 126 to 82 in 1981. Ever since the UK has been a member of the Common Agricultural Policy real price falls have occurred. In Table 13.1 the real price of all the agricultural products listed had fallen between 1970 and 1982, the all-products weighted average price to 76 per cent of its 1970 level. On the other hand, the costs of the inputs that farmers use (fuel, chemicals and labour) are bought in competition with other industries and do not exhibit such a trend. Consequently farmers are caught in a cost-price squeeze, which has the end result of forcing farmers at the small end of the size spectrum out of the industry. Another way of expressing the changing relationship between input and output prices is to describe the ratio of the two sets of prices as the 'agricultural terms of trade' with the rest of the economy.

The 'agricultural terms of trade' is the result of dividing the prices of farm products (taken as a weighted average) by the prices of farm inputs.

**Table 13.1**
**Change in the Real Price of Agricultural Outputs and Inputs, 1970–82**

|  | 1982 (Index, 1970 = 100) |
|---|---|
| *Agricultural outputs* | |
| All farm and horticultural crops (weighted) | 76 |
| Cereals | 83 |
| Potatoes (main crop) | 78 |
| Sugar beet | 95 |
| Fresh vegetables | 67 |
| Animals for slaughter | 88 |
| Milk | 80 |
| Eggs | 56 |
| *Agricultural inputs* | |
| All currently consumed (weighted) | 94 |
| Animal feeds | 107 |
| Fertilisers | 90 |
| Fuel, oil, electricity | 90 |
| Plant protection | 130 |
| Investment inputs (weighted) | 111 |
| Tractors | 120 |
| Buildings | 110 |

SOURCE    Derived from MAFF, *Annual Review of Agriculture* (London: HMSO). Changes in nominal prices have been deflated by the Retail Price Index (all items).

Normally, all current inputs are used, that is seeds, fertilisers, fuel, foodstuffs and other inputs that are completely used up in one production cycle, although indices are also available which include all inputs. Figure 13.3 shows changes in product and current input prices in the EC from 1975 to 1982; Figure 13.4 shows the result of dividing the one series by the other to construct a terms of trade graph. These are indicators of a much longer

**FIGURE 13.3**
**Input and Output Price Indices, EC (10) (1975=100)**

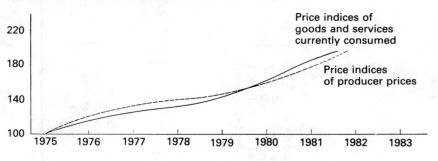

SOURCE EC (1984) *The Agricultural Situation in the EC* (Brussels) pp. 278, 288.

trend which is nevertheless subject to short-term variation. In nominal money terms (that is, before allowing for inflation) both sets of prices were rising over the period. Despite the attempts by policymakers to support the prices which farmers in the EC receive for their products the terms of trade marginally turned against agriculture over the period.

Turning to the UK, it is evident in Figure 13.4 that the terms of trade moved against agriculture to a greater extent than was the case for the EC as a whole, reflecting the greater rate of inflation in the UK and the reluctance of the government to take options open to it (such as devaluing the **green pound**, – see appendix to this chapter – which would have had the result of raising farm prices in the UK) for fear of fuelling still more inflation. After 1980 the ratio between output prices and current inputs seems to have remained fairly stable, but these are of course only part of the input mix used by farmers. The index of *all* input prices divided into product prices (available since 1978 for the UK and shown in Figure 13.4) continued to decline after 1980 (the dotted line in Figure 13.4) and this seems to have resulted from changing costs of labour, machinery and other items not consumed by agriculture in current production.

The cost-price squeeze says nothing about the volume of farm output or volume of inputs purchased. So long as productivity (in the form of output per farm) increases faster than the terms of trade drift down, average farm incomes will increase. Not all types of farming will experience the same change in costs relative to prices, and productivity improvements will vary from product to product. These differences are reflected in the marked divergence in income patterns over the 1980s seen in Chapter 12.

**FIGURE 13.4**
**The Changing Terms of Trade (1975=100)**

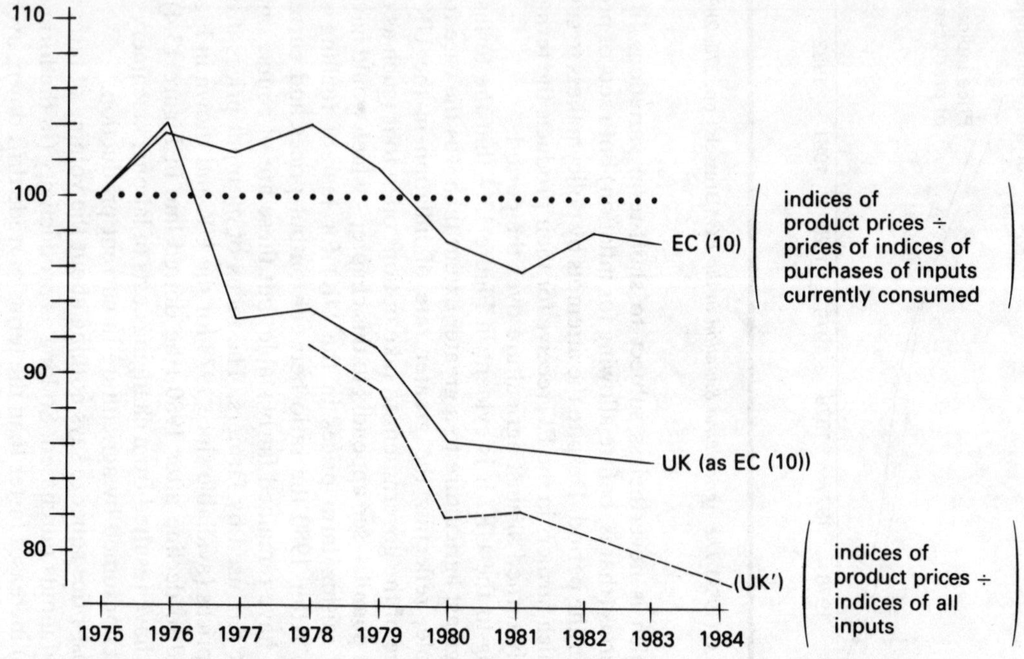

indices of
product prices ÷
prices of indices of
purchases of inputs
currently consumed

EC (10)

indices of
product prices ÷
indices of
all
inputs

UK (as EC (10))

(UK')

1975 1976 1977 1978 1979 1980 1981 1982 1983 1984

110
100
90
80

## 13.5 Government Interference in the Market for Farm Products

Bearing in mind the basic nature of the price system and the characteristics of the demand and supply experienced in agriculture, together with the price movements noted above, we next consider the ways in which the market system is manipulated by the state. Chapter 2 has already covered the food chain beyond the farm gate, so we will not repeat material on the structure of food marketing and how institutions such as state-backed marketing boards may manipulate the food market to the benefit of farmers. Here we are concerned with market interventions which impinge more directly on farmers, such as subsidies they receive and quotas that constrain them. At this stage it is not necessary to describe in detail the policy objectives of the UK government or how they relate to the EC's Common Agricultural Policy. These matters will be reviewed in Chapter 14. All we need note here is that market intervention is used in various forms to achieve certain policy aims, such as raising the incomes of farmers through giving them higher prices for their products or lowering their costs of production, or increasing the level of self-sufficiency by raising the volume of production from home agriculture. Because of the complexity of the market intervention system as applied under the CAP, and because it is constantly under modification, this chapter deals primarily with the principles of intervention mechanisms. For those readers needing to know which applies to which products, a list is given in Table 13.2 for the situation in early 1986.

Some idea of the importance of market intervention as a tool of policy is given by figures from the Annual Review of Agriculture White Paper which details the amount of public spending on this industry. In 1983–4 the UK government spent an estimated £1.7 billion on agricultural support of which £1.3 billion was spent on price support measures.[2] About 80 per cent of payments on price support were refunded to the UK by the European Agriculture Guarantee and Guidance Fund (EAGGF, also known as FEOGA). Within the agricultural budget of the EC price support in various forms takes all but three or four per cent of the total expenditure. Price support in Britain, the EC and the USA extends to all the main agricultural products with very few exceptions. Where there is no price support, other measures operate – in Britain, for instance, potato growers finance the Potato Marketing Board, a statutory cartel which restricts supply and plays a price supportive role.

## 13.6 Methods of Market Intervention

Governments use a range of ways of manipulating the supply and demand of farm products. There are three main policy options for government to choose between if they wish to intervene in agricultural markets. They are to:

1. manipulate (i.e., raise) product prices with or without controlling supply from home or foreign suppliers
2. change the level of costs faced by farmers by the use of subsidies and taxes linked to inputs
3. make payments to farmers not linked to production (though possibly linked to non-production) so that their profit-income objective is approached in a way that does not directly flow from agricultural production.

These three options are considered in turn, though inevitably actual policies contain aspects of more than one of the approaches. Most attention will focus on the first two, since in practice they have dominated the pattern of intervention for policy purposes. We discuss the policies with the aid of simple linear supply and demand diagrams, with the usual assumptions concerning independence of supply and demand and smooth, continuous functions. Throughout we assume constant technology, incomes and tastes, no uncertainties about the future, no lags in adjusting to market conditions and no significant elements of monopoly or monopsony power.

## 13.7 Manipulating Farm Prices

In a free market the interaction of demand with supply ensures that the quantity that enters the market from suppliers is all taken from the market by purchasers, and prices will adjust by the interaction of buyers and sellers to ensure that this balance is achieved. Both buyers and sellers may make adjustments to the stocks which they hold, but there will not be an accumulation of farm products in the market itself. Nor does the market hold stocks which can be released to buyers to counter a shortfall if the quantity supplied is curtailed by outside influences, such as adverse weather. The term market refers to a set of interactions or relationships between buyers and sellers and, though these may be institutionalised to a specific building or site (such as a cattle market) and need facilities and staff to operate (auctioneers for example), the market institutions do not themselves normally act as either buyers or sellers. Adjustments to the balance between demand and supply are made solely by changes in price.

It follows that a government that attempts to manipulate farm prices and distort the price from its equilibrium level is also faced with the consequence that the quantity demanded will not equate with the quantity supplied at the new price level. The government must have a mechanism which removes surpluses and supplies deficits at the new price if it wishes to allow market forces to continue to operate.

### 13.7.1 Buffer stocks

The simplest solution is the buffer stock principle, where intervention is designed to stabilise prices of those agricultural products that are subject to wide price variation brought about by weather influences on supply. The buffer consists of the state buying farm products when prices are low, taking them into storage, and releasing them onto the market when prices rise to unusually high levels. In reality no state agency will try to maintain a single, unfluctuating price. More likely it will set a minimum and a maximum price it is willing to see in the market, buying if the actual price tends to slide below the permitted range and selling if it appears to be likely to exceed the desired maximum. Figure 13.5 shows such an arrangement.

It is unlikely that the buffer stock managers will hit on exactly the right average price or range to maintain. What if they experience a run of low-yield, high-price years soon after setting up the buffer stock? The market deficit cannot be met (without buying in stocks from elsewhere) and the price rises above the desired maximum. What if there is a run of glut years? Unwanted stocks build up, which must be destroyed, exported or otherwise got rid of. Attempts to give them away free to the poor and needy may be a socially and politically attractive solution, so long as the poor are a long way away; otherwise the minimum desired price will be undermined.

The buffer may operate not simply to stabilise prices, but rather to stabilise revenues, that is price *times* quantity sold by producers. There may also be income implications of stabilisation programmes. These complications are beyond our treatment here. In practice, buffer stocks are not used in their pure form as UK or EC agricultural policy instruments. Rather they become enmeshed with support buying, considered next.

### 13.7.2 Support buying

The key to a buffer stock is the variation in supply that means a government (or its agent) is sometimes a buyer and sometimes a seller. In the case

**FIGURE 13.5**
**A Buffer Stock**

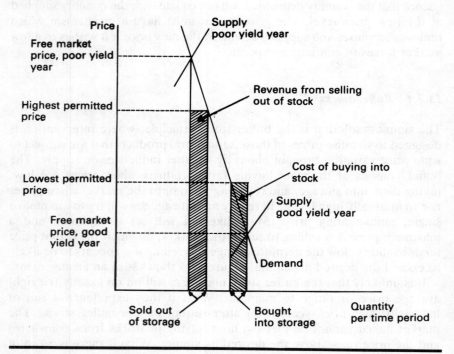

NOTES   1. Over a period of time the quantity bought into storage should be balance by that sold out of storage.
2. The choice of permitted maximum and minimum price is crucial to achieving this balance.

of support buying, the opportunity to be a seller seldom if ever exists. The government is providing farmers with a higher price than a free market would permit. To do this, it must operate on the internal and on the external markets. Figure 13.6 illustrates the situation when an internal price is fixed above the free market price. Farmers can sell as much as they like at this fixed price and the state buys up any excess. It is assumed that no imports or exports take place and that the government places a block on the entry of imports which might otherwise be attracted to the UK to take advantage of the raised prices.

The 'without-intervention' equilibrium situation would produce a price at which the quantity supplied to the market exactly balanced the amount

**FIGURE 13.6**
**Support Buying**

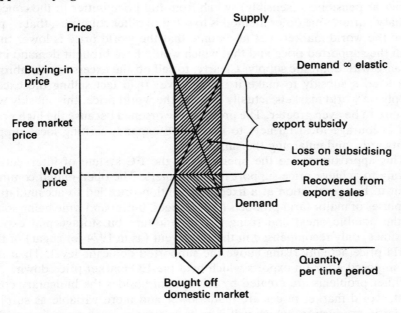

NOTES 1. A government agency which buys a commodity from the market creates a new 'kinked' Demand curve which is infinitely elastic at the buying-in price. The higher price causes domestic output to rise.
2. Accumulating stores have to be disposed of. Selling them on the world market will involve subsidising the export, but this may be more acceptable than destroying the food surplus.
3. Support buying will only be effective if imports are controlled, usually by import levies under the CAP.

demanded. If, to support farm incomes, the government fixes price at a higher level, local farmers will produce an excess of supply over demand. This will accumulate in the state's stores. Clearly this accumulation cannot continue indefinitely and some way of disposal will need to be found. It could be destroyed by fire, changed in some way such as making wine into industrial alcohol or butter into oil for which a market might exist, or simply left to rot, but this type of action is politically sensitive when there are members of the Community in need of an improved diet. Such action has also proved increasingly unacceptable when viewed against recurrent famines elsewhere in the world. To sell it cheaply to certain disadvantaged groups, such as the sales of butter stocks to old-age pensioners at Christmas, carries the danger of undermining the normal market demand for butter as pensioners (sensibly) switch from full-price butter to the concessionary butter. One other method is to sell it to other countries, that is, put it on the world market. Let us assume that the world price is lower than both the supported price and that which would have brought demand into balance with domestic supply. Clearly, to sell off the excess supply abroad will need a subsidy to make it competitive. If in fact selling the excess supply on world markets actually reduces the world price, this subsidy will require to be even greater. The problem is worsened because the high price will encourage local farmers to invest and expand, shifting local supply rightwards, and enlarging the surplus.

This approximates to the operation of the EC system of intervention buying to achieve price support for its farmers. The open-ended commitment to buy production at a fixed intervention price led to accumulating surpluses of major farm products (wheat, beef, butter and wine being some of the notable ones) and rising EC expenditure on storage and export subsidies, only recompensed in the rare event (as in 1974 for sugar) of the world price actually rising above the supported domestic level. Then the EC imposed a tax on exports which held the EC market price down.

Other problems are created by this system besides the budgetary cost. First, world market prices are made lower and more variable as surplus domestic production, which will vary from year to year according to the weather and other short-term supply factors, is put on the world market at subsidised prices, a cause of complaint by traditional farm exporting economies in Australia and the Americas. Second, farmers lose the discipline of market forces and are unable to relate their costs to costs elsewhere or to the strength of consumer demand. Third, consumers pay higher prices than they need and there are the trade distorting effects already noted in Chapter 4.

### 13.7.3 Restricting supply from abroad

Supported prices for domestic farmers can only be maintained if world market supplies at lower prices are kept off the home market. Where a significant proportion of total supplies normally come from abroad, it is possible to engineer price rises for domestic farmers by restricting this foreign supply without any domestic intervention buying or other government schemes. The CAP price support system was initially of the import-restricting type but has been forced increasingly to adopt support buying and other measures as Community farmers have responded to the higher prices by expanding their supply, displacing imports to the extent that the EC is a net exporter of many farm products (see Chapter 4).

## FIGURE 13.7
## A Levy on Imports

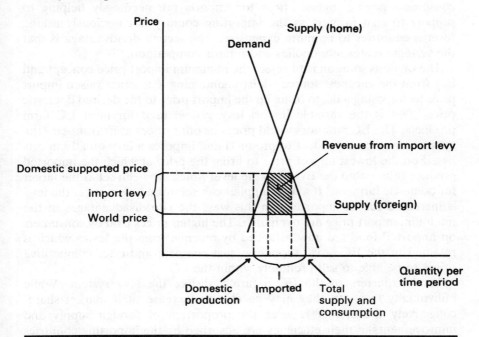

NOTES  1. If prices are raised to home suppliers by using an import levy, home suppliers expand but imports are reduced by more, so that consumers take less from the market and pay higher prices for it.
2. If a levy is used there will be revenue to the state.

To explain the workings of such methods of support it is necessary to go back to the idea of the world price, that is the price at which a country could import the commodity from foreign suppliers. For simplicity we will assume that this foreign supply is infinitely elastic, and the importing country could buy as much as it wished without affecting the world price. From Figure 13.7 it can be seen that in a free market situation some food would be imported at the world price and home farmers would produce some too.

Governments can attempt to establish higher home prices by minimum import prices. The EC uses these for fruit and vegetables (and the UK in the 1960s used them for cereals). Imports are only allowed into the country at this price, not below. This is an easy system to negotiate so long as supply is restricted to a few well-established countries because they can see a gain in prices they receive to offset their loss of sales in the country. The higher price for home farmers increases local supply and cuts local demand for the product and thus reduces the quantity of imports.

The minimum import price system has two major disadvantages. First, consumers paying higher prices for imports are needlessly helping to support foreign farmers as the importing country is consciously asking foreign countries to be more expensive. The second disadvantage is that the system creates monopolies and retards competition.

The obvious solution is to reject the minimum import price concept and buy from the cheapest source whilst maintaining a constant raised import price by imposing a tax to bring up the import price to the desired domestic price. This is the variable import levy system used for most EC farm products. The EC monitors world prices or offer prices at its frontiers (the system varies from product to product) and imposes a levy on all imports based on the lowest import price to bring the price at which the imported produce sells within the EC up to the level which has been set as the target for domestic farmers.[3] If a new supplier can sell for a lower price, the levy is increased for all importers. In this way, the two disadvantages of the minimum import price are overcome. The higher prices paid by consumers on imported food are now balanced by revenue from the levies which is retained in the EC. Also suppliers must compete and offer competitive prices to be able to sell profitably within the EC.

Food exporters to the EC naturally dislike the levy system. While individually they can they may be able to increase their market share, collectively the system reduces the proportion of foreign supply and improvements in their efficiency are absorbed by the importing countries through charging variable levies. Exporters also dislike the uncertainty which is introduced by the variable nature of the levy. Taxes (termed tariffs) on imports have long been used in agriculture to raise the prices for home producers, but these have generally been of a non-variable nature, at least within budget years, and have not been linked to specific target prices for domestic farmers. Tariffs either relate to a physical amount of import, such as a tax of £50 per ton of a good imported (a specific tariff) or to value of imports, such as a 16 per cent import tax (an *ad valorem* tariff). While having similar effects of reducing the level of imports and giving the home supplier a higher price, foreign suppliers who become more efficient and capable of selling at lower prices can gain a greater share of the market by undercutting the less efficient home producers. This opportunity is denied them by the variable nature of the current EC variable import levy system.

The EC combination of support buying at home and import levies is attacked from three quarters – domestic consumers, foreign exporters and, as imports disappear and therefore levy receipts, and the cost of disposing of surplus home production grows, ministers of finance and governments concerned with public expenditure. The beneficiaries are local farmers and, less obviously, other importing countries around the world. The system has encouraged cheap food on world markets to the good fortune of countries such as Bangladesh, Nigeria and Egypt. They might prefer help in other ways besides cheap food, but they can see the benefit in lower foreign exhange costs and a greater availability of food within their country.

A major problem with market intervention which raises the prices of farm products is that, not unreasonably, farmers interpret the higher prices as a signal to expand production. This may not be the intention of policymakers, who may really be aiming at supporting incomes through product prices. Consequently the cost of price support may be much greater than is at first envisaged. From experience it has been found necessary to limit the amount of spending on market intervention by making farmers aware through the price system of the consequences of their expansion. The UK has used the notion of **standard quantity**, that is, to link the amount of support going to the producers of a commodity to a predetermined quantity of output; exceeding that output would result in a dilution of support as the available spending is spread over the greater volume of production. A similar notion under the CAP is the **guarantee threshold**. Introduced in 1982, by 1985 there were guarantee thresholds for cereals, rape, sunflower and tomatoes. The threshold is usually the average EC production in the previous three marketing years. For cereals, the intervention price was reduced by 1 per cent for each 1.0 million tonnes of EC cereal production beyond the threshold level.

The advantage of such a system is that it is automatic. In an EC where political disagreement makes price-cutting difficult to achieve, measures are most likely to work when they are independent of the Council of Ministers and European Parliament, both open to strong pressure from the farming lobby. The problems with the guarantee thresholds as presently conceived are that they are far too timid in the amount of price penalty imposed for exceeding the threshold quantities, and they are still open to political manipulation. **Co-responsibility levies** are a further variant on this theme, with a tax being imposed on producers of commodities in surplus (really another form of price reduction) which is intended to assist with the disposal costs of excess production.

Other ways of achieving higher prices for farmers have been and are being used in the UK and EC. While of less general application than the two measures given above, they are nevertheless important for particular commodities.

### 13.7.4 Domestic quotas

Quotas are physical limits on the amount of output that farmers can produce. Potatoes in the UK have long been the subject of acreage quotas, with the individual farmer restricted to the area he can grow. The Potato Marketing Board has thereby attempted to regulate the quantity of crop

## FIGURE 13.8
### Quota on Home Supply (no imports)

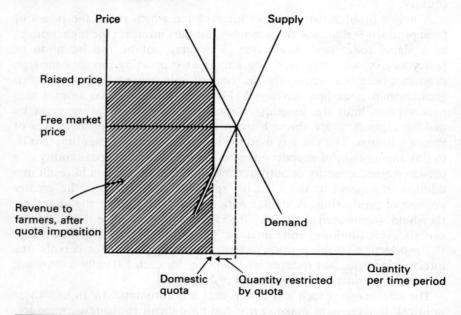

NOTES  1. By imposing a quota a new kinked supply curve is created.
2. Output is reduced, market prices rise.
3. Revenue to the farmer is raised (assuming demand is inelastic) – indicated by the price x quantity rectangles.

entering the market. Quotas were thrown into particular prominence by the 1984 decision of the CAP to impose milk quotas at the farm level at very short notice in answer to the potentially disastrous budgetary consequences of rising milk production which other attempts at constraint had done little to curtail.

There has been much discussion in the mid-1980s as to whether quotas should be imposed on cereals which display a static consumption and explosive production trend similar to that of milk.[4] Figure 13.8 illustrates the effect of a quota scheme using a supply and demand diagram, taking a case where there are no imports. The effect of restraining supply to the quota quantity is to raise market price paid by consumers. We have seen in Chapter 2 that demand for milk is gradually falling; this means that the demand curve is shifting to the left. As this happens so the quota must be reduced or prices will fall. In reality the regime for milk under the CAP fixes the price farmers receive above that which quota-restricted supply and demand would result in, so there is surplus on the market which has to be mopped up by support buying.

### 13.7.5 Physical constrictions on foreign supply

When imports form a significant portion of total supply, a physical limit on the volume of material that can be imported can raise prices for domestic farmers. Supply restriction from abroad can be done in several ways. The first is by some variant of voluntary restraint; foreign suppliers of apples for instance are simply asked not to send more than a certain tonnage of produce at times of the year when the EC price would be sensitive to such supplies (i.e., July/August). The UK used this approach widely in the 1960s to limit imports (a bacon market-sharing arrangement with Denmark, a beef importing 'gentleman's agreement' with Argentina and import quotas for sugar and butter). Currently it can be seen operating in the market for cars, in which Japan has agreed voluntarily to restrain its exports to the UK.

Other barriers to trade may be used which purport not to be formal ways of raising the prices enjoyed by home producers but have this effect. The zealous application of health and hygiene controls may enable foreign food to be kept out (this has been used both by the UK and France in the 1980s), the retention of cumbersome import documentation and slow border customs, the insistance that imports travel in ships or lorries of the importing nation only and other trade impediments result in enhanced prices for home producers.

More formal limits on foreign supply could be imposed as quotas, illustrated in Figure 13.9. Prices on the home market are increased as total supply contracts up the inelastic demand curve. Such a system has all the characteristics of any price raising measure, benefiting home producers through higher prices and easing the competitive pressure on them at the expense of consumers (particularly the poorer members of society and foreign suppliers), plus the problem of allocating the quotas and reallocating it as new producers arise.

### 13.7.6 Deficiency payments

One of the problems with price fixing and support buying is the effect this has on consumer demand. Food prices are increased and the quantity

**FIGURE 13.9**
**Quota on Imports**

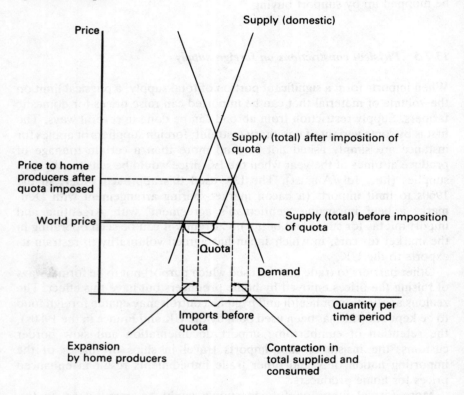

NOTES   1. A quota on imports which is less than the initial level of imports gives the total
        supply curve a kink. The new supply curve intersects demand at a higher price.
       2. At the higher price domestic producers expand production. Foreign suppliers sell
        a reduced quantity but receive a higher price.

demanded is reduced. This has both political and economic ramifications, although as far as individual voters are concerned, food prices are less politicallly sensitive than might be supposed.

Raising product prices by deficiency payments avoids this undesirable effect and was the typical method of government intervention in the market for farm products before the UK entered the EC and adopted its CAP. Although now largely displaced by the system of import levies and support buying, the idea of the deficiency payment is retained in the beef

**FIGURE 13.10**
**Guaranteed Price and Deficiency Payment (No Foreign Supply)**

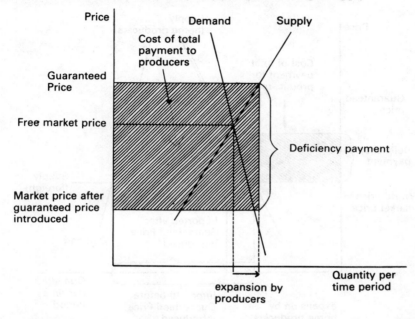

NOTES   1. A guaranteed price higher than the market price causes producers to expand output. They will never produce less than the quantity corresponding to the Guaranteed Price; a new kinked supply curve is the result.
   2. Market price is pushed down by the expanded output.
   3. Farmers' revenue (including the deficiency payment) is increased although revenue from the market alone is reduced (market price x output).
   4. Government spending, much greater than the difference between the original market price and the guarantee *times* the original quantity, may be limited by the use of standard quantities or other measures (see text).

variable premium (a deficiency payment in all but name) and under the CAP, tobacco, rapeseed and other oilseeds and olive oil all have deficiency payments.

The basic idea of a deficiency payment system is to allow free market forces to determine the level of market prices, which with an internationally traded commodity will be the world price. Farmers sell their output for this market price and the government agrees to make up to farmers the difference between this market price and some pre-determined guaranteed price by means of a direct payment, termed the deficiency

**FIGURE 13.11**
**Guaranteed Price and Deficiency Payment (With Imports)**

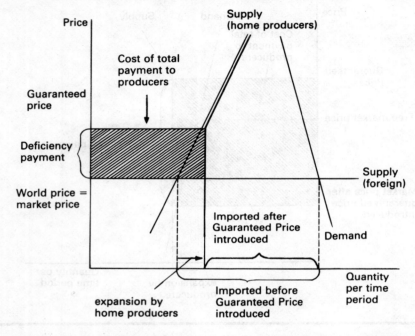

NOTES 1. A guaranteed price causes home producers to expand, displacing imports. The deficiency payment has to be paid on the higher, expanded output.
   2. If foreign supply is infinitely elastic, as shown here, the market price, which is dictated by the world price, will not alter. If, however, foreign supply is less than infinitely elastic, market price will fall as less is imported.

payment. The use of an average market price in order to calculate the deficiency payment means that there is still an incentive for individual farmers to seek the best prices for their output, that is marketing efficiency is still promoted. Figure 13.10 illustrates what happens when there are no imports, Figure 13.11 when there are imports. Whatever the market price, farmers receive the guaranteed price and produce the quantity corresponding to this. Thus their supply is unresponsive to market prices below the guarantee level. Farmers respond to the higher guaranteed price by producing more, which can only be sold if market price falls. The deficiency payment required is thus greater than the difference between the original free market price and the guaranteed price because of this depression of the market. There is no surplus because consumers take all the supply at the lower market prices. Such a system is therefore counter-inflationary in that it leads to lower food prices to consumers and is progressive in its income redistribution effects; lower food prices benefit the poorer members of society proportionally more and the cost of the deficiency payments to farmers is borne by taxpayers, who tend to be the relatively better off members of the community.

The disadvantages of the deficiency payment system are the administrative problem of paying farmers individually according to their production levels, the high cost to the public budget of the payments and the inevitable political attention that they attract (whereas price support through the market may be far less transparent though just as real and, indeed, may imply a less desirable burden of support than one financed through taxation) and the vulnerability of expenditure to influences outside the control of the government, such as a movement in world prices because of good yields in foreign countries.

In the UK, deficiency payments became too expensive to leave open-ended. First milk, then cereals and meats were linked to devices such as guarantee thresholds or import restrictions in order to prevent increasing the burden on taxpayers. Minimum import prices also held up to minimise the gap in the case of cereals. In short, a system devised in the 1950s for a 50 per cent self-supply situation proved too expensive by 1970 for a 70 per cent self-supply UK. Plainly it would be politically and fiscally highly unlikely to gain favour in the EC of the 1980s with even higher self-supply.

### 13.7.7  Production bounties and taxes

Finally in this section we turn to production bounties, such as a payment for every calf sold from a farm or the closely related headage payment given to farmers keeping livestock in hill areas. The effect is to shift the supply curve for these enterprises downwards – farmers are willing to supply the same quantities to the market at a market price lower by the extent of the bounty than before. In terms of market price and output it is evident from Figure 13.12 that output rises and the market price falls somewhat, partly cancelling out the benefit given to the farmer through the bounty, although this effect will depend on the elasticity of demand for the commodity in question. A tax on production would exert the opposite effect (Figure 13.13); not all the impact of the tax would fall on the farmer as the contraction in supply that would result would in part pass the burden to the consumer. One instance where such a tax is used is on farmers who exceed quota levels of output; by penalising over-quota production consumers have to pay more for their purchases.

We move from considering the intervention of the state in the markets for agricultural products to the ways in which it attempts to manipulate the markets for the inputs which farmers use.

**FIGURE 13.12**
**A Bounty (Subsidy) on Production**

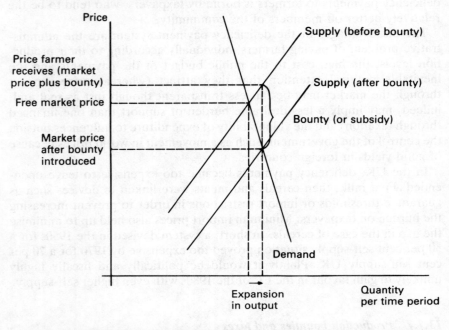

NOTES   1. A bounty (subsidy) lowers the supply curve by the extent of the bounty.
        2. The producer expands in response to the bounty but this pushes down market
           price. Consequently, the final price the farmer receives is less than the *original*
           market price plus bounty; it is the new lower market price plus bounty.
        3. The benefit of the bounty is thus in part passed on to the consumer, the extent
           depending on the elasticity of demand.

## 13.8   Intervention in the Markets for Farming Inputs

The markets for agricultural land, labour and capital have been described
in Chapter 6, 7 and 8. From these it emerges that although agriculture is
the majority user of land, overwhelmingly so in rural areas, agriculture
otherwise plays a very minor role in the national markets for resources.
These factor markets are heavily influenced by distortions in competition
and by institutional controls, of which the role of government in influencing farm demand for inputs is only one. In the case of managerial talent,
for which inevitably a market exists, the market is particularly constrained
by the social structure and legal framework which ensure that the supply of

**FIGURE 13.13**
**A Tax on Production**

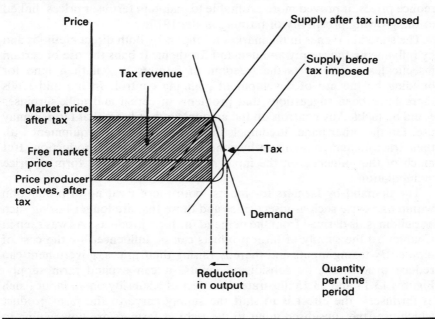

NOTES
1. A tax on output raises the supply curve by the extent of the tax. Producers reduce output.
2. The market price is raised, but by less than the tax. The producer receives a lower price than before, but the fall is less than the size of the tax.
3. The division of the burden of the tax between consumer and producer will depend on the elasticity of demand.

talent is restricted almost entirely to the members of farming families or others with access to large amounts of land and capital.

Of the purely industrial inputs, fertilisers are the most important, approaching £1 billion in 1984, about one-sixth of total expenditure on purchased inputs, and it is salutory to note about half are purchased from one company (ICI). Pesticides represent the fastest growing input purchased, linked to new techniques of cereal and oilseed rape growing, and amounted to about £300 million in 1984. A small number of firms dominate the supply of fertilisers, chemical sprays and farm machinery. Further, the agricultural business is often a small part of their total business (i.e., ICI, Ford, Shell Chemicals). Consequently, their supply to farmers depends on factors in addition to the level of price, such as when a firm wants a long production run or uses wastes from other processes. Nor do input prices necessarily reflect marginal costs. For instance some fertilizer manufacturers obtained cheap North Sea gas on long-term contract in 1973, which helped constrain cost of production for ten years. However, rather than reduce prices, it proved more profitable to maintain fertiliser prices, linked to the growing prosperity of farming in the 1970s.

The state intervenes in the market for inputs by both direct methods and by influencing the supply and demand for them. It bans the use of certain insecticides and restricts the freedom of farmers to sell their land for building by the use of development planning control. In the mid-1980s there have been suggestions that problems of cereal and milk surpluses could be tackled by controls on the amounts of fertilisers that farmers may use. On the other hand, it compels farmers to buy safety equipment with their tractors and to maintain hygiene standards in milk production. But much of the influence on the input markets comes in the form of price manipulation.

The demand by farmers for inputs, both those used up in production within one cycle such as insecticides and those that are longer-lasting such as buildings, is derived from the demand for farm products. As was seen in Chapter 10 the supply of farm products can be influenced by the cost of inputs. By taxing inputs and thereby raising their prices, government can reduce production; by subsidising inputs, it can expand farm supply. Figures 13.14 and 13.15 illustrate the effect of a subsidy on an input (such as fertiliser); the effect is to shift the supply curve of the farm product which uses the subsidised input to the right as farmers are now willing to supply a greater volume of ouput at each market price than they were before. This will have an effect on the price of the product which depends on the market situation. In Figure 13.14 the rising supply meets inelastic demand and depresses price and revenue. In Figure 13.15, where there are imports and an infinitely elastic supply from the world market, the effect is to increase the amount supplied by home farmers whose rising output displaces imports. The subsidy will also be instrumental in helping the balance of payments (so long as the extra exports outweigh any extra imports of inputs).

It has been assumed so far that the increased demand for the input does not raise the price of the input. However, this may well happen if the supply of the input cannot easily respond to the increased demand for it; in other words if the supply of the input is less than infinitely elastic. Part of the subsidy goes then not to the farmer but to the supplier of inputs (see Figure 13.16); in the extreme case of totally inelastic input supply, an input subsidy simply results in a rise in market price of the input which is of the same size as the subsidy and the farmer ends up paying the same price for the input as before. No more input will be used and the supply curve of the product will not be shifted at all. Government is unlikely to allow such a

**FIGURE 13.14**
**A Subsidy on Inputs (No Imports)**

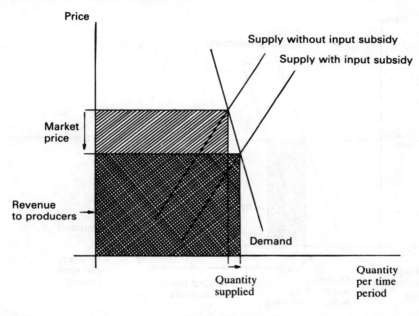

NOTES 1. The effect on the market for the product of a subsidy on inputs is to move the supply curve to the right. Output rises.
2. The revenue to producers will *fall* if demand is inelastic (as shown here). Whether producers are better or worse off will depend on the relative size of the revenue fall and subsidy on inputs.

crude response by input suppliers but some price increase seems likely to cover increasing costs of input suppliers.

The UK has operated numerous input subsidies at one time or another including subsidies on labour (briefly), machinery, land improvement, building, breeding livestock, fertilisers and lime, exempting farms from local taxes (and therefore subsidising local services to farmers), subsidising fuel for heating and reducing taxes on vehicles. The capital input subsidies appear to have been particularly output-increasing and have also led many to question how far they have accelerated the shedding of labour by farming (and reduction in rural jobs), something that is less desirable in times of significant unemployment than in the 1950s to 1970s when such grant schemes were initiated.

**FIGURE 13.15**
**A Subsidy on Inputs (with Imports of Product)**

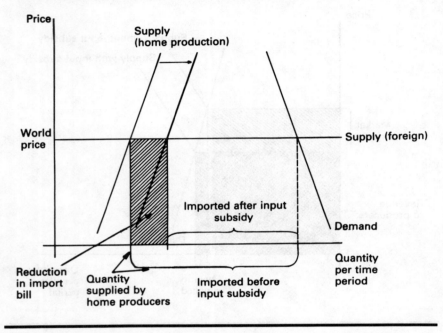

NOTES  1. The subsidy on input costs shifts the product supply curve to the right.
  2. More home production occurs, displacing imports. Prices does not alter, as this is determined by the cost of imports. It is assumed that foreign supply is completely elastic.

## 13.9  Direct Payments to Farmers

So far intervention in the market has been discussed in terms of altering the prices farmers receive for their produce or the costs they pay for their inputs. However, one way in which it is felt possible to influence the market for products is to make payments to farmers which can be enjoyed as income; this may relieve them from the necessity to derive a livelihood from agricultural activity. The implication is that they will then not produce as much, thereby relieving in part the problem of over-production in the EC. From Chapter 10 it would be easy to deduce that a direct income payment without any conditions attached would affect neither the level of output nor the use of inputs; farmers would still behave as profit maximisers in the way they arranged their farms, although perhaps a modest reduction in activity might be exhibited by those who preferred to take more of their reward in the form of leisure.

**FIGURE 13.16**
**The Effect of An Input Subsidy on the Market for the Input**

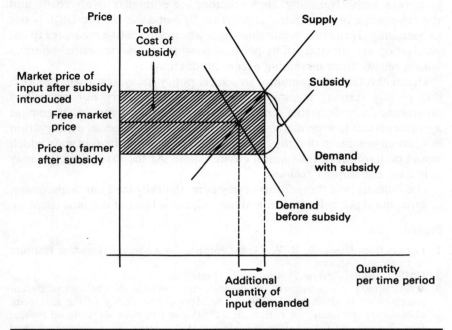

NOTES 1. A subsidy per unit raises the demand curve of the input by the extent of the subsidy.
2. The market price of the input rises.
3. The price paid by the farmer (after subsidy) falls, and he uses more of the input.
4. Although part of the subsidy benefits the farmer, some is passed on to the input supplier.
5. The more inelastic the supply of the input (the steeper the supply curve) the more the subsidy results in a price rise for the input supplied. In the extreme, when an input has totally inelastic supply (land) the subsidy simply results in an equivalent rise in market price. The farmer ends up paying the same net price.

On the other hand, direct payments could affect the markets for farm products and of inputs (especially land) if they were linked to specific actions by farmers. Under the EC policy towards disadvantaged areas, farmers on hill land are paid an income supplement linked to the number of livestock to encourage them not to follow the dictates of the market and abandon the land, but to remain; the objective is social and environmental rather than agricultural. While these payments could be regarded also as a production subsidy, it is clear that the objective is not concerned at all with encouraging output.

Schemes designed to cut production by limiting the use of land are called 'set-aside', after the long-running programme adopted in the USA. Access to price support, deficiency payments and production quotas can be linked to an agreement by farmers to leave some land uncultivated, temporarily or permanently. Naturally, such schemes are administratively costly and the experience of controlling production by 'set-aside' in the USA is not encouraging. However, in circumstances where alternative measures to cut production are constrained by political considerations, 'set-aside' offers at least a respite from increasing excess production.

Under the latest Community structural policy measures it will be possible to pay farmers income supplements if they carry out prescribed environmentally desirable practices. On the other hand, management agreements can now be made with occupiers of land of special conservation interest on condition that they give up plans to use their land in ways which would be harmful to the natural environment. At the same time this may help ease the surplus problem.

The objectives of these forms of payment naturally lead into a discussion of agricultural policy in the UK and the EC, the subject of the next chapter.

## Notes

1. Figures from Howarth, R. W. (1985) *Farming for Farmers?* (London: Institute of Economic Affairs).
2. HMSO *Annual Review of Agriculture* (London).
3. A description of target, guide, sluice-gate, etc., prices as they relate to different commodities is given in EC (1982) *The Agricultural Policy of the European Community* (Brussels), or Hill, B. E. (1984) *The Common Agricultural Policy: Past, Present and Future* (London: Methuen), or Harris, S., Swinbank, A. and Wilkinson, G. (1983) *The Food and Farm Policies of the European Community* (Chichester: Wiley).
4. A description of the milk quota arrangements can be obtained from Milk Marketing Board (annually), *Dairy Facts and Figures*. See also the monthly magazine of the Board, *Milk Producer*.

The 'green pound' is the informal name for the representative rate of exchange between the European Currency Unit (ecu) and the pound sterling. Each member state of the EC has a green rate. Common EC prices in the Community are set in terms of ecus and translated at the green rates into national currencies. Thus, if the EC price is 100 ecu per tonne and the green pound is 1.0 ecu: £0.4, then the EC price in Britain will be 100 x 0.4 = £40 per tonne. Green rates have been used since 1969 as a way of handling changes in market rates of exchange between currencies. For instance, in 1986 the pound depreciated against most other EC currencies by about 20%. The fixed green rate ensured common prices remained stable in Britain despite this depreciation. In order to prevent intra-EC trade distortions, a series of border taxes and subsidies are instituted, called monetary compensatory amounts (MCAs). Thus, in 1986 imports of most agricultural products into the UK from other EC countries received an MCA subsidy of around 20% and exports from Britain to other EC countries of similar farm products paid an MCA tax of around 20%. The system has proved awkward and at times clumsy. However, the EC has preferred this to the alternative of abandoning the concept of common prices in the face of fluctuating exchange rates. The green rates have also proved useful to member countries by giving them some national control over the EC prices their farmers receive. In 1985 for instance, the West German government devalued the 'green mark', so that the deutschmark price of wheat paid to German farmers would not fall despite a cut in the ecu price of wheat agreed in Brussels. The greater the degree of exchange rate stability between EC member states, the more likely it is that green rates can be adjusted to equal market rates and remain stable. In recent years, Britain, and to a lesser extent, Ireland, Italy and France have found such stability elusive compared with Germany, the Benelux and Denmark. For most of these countries, the creation of the European Monetary System in 1978 has subsequently encouraged greater exchange rate stability. For a full account of the system of green rates of exchange, see the references listed in note 3 above.

**Table 13.2**
**Support Arrangements for Agricultural Products, 1986**

| | Prices and production supported: | imports controlled (Export subsidies) | Special measures |
|---|---|---|---|
| Milk and Milk Products | Target prices and intervention prices backed up by buying up at intervention prices | Threshold prices backed up by Variable Import Levy (Export subsidies important) | Quotas at farm level and national level; various attempts to stimulate increased consumption especially of butter; co-responsibility levy omitted for small farms. |
| Beef and Veal | Guide prices backed by intervention at buying in price at given % of guide price; Hill livestock Compensatory Amounts (production bounties) | Customs duties, Basic levies on imports (Export subsidies important) | Minor variations between UK and rest of EEC for price support; special arrangements with some third world countries for imports. |
| Sheepmeat | Basic guide prices backed by storage subsidy and intervention buying; annual compensatory premia in first 4 years of scheme; Hill Livestock Compensatory Amount | Customs duties and Import Quotas | Variations between UK and rest of EEC; variable premia paid to sheepmeat producers. |

*continued on p. 358*

*Table 13.2 continued*

| | Prices and production supported: | imports controlled (Export subsidies) | Special measures |
|---|---|---|---|
| Pigmeat | Basic price backed by storage subsidy and possibility of intervention buying | Sluicegate prices, basic levies and Supplementary levies on imports. Partly related to EC cereal price | |
| Cereals | Target prices and intervention prices backed up by buying up at intervention prices for 'quality' products: intervention limited | Variable Import Levy (Export subsidies important) | Export levies charged if EEC price is below world price; co-responsibility levy omitted for small farms |
| Sugar | Target prices and intervention with the addition of production quotas | Variable Import Levy (Export subsidies important) | Special arrangements for third country imports from certain countries |
| Oilseeds | Production subsidies to raise EEC production and prices | Few controls, except olive oil | Processing grants |
| Fruit and Vegetables | Basic prices used as guide; producer organisations control marketing | Customs duties, countervailing charges (effectively a variable import levy) related to reference prices | Voluntary import restrictions |

SOURCE  Brassley, P. W. (1982) in McConnell, Primrose, *The Agricultural Notebook*, 17th edn. Edited by R. J. Halley. London, Butterworth. *Green Europe* (1986), 1986/7 Agricultural Prices Commission Proposals. No. 35 (Brussels).

# Public Policy Towards Agriculture **14**

## 14.1 Policies and Agriculture

Earlier chapters in this book have established, first, that agriculture has some special problems, such as income instability because of farming's reliance on weather and the necessity for adjustments in the structure of farming as an expanding output runs up against a static demand for food. Second, it has been shown that governments have become involved with their agricultures, principally through intervention in the markets for farm products and inputs. The first characteristic is often the key to the second. However there are also many instances where government involvement has arisen from historical accident, such as from the urgent necessity to expand domestic food production in times of war, or from a desire to influence other sectors of the economy, such as to maintain a socially viable rural population in remote areas. This attempt to influence the size or shape of the farming industry, for a range of reasons, is broadly labelled 'agricultural policy'.

The term 'policy' implies that there is some attempt by government to alter the outcome which would otherwise result. A decision *not* to interfere, while clearly forming one possible option, is not usually graced with the description of a policy. If a definition of agricultural policy is required it might be 'That set of measures taken by central governments which are aimed at influencing, directly or indirectly, agricultural factor and product markets'.[1]

Agricultural policy is only one aspect of a government's attempt to shape its country's economic, social, physical and political character. Consequently, agriculture will be affected to various degrees by policies aimed primarily at other parts of the nation's fabric. For example, capital taxes, aimed at breaking up large accretions of wealth for reasons of greater social equity, have had a major influence on the way that farmers arrange their businesses, and tax concessions given to the owners of farmland have been blamed for making land an attractive investment for non-farmers, so bidding up its price with important but diverse implications for the farming community. Taking another example, public policy towards promoting a healthier national diet might alter consumers' buying habits and hence change the demand pattern for food. Other policies which have obvious relevance to agriculture are those on trade, land use and planning, pollution and nature conservation, rural employment and housing, and animal welfare. The influence of some of these on farming has been greater than those policies more directly aimed at the agricultural industry.

As a corollary, agricultural policies can carry non-agricultural implications. The stimulation of agricultural output as a way of improving the incomes of farmers has tended to reduce the amount of food imported, which in turn may restrict the abilities of our exporting industries to sell abroad. Higher prices for farmers will mean that consumers and/or taxpayers will find themselves paying more for food. Capital grants which subsidise the cost of farm buildings and machinery as a way of promoting productivity improvements may displace labour and so aggravate rural unemployment and depopulation from some remote areas. And, perhaps of most topical interest in the 1980s, modern high-output farming encouraged by enhanced product prices may bring with it the destruction or degradation of the natural environment.

In short, agricultural policy must be viewed against the background of a whole array of general government policies. Within this array there will be both policies which work in sympathy with each other and examples of stark conflicts, and many oblique relationships. Even under the label of agricultural policy there will be contradictions and inconsistencies, as different aspects of policy strive to achieve different ends.

### 14.1.2  The scope of agricultural policy

Agricultural policy consists of a bundle of measures which are more or less integrated with each other but, by definition, are aimed primarily at the farming industry. In recent times the scope of what is embraced by the term 'agricultural' has both broadened and lengthened. In the vertical sense, the links between farming and the sale of produce have meant that agricultural policy covers marketing. Some of the earliest UK policy measures (in the 1930s) which marked the start of active agricultural policy were of this form – the setting up of marketing boards for major products (see Chapter 3). More modern developments have extended this concern with marketing to the wholesale and retail food chain levels, with advertising of UK dairy products and meats and the promotion of schemes such as

Food from Britain's Quality Mark and the Kingdom Scheme for apples. On the input side, government control of the quality of seeds has a long history, and labour training through publicly financed colleges and the Agricultural Training Board has served to increase the quality of this input. The approval scheme for agricultural chemicals could also be seen as part of agricultural policy.

Agricultural policy has more recently broadened to merge at its edge with land use planning, pollution control, environmental protection, tourism and policy on access to the countryside. It is increasingly thought of as a subset of rural development policy, with farming considered as just one alternative way of improving the incomes and living standards of all those who live in the countryside, not just farmers. Although most frequently thought of in national terms, special agricultural policies applied to certain areas, such as the payments made to farmers in upland areas (Hill Livestock Compensatory Payments) can reinforce the regional policy aims of maintaining a viable population in these areas. This broadening has provoked comments that the objectives of many important agricultural policies (in terms of government spending) are not really agricultural at all in the sense of being primarily directed at the production processes, but that they are aimed at social goals, such as improving the incomes of the farming community or maintaining people in remote districts.

## 14.2 The Policy Process

Any policy consists not only of the government measures which attempt to effect a change (such as grants or taxes or import bans) but the steps leading up to those measures and the monitoring of their performance. Policy can be thought of as a process of which the practical implementation is only a part. The components are shown in Figure 14.1

### 14.2.1 Social values

The background to policy is society's set of **values**. Examples of these include the feeling that a reasonably stable society is something which is desirable, so that political, economic or social stability might be thought 'good' and instability 'bad'. Peacetime is thought preferable to hostilities. On a less dramatic scale, agricultural slumps and booms would be thought less desirable than a more equable prosperity. Economic growth is generally considered 'good' because it enables people to enjoy a greater consumption of goods and, given the choice, they will usually prefer more to less. Equality of opportunity is also thought desirable, and this has been

**FIGURE 14.1**
**The Policy Process**

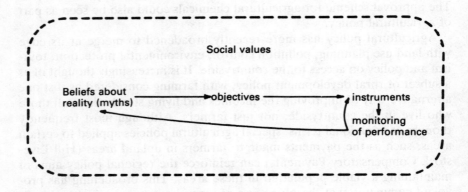

partly achieved in education, although in agriculture if anything there is less opportunity to become a farmer in the 1980s, unless one comes from a farming family, than there was fifty years ago. A more efficient use of national resources is usually thought preferable to a less efficient use, and freedom to work at an occupation of one's choice is preferable to coercion.

Different societies will attach different weights to different values, so that for example personal economic freedom is valued more highly in the USA than in China.[2] Even within countries a wide range of views may be held, and the political system exists to enable a compromise in settling national values to be reached. Journalists, economists, philosophers, professional politicians, divines and countless others will help shape the nation's set of values. In the UK there is a regular appeal to the public, with different parties presenting alternative sets of values. The political process can be thought of as indicating 'whose values count' and there must be serious misgivings if the process is heavily distorted in favour of one sector of society. Those of us living in democracies usually claim that this system is relatively free from the imposed values of a small ruling autocracy.

### 14.2.2  Myths about reality

Against this background arise **beliefs about how society works in reality**, beliefs which suggest ways in which society can be changed to a 'better' state. Examples abound:

that a large rural population leads to social stability

that unregulated markets for agricultural products lead to low and unstable farm incomes

that high prices for farmers cause inflation

that a prosperous farm population is necessary for adequate nature conservation

that prosperous farming will ensure a generally prosperous rural population

that farming is the best employment creator in the countryside

that a viable landlord-and-tenant sector is necessary for an efficient farming industry

that expanding home food production will improve the nation's balance of payments

that rewards in agriculture are lower than can be achieved in other industries

that land lost annually by farming is a threat to our ability to feed ourselves

that a more highly capitalised industry is more efficient

that the sons of farmers make the most efficient farmers.

All of these are statements about how the world is believed to function. On these sorts of beliefs policy is based. All are testable, or at least potentially so, although to test them may necessitate reshaping the statements in a more quantifiable form, such as giving the criterion for the 'most efficient' farmer in the last of the list above. Yet many of these statements will not have actually undergone testing and policy will frequently be based on relationships which are more mythical than proved. For this reason the term 'myths' is often used about these believed relationships, not in the 'fairly tale' sense but to indicate that their validity has yet to be established. It is the role of the economist to examine these myths and thereby to provide a more secure basis by which the values of society can be implemented through policy measures.

### 14.2.3 *Objectives*

The next stage in the policy process is the setting of *objectives* or goals, which are descriptions of a preferred state of things towards which the agricultural industry is to be steered. Examples are:

higher prices for farm produce
stable prices for farmers and consumers
stable supplies of food for the community
stable farm incomes

Economics for Agriculture

the elimination of very low farm incomes
achieving parity of income between farming and other occupations
improving the efficiency with which the nation's agricultural resources
are used
low food prices for consumers
increasing the degree of self-sufficiency in food supply
eliminating certain diseases from UK farming
maintaining farming in the hill areas.

At any one time there may be a whole array of policy objectives which a government is attempting to pursue for its agriculture. Some will be consonant with each other – such as stable prices for farm products and stable incomes to farmers – while some conflict: higher prices for farm produce will run directly counter to lower food prices to consumers. Agricultural objectives may conflict with the objectives of non-agricultural policies: promoting greater efficiency may result in less labour being employed and hence exacerbate rural depopulation in upland areas in which social policy is trying to maintain the numbers of inhabitants. In turn, agriculture may be affected by non-agricultural policy objectives: capital taxation aimed at redistributing wealth might break up farms which could result in a less efficient use of national resources, and the conservation of wildlife in the wetlands may conflict with agriculture's policy of making land more productive by drainage. A matrix of interactions of agricultural and non-agricultural policies and objectives is given in Figure 14.2

In order to avoid conflicts between individual policies it would seem paramount that objectives should be clearly set out. Areas of inconsistency could then be identified and the individual parts be formed into an integrated overall policy towards food, farming and the rural economy. However the mechanism by which objectives are set militates against such an integrated approach. History is full of examples of agricultural policies which have been brought in rapidly to deal with particular problems without adequate appreciation of the longer-term consequences or the impacts on other policies. Two examples will suffice: first, raising cereal prices in the 1980s has resulted in their extension into counties not normally associated with them with consequential concern caused to nature conservationists (see Chapter 6); and, second, legislation which enabled the sons of tenant farmers to succeed their fathers resulted in a large reduction in the number of farms coming available to rent at a time when there was rising concern about the contraction of the let sector in the UK. As will be shown later, the objectives of agricultural policy have rarely been expressed in other than vague terms. While this permits some useful flexibility of interpretation (that is, useful primarily to politicians and administrators to enable them to adapt policy to changing conditions) it has also resulted in serious inconsistencies and conflicts.

# FIGURE 14.2
## Interactions Between Policies and Objectives

### A *Principles*

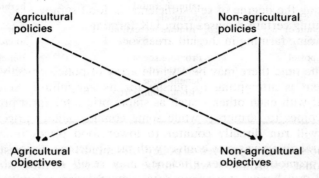

Agricultural policies     Non-agricultural policies

Agricultural objectives     Non-agricultural objectives

NOTE    The implications can be beneficial or harmful.

### B *Examples*

1. | *Agricultural policies* | *Agricultural objective(s)* | *Non-agricultural implication* |
   |---|---|---|
   | Raising price of farm products | Increased farm output and higher incomes | Less resources available for other industries |
   | Grant aid to investment in buildings and machinery | Improved output and efficiency | Buildings change appearance of countryside; machinery displaces labour |
   | Grants for drainage | Improved output | Loss of wetland environments |

2. | *Non-agricultural policies* | *Non-agricultural objectives* | *Agricultural implications* |
   |---|---|---|
   | Capital taxation | Redistribution of wealth and break-up of large wealth holdings | Potential break-up of large farms, and taking of steps to prevent it |
   | Planning of urban development | Restriction of housing and other building to designated areas | Some farmers near towns become very wealthy if they gain planning permission and invest heavily in farming elsewhere |

*continued on p. 366*

*continued from p. 365*

| 2. Non-agricultural policies | Non-agricultural objectives | Agricultural implications |
|---|---|---|
| Food quality control | Prevent the population eating harmful chemicals | Banning of certain pesticides and herbicides |
| Protection of footpaths | Promote recreation in the countryside | Some damage to crops |
| Rural postal services | Promote social contacts and benefit those without transport | Benefits the better-off too |

But who sets the objectives of policy? Essentially the responsibility rests with national governments who, in the case of the member states of the European Community, may agree on a collective policy for many important aspects – the Common Agricultural Policy. However, policies are not drawn up in a political vacuum, and pressure groups will endeavour to ensure that their particular interests are well-served. This they do by, first, promoting certain myths about reality (for example, that a prosperous farming sector is necessary for a beautiful countryside and a viable rural population) and by ensuring that workable policies are brought to the attention of politicians and policy-shaping civil servants. 'If you want British agriculture to expand to produce at home the food we currently import from abroad, then you must give farmers the resources to do so' is the sort of statement that was frequently put forward in the 1950s, 1960s and early 1970s. It assumes that it is desirable to displace food imports (as we saw from Chapter 4, a questionable assumption) and suggests that the only way to achieve this end is by higher rewards, meaning higher prices for farm produce, greater levels of grant aid and other methods of subsidising farming.

In the UK the National Farmers' Union has been frequently described as the country's most effective pressure group, securing for itself an influential role in agricultural policymaking through taking part in the discussions leading to the government's Annual Review of Agriculture White Paper. This role was initiated during the Second World War and consolidated in the period which followed, when great expansion of output was required; it was largely maintained until the UK joined the EC, since when the NFU's role has in part changed to influencing CAP development through the Community association of farmer's unions (COPA) although there are still significant national aspects of policy it can influence directly.

The NFU does not sponsor Members of Parliament (although it did so when first established in 1908), but the agricultural interest is heavily represented by MPs who happen also to be farmers or owners of agricultural land. In the 1983 House of Commons 59 MPs were farmers or had strong farming connections; had the House mirrored society at large there would have been around 10. The proportion of farming MPs seems to have risen since the last War although the latest House is less heavily stocked with farmers than in 1970, when there were 90.[3] The Minister of Agriculture has generally been a farmer, an odd position when it is seen that Ministers of Education or Health, for example, have not usually been practising teachers or medical practitioners. Agriculture tends also to be disproportionately represented in the Cabinet, where in 1983 10 Ministers had close connections with agriculture and the House of Lords, particularly the hereditary members.

These personal interests, and the association between the NFU and MAFF, have according to some commentators resulted in the interests of farmers being given undue political prominence, with the Minister and Ministry acting *for* agriculture, and ensuring that agricultural policy was conducted primarily for the benefit of farmers and landowners. This has occurred under both Conservative and Labour administrations, with the suggestion that a political stance towards greater or less government involvement has only emerged in respect to agriculture when parliamentary majorities have been particularly large. However, since about 1980, the awareness of a politically important and growing block of voters who are concerned with the rural environment, coinciding with a big Conservative majority, seems to have turned the attention of government and MAFF, and they have recently adopted a broader range of rural objectives.

During the 1970s and 1980s it has become noticeable in the UK that the initiative in drawing up the policy agenda has been slipping away from the agricultural 'establishment', broadly consisting of the MAFF, the NFU, the Country Landowners' Association (CLA), the Economic Development Committee for Agriculture, the Marketing Boards and Authorities and so on. The environmental lobby in particular is much more vocal and effective in pointing out the implications which agricultural policy has had for land use, loss of flora and fauna, changes in the appearance of the countryside and so on. Their criticisms have been increasingly pertinent when, within the EC, there is an embarrassing over production of many major farm products which are costly to dispose of. Two arms of government, the Countryside Commission and the Nature Conservancy Council, are particularly influential propagators of the environmental case, but they are augmented by a wide range of semi-public and voluntary bodies (National Trust, Royal Society for the Protection of Birds, etc.).[4] As will be seen later, recent agricultural policy objectives reflect the growing political importance of the environmental lobby. Britain seems to have gone further than other EC member states in shifting the control of the policy agenda.

## 14.2.4 Instruments of policy

Policies get put into practice in the form of **measures** or **instruments** which set out to achieve the objectives of policy. Agricultural policy measures take many forms, ranging from the buying by public intervention agencies of cereals and butter into storage so as to keep up market prices, taxes on imported foods to achieve the same result, schemes which subsidise the cost of buildings or machinery (a description of the workings of these forms of market intervention was given in Chapter 12), the government-run agricultural advisory service, the state veterinary service, the Agricultural Training Board and agricultural colleges, and the whole body of agricultural legislation, covering items such as landlord and tenant relations, rights of succession, and countryside protection. Non-agricultural measures, which nevertheless are important to agriculture, are the tax system, health and safety regulations, and planning controls. Some measures, such as legislation on rights of access to the countryside and on protection of the natural environment, straddle the farm/non-farm divide.

Civil servants are responsible for the design of these measures, and they will be on the look-out for instruments which are considered efficient in terms of achieving their objectives yet are consistent with other objectives (or at least do not run conspicuously against them) and which are administratively feasible. Thus, a measure which attacked the problem of low incomes in the most direct way – by direct payments to those farmers found to have unacceptably low incomes – might founder not on the grounds of effectiveness or conflict with other objectives, but because administratively it might prove costly. A less direct scheme might be preferred – say, by giving *all* farmers higher prices and thus larger incomes as a way of improving the lot of the needy – if it were administratively simple and cheap. A compromise is therefore struck, with a natural preference for introducing measures which are extensions of existing ones which have proved themselves administratively feasible. Problems have arisen when too great a change has been introduced, such as the unforeseen consequences of the 1976 Agriculture (Miscellaneous Provisions) Act which granted rights of succession to the family of tenants, contributing to the drying up of the supply of farms available to rent, and the as yet unresolved difficulties over the financing of management agreements made with farmers under the 1981 Wildlife and Countryside Act.

In order to implement policies a range of institutions has grown up. Indeed, a guide to agricultural policy might conveniently take as its starting point the existing institutions, the reasons why they were set up and their current functions. Figure 14.3 illustrates the more important extant ones.

Instruments of policy affect agriculture at different stages and levels. The best known tend to be those which affect the **prices of farm products** (such as the CAP's intervention buying, taxes placed on food imports or subsi-

## FIGURE 14.3
## Public Institutions of Special Relevance to Agriculture in England and Wales[1]

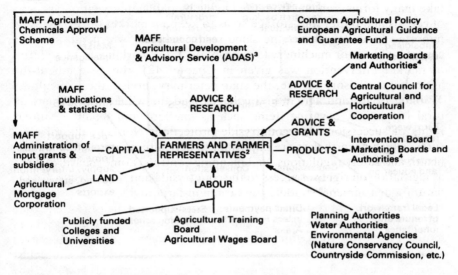

dies given on exports) or the **cost of farming inputs** (such as national or Community grants on buildings and machinery, and, in other member states, subsidised interest rates). Others aid production in less obvious ways, such as the MAFF's advisory service to farmers (mostly free in the UK, at least until recently), the Agricultural Training Board to which farmers make no direct contribution, the education provided by agricultural colleges and universities, and the research carried out by these and other publicly funded institutions. These also can be viewed mainly as ways of lowering the costs of production faced by farmers.

**FIGURE 14.4**
**Directions From Which Policy Measures Impinge on Farming**

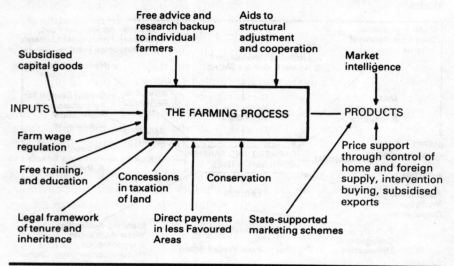

Other measures have attempted to accelerate **structural change** in this industry by encouraging elderly farmers to retire, or small farms to amalgamate, by offering financial incentives (lump sums or pension for the former and grants for the latter). In upland areas where falling farm numbers are thought undesirable, special aids are available to try to slow change. And perhaps the most pervasive aspect of agricultural policy is the body of agricultural legislation, covering items such as the setting up of Marketing Boards and Authorities, minimum wages through the Agricultural Wages Board, tenancy agreements and rent arbitration by the Agricultural Land Tribunal, animal health controls and so on.[5]

There is a danger that an analysis of agricultural policy can become bogged down in the detail of instruments and legislation. These are, however, only one aspect of the policy process, albeit that part which is most easily visible. When encountering Acts of Parliament or Community Directives, price support mechanisms or forms of grant aid, it is necessary to go further and ask what the objectives are that these measures are supposed to be approaching. Frequently schemes become divorced from their aims or outdated and yet continue, developing a life of their own. With this warning, an indication of the directions from which policy bears on UK agriculture is given in Figure 14.4; the multiplicity of measures means that this cannot pretend to be exhaustive.

### *14.2.5 Assessing the performance of policy*

The fourth and final stage in the policy process is the assessment of the **performance** of measures and instruments. Do they achieve the objectives for which they were designed, completely or partially? In monitoring their performance modifications can come to light by which their effectiveness can be improved; for example, schemes to encourage the amalgamation of small farms into viable units by offering financial incentives were soon found to contain conditions on the resale of land which proved unacceptably restrictive to potential amalgamators, and the length of time before which parts of an amalgamated farm could be resold (without repaying grants) was progressively shortened in an attempt to promote amalgamations. In the event, this whole scheme proved ineffective, and it was eventually withdrawn.[6]

The increased knowledge which comes from monitoring can change basic beliefs about reality on which the policy itself is based. For example, the commonly held view that more capital goods on farms leads to greater productivity receives a hard knock if monitoring the performance of farms which have received grant-aid towards new buildings fails to reveal the looked-for productivity gains. In order that performance can be assessed, however, not only is monitoring necessary (and important instances can be given where this has been lacking) but also the objectives of policy must be clearly set out. The failure to specify what major measures are supposed to achieve in anything other than broad, virtually untestable generalities is a characteristic not only of much UK agricultural policy but of the European Community's Common Agricultural Policy as well.

## 14.3 Is There a UK or EC Agricultural Policy?

In the UK there is no *one* integrated policy relating to agriculture. First, as was noted earlier, agriculture is affected not only by policies aimed directly at it but also by general policies (on income and wealth distribution, inflation, level of economic activity, competition, recreation, the environment, etc.). As was demonstrated earlier, the objectives of these non-agricultural policies can conflict with agricultural objectives.

Second, agricultural policy has not been designed as a whole. Rather, policy has evolved as a series of *ad hoc* measures introduced to solve specific problems as and when they arose. While care was no doubt taken by civil servants to avoid the most glaring inconsistencies when drawing up new measures, there is always the danger of a ridiculous outcome. For example, for a time in the 1970s there was in the UK a scheme which encouraged dairy farmers to cease production (as part of an attempt to lower milk supplies) yet their neighbours could expand their herds with the help of capital grants for new dairying buildings. Conflicts can (and do) therefore exist between individual policies.

Third, since the UK joined the European Economic Community and adopted the Common Agricultural Policy British farming has been influenced by both a set of Community-wide policies and measures and by a set of national (UK) measures. Dual-level policymaking is another potential source of inconsistency with Community and UK policymakers setting rather different objectives, although within EC policy regulations there is usually some room for interpretation and implementation according to national needs.[7] Policies appropriate to countries whose agricultures are dominated by small-scale farmers have at times proved inappropriate when applied in the UK; for example, an EC scheme to assist milk producers to leave the industry[8] was designed primarily for the small Continental farmer but was seized on profitably by the large UK producer.

## 14.4  The Historical Development of Policy

History and policy are impossible to separate. A more detailed chronological account of policy history is given in an Appendix to this chapter. Here we need only describe the broad pattern of development since the mid-nineteenth century in order to set the present policy in context.

Following the repeal of the Corn Laws (1846) the government policy towards agriculture was largely one of *laissez-faire*, allowing UK farming to compete openly with food supplied from other countries. The third quarter of the nineteenth century was a period of general agricultural prosperity – Lord Ernle has described it as the 'Golden Age' of agriculture – with a strong demand for food from a rising and prosperous population, improved transport by road and, especially, railways, and heavy investment in field drainage and buildings, many of which are still in use. However, from the late 1870s onwards to the First World War agriculture fell into acute depression, largely the result of severe competition in cereals coming from the opening up of Canada and the USA by railroads coupled with better shipping; soon after, refrigerated transport permitted the importing of meat and dairy products, especially from Australasia. In the UK there was no attempt to protect farming from this competition, as food imports permitted the growing population to be fed cheaply, keeping wage costs down, and encouraged UK exports of manufactured goods. While livestock farmers benefited from cheaper cereal feed, and the output of cattle, pigs, horses, fruit and vegetables all rose, overall agricultural income fell.

Abroad, Germany and France responded differently; with a much higher proportion of their populations still in agriculture they chose to protect their farmers by laws restricting food imports from the new low-cost sources. This chosen path carried implications much later for the problems of integrating UK agriculture into the CAP dominated by these two countries. Holland, Denmark and Belgium responded by a switch from arable to specialist livestock production, exporting to Germany and the UK.[9]

The First World War provided an interruption to this depression, the sea war resulting in reduced food supplies, rationing and price control. In an attempt to expand UK output (half the nation's food was imported) a campaign to plough up more land was launched. Late in the war (1917) the Corn Production Act introduced guaranteed prices and deficiency payments for wheat and oats to encourage expansion, with the intention of maintaining the system after the war as a way of protecting supplies in the climate of a continued shipping shortage. However by 1921 falling world market prices for these cereals made the scheme unacceptably expensive and it was dropped, an action frequently labelled the 'Great Betrayal'. The 1920s were marked by further depression, heightened towards the end of the decade by a slide in cereal prices associated with the rapid spread of the tractor in North America which cheapened field operations and released land previously used to grow food for horses.

From 1929 into the 1930s the general worldwide depression led to countries protecting their own economies and a rundown of trade. Obligations to members of the British Empire limited the UK's ability to protect its farmers. Perhaps the most notable measures taken were the Marketing Acts of 1931 and 1933 which enabled the establishment of Marketing Boards for Milk, Potatoes, Hops, Pigs and Bacon (see also Chapter 3). Later in the 1930s (1936–9) developments were dominated by the imminence of another war, although explanations offered at the time for measures such as the ploughing-up grant, subsidies on liming or acreage payments for barley and oats were related to combating the remaining causes of poverty and of restoring the physical condition of the soil.

During the Second World War UK agriculture became virtually a nationalised industry, or at least a ward of the state, with the objective of expanding and altering output in order to maintain the calorific value of the national diet and to provide a nearly equal distribution of food by allocation and rationing. UK prices for farm produce were set by the government in line with their priorities, and inputs such as animal feeding stuffs and barbed wire were rationed. Provisions existed by which inefficient farmers could be displaced. All food had to be sold to the Ministry of Food, the state monopsony buyer which was (until 1955) separate from the Ministry of Agriculture. During this period it became customary to conduct an annual review of the conditions and prospects of agriculture as part of the planning procedure, including a calculation of the industry's aggregate income; and in order to assess the progress of agriculture many statistics were collected.

In the postwar period the continued expansion of agriculture was vital both to food supplies and to the nation's trade position. The 1947 Agriculture Act permitted many of the war-time controls and incentives to continue in peace-time; it also contained a description of agricultural policy objectives (described below) which laid the foundation for the next twenty-five years. As farm output responded to higher prices and input subsidies the role of the market was gradually restored; all foods were de-rationed by 1955 and fixed prices for farm products were replaced by guaranteed prices and deficiency payments. (1952) The policy of agricultural expansion at *any* cost was succeeded by a more selective one.

The late 1950s and 1960s were marked by continued agricultural expansion, broadly welcomed particularly as a way of assisting the country's balance of payments, and by governmental attempts to limit the cost to the exchequer of agricultural support. The chief measure used was the guaranteed price and deficiency payment; the mechanics of this measure were discussed in Chapter 13. Basically, under this system the government agreed to make up to farmers, using money raised from taxpayers, the difference between the price farmers received from selling their produce on the open market and a pre-set guaranteed price. Imported agricultural products were generally allowed into the country without restriction if they could be bought more cheaply than UK produced foods, to the benefit of consumers and thereby encouraging international trade. However this meant that a fall in the price of foreign food could result in an increased support bill for the government, and this would also happen if farmers expanded their output, depressing prices on the domestic market and increasing the size of the deficiency payment. Various ploys were used, such as limiting the total value of support for a commodity to a specific (standard) quantity, going beyond which diluted the guaranteed price; switching spending to more predictable grants on capital and thereby helping farm incomes through promoting productivity; setting minimum import prices and encouraging effective marketing.

From 1973 the UK moved towards the CAP, adopting both its policy objectives (see below) and its system of support. This involved a shift in the burden of product price support away from the taxpayer (who financed deficiency payments) towards the consumer who gave farmers direct support in the form of higher market prices. These were (and are) engineered by a combination of taxes on imports, minimum import prices, intervention buying by Community agencies, export subsidies and some quota arrangements (see Chapter 13). These forms of expenditure are financed out of the Guarantee Section of the Community's agricultural fund (known as FEOGA, the acronym of its French title, the Fonds Européen d'Orientation et de Garantie Agricole). In addition there are subsidies on capital expenditure and other spending designed to modernise the structure of agriculture and assist the elderly or occupiers of unviably-small farms to leave agriculture, financed from the Guidance Section of FEOGA. Although this latter approach aims to provide long-term solutions to problems in the farm sector, its spending is only about one-twentieth of that by the Guarantee Section. To a major extent the two are counter-active in that product price support enables farms which are fundamentally not viable at market prices which would bring supply and demand of farm products into balance within the Community to keep in production and relieves them of some of the pressure to leave farming. Price support, although offering a short-term solution to declining farm incomes, exacerbates the long-term problem of adapting the structure of agriculture to the changing technology of farming when faced by a demand which is virtually static.

In this most recent policy phase UK (and the rest of Community) agricultural output has responded to the level of prices offered to farmers to an embarrassingly great extent, generating problems of costs of support buying, and storage and disposal of surpluses. Additionally, in the UK in particular attention has been focused on the environmental damage done as a by-product of producing high levels of output which no one in the Community is willing to consume at the prices set and which can only be exported if heavily subsidised. The Commission of the EC has declared that the existing pattern of support cannot continue because of the prospects of an ever-increasing burden of maintaining an agriculture which is fundamentally out of equilibrium with its market. It has outlined various policy options (see the Appendix to this Chapter) all of which include lowering product prices but as yet no clear avenue of progress is apparent.[10] Judged by earlier attempts to change the direction of policy, the reform as implemented after political pressures have been brought to bear is not likely to be radical.

## 14.5 The Stated Objectives of UK and EC Agricultural Policy

From what has been stated earlier, it is evident that the objectives of agricultural policy are many, diffuse and constantly evolving. The present set of objectives, in so far as they are described in UK government and Community statements, stem from two sources, the UK's Agriculture Act of 1947 which was to guide the development of the industry in the country's postwar recovery and beyond, and Article 39 of the Treaty of Rome (1957) which performed a similar role in shaping the European Economic Community's Common Agricultural Policy. When the UK joined the EC in

1973 it came increasingly under the influence of the CAP, with many of its national policy measures being replaced by CAP ones, but there remain major areas of policy in which the UK still operates on a national basis. For example, certain capital grant schemes, the MAFF's Advisory Service, agricultural education, and Marketing Boards and Authorities are essentially UK controlled and financed, although there are Community imposed limits on the extent of some of these activities: national grant schemes, for example, must carry lower rates of grant than equivalent Community schemes. These two pieces of legislation are the fundamentals from which the present set of policy objectives have evolved.

### 14.5.1 The 1947 Agriculture Act and the 1957 Treaty of Rome

The stated objective of agricultural policy in the UK 1947 Agriculture Act was that of 'promoting and maintaining. . .a stable and efficient agricultural industry capable of providing such part of the nation's food and other agricultural produce as in the national interest it is desirable to produce in the United Kingdom, and of producing it at minimum prices consistently with proper remuneration and living conditions for farmers and workers in agriculture and an adequate return on capital invested in the industry.'

This masterly piece of drafting, with its obvious delight in the unspecific and untestable ('proper' remuneration, 'adequate' return and 'in the national interest') and its inner contradictions proved itself capable of a wide range of interpretations. However, it bears a close resemblance to both the terminology and the content of the stated objectives of the CAP, as given in Article 39 of the Treaty of Rome.

The Common Agricultural Policy shall have as its objectives:
a) To increase agricultural productivity by promoting technical progress and by ensuring the rational development of agricultural production and the optimum utilisation of the factors of production, particularly labour.
b) Thus to ensure a fair standard of living for the agricultural community, particularly by increasing the individual earnings of persons engaged in agriculture.
c) To stabilise markets.
d) To ensure the availability of supplies.
e) To ensure that supplies reach consumers at reasonable prices.

While the Treaty is elsewhere also concerned with the harmonious development of trade and the protection of the environment, these items do not feature among this most discussed list of agricultural objectives.[11] One important feature of the list is the linkage (made by the word 'thus', sometimes translated 'thereby') between the first two objectives; the assurance of a 'fair' standard of living was envisaged as being obtainable through an improved agricultural productivity, especially flowing from an 'optimum' use of labour. While again this does enable a wide range of interpretations, the linkage suggests that the income assurance is not absolute but is conditional on productivity rises.

Both of these statements can only be understood in their historical context. The practice of policy has evolved from them, and it is interesting to note the difference in flavour of the latest statement of policy objectives by the UK's MAFF, as given in 1984 (Table 14.1). While the emphasis on marketing is not surprising, a new flavour is given by paragraphs 5 and 6 which refer to environmental protection and animal welfare and the recognition that agriculture is only one part of the rural economy. The none-too-clear wording of paragraph 6 can also be interpreted as stating that agriculture is now recognised as being able to contribute to wider economic and social objectives, within which might be included the public support of agriculture in upland areas not for farming reasons but for the conservation of countryside for purposes of recreation and wildlife protection. This new area of policy objectives seems to have been only recently adopted by MAFF in response to the great environmentalist pressure of the early 1980s.

## 14.6  Policy As It Has Been Practised

Broad statements about policy objectives are only a guide to what forms policy actually takes. In practice the postwar period has seen a number of interwoven strands, concerned with the following six topics.

### 14.6.1  Securing a reliable food supply

Of obvious importance during the Second World War, a secure food supply was a prime reason for encouraging the expansion of agriculture in the immediate postwar years. An unfortunate combination of economic and natural disasters caused the British food supply per head to be lower in 1947 than in the war years. However the food supply position has now greatly altered; not only has the degree of self-sufficiency of the UK risen to record levels (see Chapter 4) but the Community is oversufficient in many products and has become a significant agricultural exporter. In consequence, little is heard of this area of concern at the moment.

---

**Table 14.1**
**Stated Policy Aims of the UK MAFF, 1984**

---

1. To foster an efficient and competitive agriculture industry,
   particularly through

     the provision and sponsorship of research,
          development and advisory service;
     the provision of financial support where appropriate;
     measures to control disease, pests and pollution;
     better marketing of its products.

2. To encourage improvements in the Common Agricultural Policy
   (CAP) which will lead to greater economic rationality and full
   recognition of UK interests; and within it to seek fair conditions of
   competition for UK producers and traders.
3. To encourage the development of a viable, efficient and market
   orientated fisheries industry within the framework of a Common
   Fisheries Policy (CFP) which pays due regard to the protection of
   renewable fish resources and other UK interests.
4. To encourage improvements in the processing, distribution and
   marketing of food; to ensure high standards of good quality, hygiene
   and safety; and to safeguard essential food supplies in times of
   emergency.
5. To encourage the farming industry to adopt high standards on animal
   welfare, and on consumer, environmental, wildlife and countryside
   protection with, where appropriate, the necessary legislative controls;
   and to assist the industry in contributing to the rural economy.
6. Generally, to assist the agriculture, fisheries and food industries to
   meet the demands of consumers in the United Kingdom and abroad
   and contribute to wider economic and social objectives.
7. To manage the Ministry's human and financial resources so as
   constantly to achieve the above aims at least cost and with good staff
   relations.
8. To keep these aims under review in the light of changing
   circumstances.

---

SOURCE    MAFF (1984) *Ministerial Information in MAFF (MINIM)* (Alnwick: MAFF).

### 14.6.2    *Trade*

Although there have been times when UK agriculture has been allowed to
decline in order to stimulate trade and the exports of other sectors of the

economy, since the Second World War the emphasis has fallen on the positive contribution which farmers can make in the short and medium term to the balance of payments by expansion. Little concern has been shown over the effects of this growth on the country's overall long term pattern of trade. The policy of general expansion with strong financial incentives was replaced in 1952 by one of selective growth. Since the late 1950s various official reports and White Papers (see the Appendix to this chapter) have called for or welcomed a rise in output, although to the disappointment of the farming community the means favoured to achieve this has been the encouragement of productivity improvements rather than financial inducements. Dispute has frequently raged over the size of any benefit to come from expanding home agriculture to displace imports; estimates are difficult because of the uncertainty about the longer-term effect on UK exports of other goods and on the prices of imports and exports. These arguments have been rather left behind by the fact that the UK now operates as part of the EC and its CAP. The rules governing each member state's contribution to the Community's Budget may make a continued expansion attractive in the UK for narrow accounting reasons, even though the Community market is in surplus. As recently as the early 1980s the UK Minister of Agriculture was still encouraging more milk output; the rapid imposition of dairy quotas in 1984 put an end to further UK growth in that commodity. Though UK self-sufficiency in food has risen significantly since joining the EC, this has not come from a deliberate policy on influencing the pattern of trade. Rather, it has been an unplanned consequence of the level of domestic support. It is clear that the current overall level of agricultural output is greater than would result under freer trade and the balance between crop and livestock production is distorted, although the cost to the country of these effects is difficult to gauge.

### 14.6.3   Food prices to consumers

Consumer food prices are mentioned specifically in both the 1947 Agriculture Act ('minimum' prices consistent with other objectives) and by the Treaty of Rome ('reasonable' prices for consumers) yet, of the strands of agricultural policy, concern with prices ranks low. One major advantage of the UK support system in use before entry to the EC was that it allowed food to be imported if it could be bought more cheaply abroad than from UK sources. The farming lobby pointed to 'cheap food' as a benefit of agricultural support by the taxpayer, a dubious claim, but at least those who were too poor to be taxpayers did not find themselves contributing to the support cost. The system was consonant with policies of income transfer to the poor and with anti-inflationary measures.

Under the CAP system in which prices are engineered upwards so that farmers receive a greater proportion of support through the market, food prices are undoubtedly higher. However they have never become a politically sensitive issue in the UK. Perhaps this can be explained by: a) the difficulty of banding together into an effective pressure group a large number of relatively poor consumers who individually might only gain marginally from lower prices; in contrast, a small number of relatively well-off farmers who individually will benefit greatly from more agricultural support can more easily be formed into an effective group; b) the low and declining proportion of incomes spent on food, with the richer, more influential members of society spending even less on food than average (see Chapter 3); and c) the effectiveness of the NFU in putting the farmers' case and the pro-farming stance of MAFF.

### 14.6.4　The marketing structure of agriculture

The marketing of farm products was a matter of great concern in the 1930s, with the belief that the individual farmer was greatly disadvantaged compared with the large firms to which he often sold his output. For example, in the early 1930s some 120 000 individual milk producers dealt with a milk distribution and processing industry dominated by only three large firms, and there was suspicion that these had used their monopsony power to force down liquid milk prices to near the world price for manufacturing. The main governmental activity has been the establishment and support of Marketing Boards, Authorities and Commissions (the latter not normally trading in their respective commodities) within functions which include the provision of market intelligence and research into marketing. (See Chapter 3) In addition the government has encouraged the establishment of farmer-cooperatives for both buying inputs and selling produce (through advice, grants and technical assistance from the Central Council for Agricultural and Horticultural Cooperation, set up under the 1967 Agriculture Act). There have been spasmodic attempts at promoting British foods on the home market, organised primarily by their respective Boards or Authorities; most recently the UK government has supported an across-the-board campaign (called 'Food From Britain' and launched 1982) both in the UK and in foreign markets. However, the marketing strand of policy cannot be expected to have much impact on the shape or size of agriculture up to the farm-gate stage because of the limited opportunities for market expansion; not only do the low demand elasticities pose a problem but, when the degree of self-sufficiency has already reached high levels, there is very limited room for import substitution (see also Chapter 3).

### 14.6.5　Improvements in productivity and efficiency

The terms 'improvement in productivity' and 'improvement in efficiency' are used interchangeably in the policy context to imply an increase in output from given resources or a reduction in the quantity of inputs required to generate a given level of output. Concern has come in two forms: first, that a more efficient agriculture could reduce the amount of food which is imported or release resources which could be used for development in other sectors of the economy; and, second, that greater productivity would enable the incomes of farmers to be higher. The Treaty of Rome specifically refered to this latter link as the way by which improvements in living standards were to be achieved. The UK has promoted improvements in productivity through establishing MAFF's advisory service arm (now called the Agricultural Development and Advisory Service: ADAS), through the Agricultural Training Board (1966) and the publicly-financed agricultural research and education institutions. A major function of the Economic Development Committee for Agriculture (part of the National Economic Development Council organisation) has been concerned with studying productivity. One controversial way of attempting to achieve productivity improvements has been by the encouragement of capital investment, principally by grant aid. A succession of variously named schemes (Farm Improvement Scheme, Farm Capital Grant Scheme, Farm and Horticulture Development Scheme, Agriculture and Horticulture Grant Scheme, and others) coupled with tax concessions have accompanied the continued build-up of capital. Since accession to the EC, Community schemes have replaced or augmented the UK array of aids. Although earlier instances exist, this encouragement stemmed largely from the 1957 Agricultural Act's shift of policy towards capital grants; these aids to investment were expected to help UK agriculture become more competitive and so reduce the costs of price support. As was described earlier, the extra capital has been associated with a major fall in the number of people working in agriculture and a modest reduction in land used. Partial productivity measures (such as output per man) have shown marked increases, but overall rises in productivity, after allowing for the extra capital, are less dramatic (see Chapter 5). If displaced labour is simply added to the pool of unemployment, then the overall benefit of the extra capital is even less clearcut.

### 14.6.6  Farm incomes

Concern with farm incomes has formed the most significant strand of agricultural policy, though the income objectives have never been precisely articulated. It was a major influence on the government changing from its *laissez-faire* policy in the 1930s. During the Second World War, and for more than a decade after, the government compensated farmers (as a whole) for any cost increases the industry suffered by raising the value of the agricultural guarantee. From the mid-1950s to the present the freedom of policymakers to lower prices of farm produce, in order to curtail

382 Economics for Agriculture

unwanted increases in output, has been constrained by the implications for such actions on farm incomes. In the late 1950s and 1960s this took the form of written guarantees (in the 1957 Agriculture Act) that prices would not be reduced by more than given amounts, which meant that in subsequent Annual Review negotiations MAFF officials went in with some desirable options already barred. In the late 1970s and 1980s the price setting mechanism of the CAP has worked in such a way that price reductions for commodities running into surplus have not occurred early enough or by a sufficient amount, the underlying reason being the political unacceptability of being seen to be reducing the income of farmers.

In view of the importance of farm incomes to understanding this industry, its changes and policy towards it, a complete chapter has already been devoted to them (Chapter 12). Here only a brief reiteration of the main points will be made. Concern with incomes forms three substrands: with stability from year to year; with poverty among farmers of certain sizes, types or areas (generally combinations of these three); and with the relative earnings of those in agriculture compared with other sectors. Income instability arises primarily from farm output instability arising from weather and price variation outside the individual's control. A good case can be made on grounds of production efficiency for public intervention to dampen such instability. Examples include the Potato Marketing Board's programme of buying the crop when prices are very low and barring it from the market for human consumption by dying and selling for cattle feed. The CAP's intervention agencies for cereals or milk products can also be used to stabilise markets by acting as buffer stocks, buying or selling according to year-to-year fluctuations.

Poverty among farmers is essentially a social matter rather than an agricultural one. There may be explanations based on the economics of agriculture for low incomes, but this does not turn farm poverty into an agricultural issue. It is because some people satisfy some chosen criterion of poverty that concern is expressed, not because they are producers of food materials. The setting of the criteria of poverty is a political decision, reflecting social and ethical values. As was seen in the discussion of incomes (Chapter 12), money income is only one of a range of possible poverty parameters.

Poor comparability of reward in farming, where it occurs, has both a social and an economic connotation. If labour is trapped in farming by factors such as housing, lack of suitable training, etc., this may produce low marginal labour productivity which will be reflected in low earnings. To improve the national use of resources an economist would advocate measures which lubricate the transfer of farm labour to other uses in accordance with the principle of equi-marginal returns. This would tend to raise the incomes both of those transferred and of those remaining in farming. However, a claim that farmer earnings are somehow 'unfairly' low in comparison with those of other groups and thereby justify income

support, is essentially one based on social values and a normative judgement of what is fair. Because some individuals who are deemed to be
unfairly rewarded will also be deemed poor it is difficult to separate the two
issues.

If it is decided that incomes of farmers are to be raised on social grounds,
then it is important to know who bears the cost of this income transfer. For
example, a tax on bread would bear more heavily on the poorer members
of society at large than on the richer. And the means by which assistance is
offered is important; one criticism of the UK's chosen policy instruments is
that they benefit the rich farmer much more than the poor, and that overall
the system of agricultural support has become one of income transfer from
the relatively *poor* consumer to the relatively *well-off* farmer.

## 14.7 Methods of Income Support

The most obvious method of solving problems of low incomes is by **direct
income supplements** to individuals. This is, after all, the way that is used for
the rest of society (Family Income Supplement, Supplementary Benefit,
etc.). Yet when it comes to farmers, for a mixture of practical, historical
and political reasons these have generally been avoided. The system of
agricultural support was set up in wartime principally to encourage output
expansion, but also worked to the benefit of farmer incomes. Product price
support then had a head-start as a system for influencing incomes. At that
time UK farmers did not generally keep accounts, a situation that still
exists in many important Continental countries, so that assessing incomes
by direct measurement has been and still is impractical in some EC
member states, made worse by the delays in measuring income from
self-employment. Farmers and their unions have traditionally opposed
subsidies which were overt welfare payments (even though much of the
support system had the same effect) and administrators have steered away
from direct income supplements as being politically unacceptable. Proposals for direct income supplements applicable generally in place of price
support have been most recently rejected in 1985 by the EC because of
political resistance and a lack of reliable information on the numbers of
farmers who might qualify for assistance; pensions for those approaching
retirement age have not been ruled out as a way of quickening their exit
from farming. Despite the farming antagonism towards explicit income
aids, the UK has for quite a while used some payments which are thinly-
veiled direct income supplements in certain areas (the Hill Livestock
Compensatory Allowance under a 1975 EC directive), and recent policy
statements from the NFU (1984) have been much less hostile to such
payments, especially if coupled with environmentally-desired activities.[12]

Another method of directly enhancing the incomes of farm families is the development of **non-farming gainful activities**, or in more familiar language, encouraging part-time farming. It has already been noted (Chapter 9) that a substantial proportion of small farms, and some larger ones, are already operated by farmers and spouses who have some other source of earned income. Many of the small farms therefore are occupied by farmers with very satisfactory total incomes. Until the 1970s part-time farmers in the UK tended to be considered outside the interest of policy-makers. However in the 1970s the awareness of the widespread nature of part-time farming in the rest of the European Community and its general growth has raised the profile of non-farming activities as a way of supplementing inadequate farm incomes. Policy-makers have adopted an attitude of mild encouragement. MAFF advice is available on catering for farm-based recreation and tourism, but this is only appropriate in the more scenically attractive areas of the UK and tends to be developed primarily by the more adventurous bigger farmers who have both the capital and necessary managerial ability. Of more significance to the small low-income farm would be the creation of alternative off-farm employment (perhaps in rural factories). Even though age and training may count against many small farmers or their spouses being able to take such jobs, younger family members may embark on off-farm employment which in the long term could make the farm viable as a supplementary income source and allow the family to remain in occupation. Overall, however, part-time farming seems, in the UK, to have only a limited role in assisting a farm income problem.

In the long term **structural policies** can assist the income problem. By reducing the number of small farms the number of low-income families is reduced and, all things being equal, the income per head of those remaining in farming is raised. The UK has a long history of such policies, but initially they were designed to increase rather than to reduce the number of small farms. In 1892 the Smallholdings Act empowered County Councils to create smallholdings if the demand existed. The objectives were in major part social: to relieve the depressed conditions of agricultural (hired) labour and to reduce the tendency of the small farmer to disappear, and with him the bottom rung of the 'farming ladder'. The smallholding movement was expanded greatly in the period after the First World War, again for social reasons. Active policy in the opposite direction started in 1967 with grants for amalgamation of small farms and payments to out-goers, and these national structural measures were subsequently replaced by similar Community ones. However, overall they must be considered ineffectual. Probably of greater significance to structural change have been the capital taxation concessions available to retiring farmers and the fact that larger expanding farmers have been willing to pay high prices for small acreages. In the opposite direction, the growth of part-time farming has enabled many small farms to continue independently, but of course they do not then constitute the only income of the occupiers.

**Cost-reducing measures** have frequently been used as a way of supporting farm incomes. Public finance of the advisory service and of agricultural training and education falls into this category. More usually the idea of cost reducing calls to mind payments such as a fertiliser subsidy (used in the immediate postwar period but not now paid) or grant on capital items such as buildings, drainage or machinery. While on first examination an input subsidy might be expected to lower costs and therefore raise incomes, the ultimate effect is far from predictable. A subsidy on an input will increase the demand for the input and, depending on the elasticity of supply, its price is likely to raise. This may partially or completely nullify the subsidy (see Chapter 13) Grant aid of drainage and land reclamation has been blamed for many changes which farmers have made to the countryside's appearance. Reducing costs of inputs will also tend to cause more to be used and hence farm output to rise; more fertiliser will be used if it is cheaper. In turn, higher farm output may depress market prices to the extent that revenue falls and hence farmers' incomes might decline. An input subsidy might therefore need to be accompanied by a restriction on farm output, if income is to benefit. And there will be distributional effects: capital grants may benefit the larger high-income farmers but may not be taken up by the small low-income producer who is the real target of income support. He may actually suffer if prices fall as the result of total output rising.

By far the most widely used method of income support in the European Community, *and* the most expensive in terms of overall direct costs, is **product price support**. Various forms are encountered; a payment (or bounty or premium or allowance) per unit produced, which may be fixed in advance, or varying in retrospect such as the UK deficiency payment system, both of which give higher prices to producers but do not directly alter the market price for farm produce; or a raising of market price by restricting the supply which is available for sale, by quotas on home production or quotas or taxes on imports from abroad; or by raising market price by buying into intervention store. These methods have already been analysed in Chapter 13. The implications are that, first, greatest benefit goes to those farmers who have the greatest output and who need income support least. Thus as a way of assisting the small low-output, low-income farmer, market price support is inefficient. Secondly, higher prices induce higher output from agriculture as a whole, and this can result in the appearance of unwanted surpluses, as has happened with the CAP, which are expensive to dispose of on the world market (requiring export subsidies to make them competitive) and politically embarrassing. Government ministers do not welcome the publicity which surrounds sales of subsidised butter to unfriendly nations at prices lower than home consumers must pay, nor the spectacle of food in stores allowed to degenerate while there are members of society who are poor and ill-nourished. In addition, the expanded agriculture in the UK has been blamed for many environmentally undesirable land-use changes. And an enlarged agriculture which displaces imports is from a trading standpoint by no means entirely in the national interest. Third, higher prices slow down the pressure on producers to become more efficient. Fourth, the fundamental and inevitable structural adjustments which economic development requires, such as the phasing out of farms too small to provide a livelihood, are slowed down, damming up for later bigger changes and hindering those other government policies designed to assist these structural changes (see Chapter 11).

## 14.8 New Strands of Agricultural Policy

As circumstances change the various strands of policy assume different levels of importance. Some strands virtually disappear, such as the concern with food self-sufficiency while others, such as the farm-income question, remain strong. Without doubt the most significant new strand to emerge in the postwar period has been the concern with farming's effect on the **environment**, especially on **wildlife** and the **countryside's appearance** and the ways in which these changes can be controlled. This strand has arisen not from the traditional origins of food and farmer lobbying (in the guise of MAFF, NFU or similar members of the agricultural establishment) but from environmental organisations whose memberships grew rapidly over the 1970s (see Table 14.2).

Key events associated with this rise in interest seem to have been, first, the outbreak of Dutch elm disease in 1974, which focused attention on changes in the appearance of the countryside in general and not just on elm trees. Second, changes in farming systems which eroded the traditional nature of Exmoor led to a well-publicised Committee of Enquiry, chaired by Lord Porchester. Third, a series of reports on the relationships between farming and the countryside culminating in a polemical book, *The Theft of the Countryside*, by Marion Shoard, published in 1980. Since then a shoal of texts have appeared almost universally severely critical of the 'productivist' path taken by agricultural policy. In 1984 both the NFU[13] and CLA[14] published policy documents which accepted the desirability of an environmentally-sensitive agricultural policy, and, as has already been seen, the 1984 official statement of MAFF objectives contained a major reorientation towards the environment. Fourth, the budgetary problems of the European Community in the early 1980s threw the costs of the CAP into great prominence.

Environmentalists have pointed to the illogicality of using large funds from the Community Budget to encourage unwanted output, causing environmental degradation in the process, and at the same time trying to put right some of this with other public spending to encourage environmentally-desirable practices by farmers. The 1981 Wildlife and

**Table 14.2**
**Membership of Some Key Farming and Environmental Organisations**

|  | 1970* | 1982* |
|---|---|---|
|  | ('000s) | |
| National Farmers Union | 200 | 140 |
| Country Landowners Association | 42.8 | 48 |
| National Trust | 226 | 1200 |
| Council for the Protection of Rural England | 21 | 30 |
| Ramblers Association | 22 | 28 |
| Royal Society for the Protection of Birds | 52 | 350 |
| RSPB Young Ornithologists | 16 | 100 |
| Friends of the Earth | 1 | 18 |

NOTE    * Nearest available year where otherwise not available.
SOURCES  Annual reports and Directory of British Associations quoted in Leonard, P. (1984)
           'A review of environmental pressures facing agriculture'. Paper to Agricultural
           Economics Conference 'UK agriculture and the environment – the economist's
           approach'.

Countryside Act enabled payments to be made to farmers in environmentally sensitive areas (such as the Somerset levels) who agree to manage their land in prescribed ways which protect the natural features. The payment is intended to compensate them for the extra income they are foregoing, although critics of such a policy point out that much of this compensation would be unnecessary but for other policies which jack up prices and subsidise inputs.

In 1985 the UK was successful in persuading other EC member states that spending on certain environmental protection measures could be financed under CAP structural policy schemes, in much the same way that new buildings were grant aided under EC's modernisation of farms structural scheme. It is too early to judge the impact of this change. However, some see land use planning as the only sure way of society protecting the natural features which it thinks valuable, with permission being required to plough established grass or to apply heavy uses of fertiliser to 'unimproved' pasture. As was seen in the chapter on Land (Chapter 6) effective restrictions on land use apply on some 17 per cent of agricultural land in Great Britain, with nature conservation having an unexpected ally in the Ministry of Defence with its hold over large tracts for military training purposes, undisturbed by intensive farming or public access.

Other contemporary strands of a lesser prominence include **animal welfare** which has curbed activities of intensive livestock producers, the strong pressure coming from a small sector of the public being reflected in the latest stated aims of MAFF, and **dietary quality**. Despite its title, MAFF's area of concern has not extended to nutritional quality of the nation's food. As was seen in Chapter 2, pressure to change the composition of diet has come from other sources and has had virtually no direct effect on the shape of agricultural policy *per se*.

The final strand of policy concerns the role of agriculture in **rural development**, that is, as a generator of income, jobs and markets in the countryside. The temptation is to assume that farming is a dominant influence on levels of economic activity in rural areas but, as will be seen in Chapter 15, this is far from the truth in most of the UK. Nevertheless the notion of supporting the rural economy through supporting farming has suited well the case of those who demanded support for farmers on grounds of income, food supply or other reasons. While the argument has not been pressed too hard in lowland areas, in the uplands agricultural policy has these rural economy aspects firmly in the forefront of its declared objectives. Special payments are available to farmers in Less Favoured Areas (recently changed in name to Severely Disadvantaged Areas) in the form of headage payments on livestock and enhanced rates of capital grant. They are also eligible for special grants for the provision of tourist facilities and for carrying out environmentally desirable practices. The reasons for this assistance are almost entirely social and environmental, particularly to prevent large-scale depopulation of farming and rural areas. While not being new in concept, the use of farming in an instrumental role for other purposes within the countryside has only recently gained acceptance among agricultural policymakers in the UK, and that only guardedly.

## 14.9 The National Interest

It will be evident from this chapter that agricultural policy in the UK and EC is in a highly unsatisfactory state. Objectives are not well defined and approaches have not been flexible to change economic conditions. In particular, the main reason for continued support of agriculture seems to be concern over the income position of farmers, yet there is little serious attempt to measure the incomes of farmers in a way appropriate to the declared objectives of policy (see also Chapter 12) or to tackle the problem of rural poverty where it is found by direct means. Instead, product prices are raised in an indirect attempt to influence incomes, with the result that production greatly exceeds demand, causing disposal problems, and a distribution of the benefit from support which tends to by-pass those farmers in greatest need. The general impression is of a policy mix more suited to the 1950s than the 1980s.

The reason why this continues is complex. At the national level, British farmers may have a vested interest in not making their income position too clear since this might lead to a reduction in support from public funds, calls for the rating of land and the removal of capital tax concessions.

At the level of the CAP the UK government is unlikely to press for reforms which will result in British farmers being put at a disadvantage. A switch from product price support to direct income supplements, for example, would very likely result in this happening. Similar consideration apply in other countries on other issues and the process of reform becomes one of trading narrow political interests.

This chapter has been concerned predominantly with public policy as it affects the farming industry. However it is realised increasingly that agriculture is only one of a number of rural land users, and that it forms only a small and declining sector of the rural economy. Consequently, a much broader view needs to be taken if problems such as unemployment or low incomes in rural areas generally are to be tackled. Policies for the whole rural economy, while involving agriculture, clearly need to see farming in perspective. This perspective is the subject of the following chapter.

# Notes

1. Josling, T. (1974) 'Agricultural policies in developed countries: a review', *J. Agric. Econ.* **35**:3, 259–64.
2. Farming in countries controlled by communist parties is predominantly arranged as large-scale collective or state farms with a small proportion of total area used as private plots, the main exceptions being Poland and Yugoslavia in which private farms dominate. The performance of these institutions has been disappointing; one way of improving this has been to introduce greater independence from the state planning system and, at the personal level, economic incentives so that greater output is reflected directly in higher standards of living. In China since 1981 agricultural land has been allotted to individual households which behave very like tenant farmers, with the commune acting as landlord.
3. In the interwar years 10 MPs were farmers or had substantial farming interests; in 1940, 16 MPs; in 1950, 25 MPs; in 1951, 33 MPs; in 1956, 36 MPs; in 1959–64, 56 MPs, plus 16 landowners and 7 with ancillary interests; in 1970, 90 MPs (79 Conservative, 10 Labour, 1 Liberal) and in 1979, 30 MPs (25 Conservative, 2 Labour, 3 Liberal). In 1983 there were 59 MPs (53 Conservative, 4 Labour, 2 Liberal/SDP), 9 per cent of all MPS. (Figures from Self, P. and Storing, H. J. (1962) *The State and the Farmer* (London: George Allen and Unwin); Wilson, G. K. (1977) *Special Interests and Policy Making* (Chichester: John Wiley); and Howarth, R. W. (1985) *Farming for Farmers?* (London:

Institute of Economic Affairs). For an account of political influence in EC policymaking see Neville-Rolfe, E. (1984) *The Politics of Agriculture in the European Community* (London: Policy Studies Institute).

4. For an account of these and other organisation in the policymaking mix see Rogers, A., Blunden, J. and Curry, N. (eds) (1985) (*The Countryside Handbook* (London: Croom Helm/Open University).

5. For an account of the various measures used in the UK since the Second World War see Bowler, I. (1979) *Government and Agriculture, a Spatial Perspective* (Harlow: Longman). For the various measures adopted by the CAP see EC (1983) *The Agricultural Policy of the European Community* and Revell, B. J. (1985) *EC Structures Policy and UK Agriculture* CAS Study No. 2 (University of Reading). For the way that schemes impinge at the farm level, see Nix, J. (1985) *Farm Management Pocketbook* (Ashford: Wye College).

6. The Farm Amalgamation Scheme lapsed in 1976.

7. European policy is implemented primarily by means of Regulations and Directives. Both carry the force of law. They differ in that the latter allows greater flexibility in the way in which a member state achieves the intended objective.

8. This was implemented in the UK as the Non-Marketing of Milk and Beef Conversion Schemes. These ceased in 1980 and 1981 respectively. A national scheme has more recently been introduced to assist dairy farmers, limited since 1984 by quotas from expansion, to cease production.

9. For an account of the response by agricultures in Europe see Tracy, M. (1983) *Agriculture in Western Europe: Challenge and Response 1880–1980* (London: Granada). For greater detail of the UK see Perry, P. J. (1973) *British Agriculture 1875–1914* (London:Methuen).

10. A detailed description of the legislation of the CAP and the development of its policies is given in Harris, S., Swinbank, A. and Wilkinson, G. (1983) *The Food and Farm Policies of the European Community* (Chichester: Wiley). Also see Hill, B. E. (1984) *The Common Agricultural Policy: Past, Present and Future* (London: Methuen).

11. See note 10 above.

12. NFU (1984) *The Way Forward* (London: NFU).

13. Ibid.

14. Country Landowners' Association Advisory Group on the Integration of Agricultural and Environmental Policies (1984) *Report and Recommendations* (London: CLA).

# Appendix: An Historical Perspective on UK Agricultural Policy

## *The golden age of British agriculture: late 1840s to late 1870s*

1846    ***Repeal of the Corn Laws*** Marked the beginning of almost a century of free trade in agricultural production – An Act of 1815 (only one in a long series of measures attempting to control the trade in cereals) fearing a post-Napoleonic War collapse in grain prices, had allowed grain to be imported duty free, but wheat could only be sold if domestic prices fell to pre-determined levels (wheat 80s/quarter, barley 40s, oats 27s). These prices were in effect *minimum import prices*. In 1828 an Act brought in a sliding scale of import duties dependent on the home price of cereals. However, the effect of protection up to 1846 was marginal. The Corn Laws were incapable of keeping up prices at home when low prices were due to heavy yields and did not keep out imports when prices were high. The Acts were repealed because (a) they were being ineffective; (b) a report by the Board of Trade thought that they raised UK wages and hence reduced our competitiveness abroad and that they caused retaliatory tariffs; (c) a tide of free-trading opinion was in Britain; (d) famine in Ireland; (e) Peel's belief that their repeal would 'discourage the desire for democratic change in the Constitution of the House of Commons' and generally preserve institutions.

Over the years following repeal the price of corn was stable at about 53s/qtr. Agriculture was generally and increasingly prosperous, and for pasture farmers product prices rose steadily and substantially. Railways and better communications benefited farming. Drainage became cheaper, with investment aided by a government loan. Agricultural output rose almost as fast as population, and 80 per cent of UK food was still produced at home in 1868.

## The first 'great depression' of British agriculture: late 1870s to First World War

Grain prices fell as the result of government *laissez-faire* policy and the development of Canada and USA – railroads and shipping improvements. Then meat and dairy produce became available from Australasia. Over the period 1875–1900 imports of grain rose 90 per cent, butter and cheese 110 per cent and wheat 300 per cent. However, the agricultural depression was *not* general as a rising population (10 million increase) presented a rapidly rising demand for pasture products. Thus, while the wheat acreage fell sharply, the output of cattle, pigs, horses, fruit and vegetables all rose. Livestock farmers benefited from cheaper feed. However, on balance, the agricultural income fell.

The Government pursued a 'laissez-faire' policy up to 1917, apart from some legislation on landlord/tenant relationships strengthening the position of tenants, on smallholdings, rural housing and animal importation.

1982 **Smallholding Act.** County Councils were empowered to create smallholdings if a demand existed. The reasoning behind this Act was (a) to relieve the depressed condition of agricultural labour; (b) to reduce the tendency of the small farmer to disappear, threatening the existence of the 'farming ladder' by removing the bottom rungs; i.e., primarily *social* reasons. It was later augmented by the Smallholdings and Allotments Acts of 1907 and 1908. The number of smallholdings was expanded greatly after the First World War by the Land Settlement (Facilities) Act 1919 and continued by the 1926 Smallholdings and Allotments Act and the 1947 Agriculture Act. The 1970 Agriculture Act at last recognised the absence of a farming ladder and enabled the rationalisation of smallholdings.

1896 **Agricultural Rates Act.** Gave partial financial relief to farming.

1902 **Finance Act.** Imposed a slight import duty on cereals and flour.

## The First World War 1914–1918 and the postwar boom

At the start of the First World War half of UK food was imported. The sea war resulted in reduced supplies, rationing and price control. To help generate output there was a ploughing-up campaign.

1917   **Corn Production Act.** Used guaranteed prices and deficiency payments for wheat and oats to encourage output. Government policy was to protect supplies in light of a continued shipping shortage and it kept prices high.

1919   **Land Settlement (Facilities) Act.** Encouraged local authorities to acquire land for smallholdings mainly for ex-servicemen. Also central government set up its own smallholdings scheme (the most notable being the Farm Settlements Estate in Lincolnshire) following the 1915 **Verney Committee Report** which recommended organised land settlement.

1920   An Act made the wartime guaranteed prices permanent. Prices reached a peak with the index for all agricultural products reaching 321, as opposed to 111 in 1914 and 100 in 1906–8.

1921   Market prices for cereals fell, and the Act giving price supports was repealed: The 'Great Betrayal'.

## The 1920s (from 1921 on)

A decade of chronic depression, becoming more acute after 1926 with many bankruptcies. The late 1920s saw a slide in cereal prices brought about by the use of tractors in North America which cheapened field operations and released a significant number of hectares previously used to feed horses. The 1920s policies were indirect in their effects on prices, costs and incomes, with the exception of measures on sugar beet and wages. However, the decade also saw a strengthening of Advisory Services and the completion of a nationwide system of farm institutes.

1922   Permanent Joint Milk Committee set up, a discriminating monopoly with reduced prices if farmers exceeded target output.

1923   **Agricultural Credits Acts.** Set up a system of credit cooperatives. The Government's first venture into agricultural credit, and an immediate and total failure.

1924   Linlithgow Committee on the Distribution and Prices of Agricultural Produce 1924 (whose findings concentrated on costs and ignored or underplayed the commercial and social values of distributors) formed the background to important marketing legislation.

1924–5   Aid was given to arable farmers, more especially to rural unemployment by setting up a British sugar industry with guaranteed prices made possible by the **British Sugar (Subsidy) Act** 1925. As a counter-balance, the Government introduced a scheme for regulating farm workers' wages (**Agricultural Wages Regulation Act** 1924).

1926   **Smallholdings and Allotments Act.** Continued pressure on local authorities to provide smallholdings to meet the demand for them, i.e., social motives were dominant in their creation.

1928   **Agricultural Produce (Grading and Marketing) Act** aimed at stabilising and raising prices with a measure of quality control.

1928–9   Agricultural Mortgage Corporation (England and Wales) and Scottish equivalent established, lending on security of land for its purchase and improvement. Their activities were slow to grow.

1929   In the local government finance reorganisations, farmland and buildings de-rated.

## The Great Depression: 1929 onward

Prices collapsed for most products and after 1931 the Government had to act. All sectors of the economy were protected, but full protection could not be given to home food producers because of responsibilities to the Empire. In the 1932 **Ottawa Agreements** the general raising of tariffs on non-food imports was not applied to food from members of the Empire but limits to imports were agreed. But agriculture in the UK was affected less than in many other countries because (a) the UK was a net importer of cereals, and (b) liquid milk, potatoes and most vegetables were naturally protected from cheap imports.

1932   **Wheat Act.** Introduced subsidies on wheat financed through tax on wheat imports (of which there were plenty). The levy was related to a 'standard' price and was extinguished if home price reached the standard. The idea of a standard quantity was used too, with a scaling down effect of the subsidy if domestic production exceeded the standard quantity. Following the introduction of this subsidy, wheat acreage rose by 60 per cent between 1931 and 1938. The Wheat Commission was set up to administer the system.

1931   **Agricultural Marketing Acts** (with ability to regulate supplies). Led to the
and    establishment of Boards for Milk, Potatoes, Hops, Pigs, Bacon. Improve-
1933   ments to be hoped for were (1) an improved bargaining position for farmers; (2) better marketing, arising from the suspicion that the distributive system was hopelessly inefficient and expensive with greedy middlemen. Powers of Marketing Boards variously included an ability (1) to trade in the regulated products; (2) to control movements and transactions other than by trading; (3) to limit production. Imports could be restricted by Minister's action on advice from Market Supply Committee set up under 1933 Act. The main impact was on imports from N. Europe. e.g., Danish bacon: 1932–391 000 tons; 1936–172 000 tons. This was probably the only effective import curb.

1934   **Milk Act.** Introduced a subsidy on butter and cheese. **Cattle Industry Act**. Introduced subsidies on fat cattle, with the Livestock Commission set up to administer them.

## Preparation for war

Agricultural Developments 1936–39 were dominated by the imminence of another war, although the explanations offered at the time were related to combating the remaining causes of poverty and of restoring the physical condition of the soil. The setting-up of a shadow Ministry of Food (separate from agriculture), formed initially as the Food (Defence Plans) Department (1936) within the Board of Trade was an important part of the preparations.

1937   **Agriculture Act 1937.** Gave a ploughing-up grant, believed to be a rejuvenating treatment. Also acreage payments were made for barley and oats.

1939    **Agricultural Development Act.** Introduced a subsidy on liming and on phosphate-rich basic slag. By the late 1930s the only major products *without* some form of assistance in the form of Marketing Board, subsidy, import tariff or quantitative restriction (some or all) were poultrymeat, wool and horticultural produce.

## Second World War: 1939–45

Policy objectives were: (1) the maintenance of calorific value of the national diet, and nearly equal distribution of this by allocation and rationing; (2) a major switch of production from animal to vegetable foods; (3) a fuller utilisation of domestic resources; (4) the replacement of imported animal feed by a smaller home-produced quantity capable of sustaining the few livestock.

Methods used were: (1) a ploughing-up campaign (as in the First World War) with some 2.5m ha of permanent pasture ploughed, changing the balance of land use from 2/3 permanent grass to 1/3 by the end of the War; (2) the establishment of County War Agricultural Executive Committees, consisting mostly of farmers nominated by NFU, to encourage ploughing up, and with power to dispossess inefficient, idle farmers; (3) creation of Women's Land Army; (4) arrangement of machinery sharing; (5) encouraging the rapid spread in use of tractors; (6) the rationing of animal feedstuff, with top priority given to milk production.

The result of these measures was that output rose by 30 per cent in 6 years, with a 70–95 per cent rise in calorific content. There was also, as a by-product of greater government monitoring, control and rationing, a great improvement in agricultural statistics over this period. For the first time a national farm income could be calculated. During the war the Ministry of Food emerged as the monopoly state trader in foodstuffs. It bought all home-grown (and foreign) food and resold at lower prices, subsidised by heavy general taxation. High prices were offered to farmers for priority foods, e.g., milk, but very low prices for others. This brought conflict with the Ministry of Agriculture who supported the NFU view that not all farmers and land could produce the priority foods.

This led to a chaos of price adjustments by different ministers and officials at different times. In 1943 the system was rationalised so that (1) price changes were put under the control of the Ministry of Agriculture; (2) prices were to be reviewed only annually; (3) prices had to be considered altogether, i.e., in relation to other products and their overall effect.

1940    Wages of farm workers increased substantially after 1940 when Ernest Bevin became Minister of Labour in Churchill's coalition government (thought to be a condition of his acceptance of the post) but the increased wage costs were fully recouped by farmers in the form of higher prices.

1943    Tax: high national farm income caused by raised product prices attracted the attention of the Inland Revenue Department. The aggregate farm income had increased fourfold since 1939. Previously, taxable income was deemed equal to rental value. Now a change was instituted towards tax on actual income. This started the necessity for farmers to keep accounts which also proved useful for management purposes.

1944    The NFU elected James Turner (later Lord Netherthorpe) as President. He managed to oust the chairmen of the Marketing Boards from the Annual

Price Review. The NFU's first objective was to secure full recoupment of cost increases (despite rapid increases in income). No allowance was to be made for increased productivity.

1945 An Act consolidated the advisory service and founded the National Agricultural Advisory Service (NAAS) by bringing together former County Council, University and College advisers (financed by the Government). Only one major group was not included, agricultural economists.

1945 The Labour government decided that wartime Price Reviews should continue.

## Postwar recovery

By the end of the war the British economy was in extreme difficulties. Two-thirds of our export trade had been lost; most overseas assets had been sold off; foreign exchange reserves were exhausted; old food-supplying partners had no longer food to sell in former quantities as a result of their enforced self-sufficiency during wartime and the establishment of their own manufacturing industries, combined with more domestic food demand from their own population (e.g., Argentina). The effect of war was aggravated in 1946 by the abrupt ending of the Lend Lease aid from the US and in 1947 by flooding in East England. Production of many food products was already heavily expanded. The result was that food supply was very short with reduced levels of rationing to below war-end levels and, lowest of all, in 1947. The government's policy was to allocate its slender foreign reserves to buy animal feed to enable the expansion of livestock production.

1946 **Hill Farming Act.** Continued wartime subsidies to hill farmers on hill sheep and cattle (headage payment) and gave grants on capital expenditure, with the objective of boosting production.

1947 A **Price Review** set as a target a 20 per cent increase in net output, with component targets for the main products. The prices of products were increased plus an additional £40m of income given to farmers for 'capital injection'. The Ministry of Food was retained as the state trader in food and rationing used to control food prices. The removal of control in the food market would have been inflationary and would have led to a worsening balance of payments; the government was particularly worried about imports from dollar countries.

1947 A new **Agricultural Act** was necessary to continue Price Review arrangements in peacetime. Section I contained the famous declaration of policy aim (quoted in Chapter 14.5.1). The terminology assumed the continued existence of a Ministry of Food as the state sole food trader, and an implied commitment to a socialist-type organisation. Method of implementing the Act comprised 'guaranteed prices' and an 'assured market', with provisions for Annual Reviews (with interim reviews if necessary) conducted between the Ministry of Agriculture and the NFU. Prices were guaranteed for 5 years and adjusted (upwards) annually and input subsidies were given (for example on fertiliser).

1947 **Town and Country Planning Act.** Imposed controls on the urban development of agricultural land. Farm buildings were exempt, as were changes of land use within agriculture. This Act nationalised the right to develop land. Followed reports by Barlow, Scott and Uthwatt.

1948     **Agricultural Holdings Act.** Consolidated material in the 1947 Agriculture Act and earlier legislation. It gave tenants virtual life-long security if they paid their rents. One reason for tenure security was the feeling that permanence would result in better and more productive farming by tenants. Arbitration on rents was to take as its definition that rent which 'should be properly payable' – interpreted by arbitrators as levels currently paid for similar land in the same district with a consequence that rents were kept down. The Act allowed lettings for less than 1 year. Lettings of 2 years or more were to revert to normal tenancies on their termination. Licences (not lettings – these must be less than one year) could be granted for special purposes (development, defence, special co-ops) only with Ministry approval.

1949     **Agricultural Marketing Act.** Revived most Boards under pressure from the industry and the Ministry of Agriculture, but with limited powers while the Ministry of Food was still the sole food trader.

1949     **National Parks and Access to the Countryside Act.** Following influential reports by Dower (1945) and Hobhouse (1947), established the National Parks Commission (later changed to the Countryside Commission, 1968) and the Nature Conservancy (later the Nature Conservancy Council, 1973). Under this legislation the Commission set up parks and Areas of Outstanding Natural Beauty (AONB), and the Conservancy the Sites of Special Scientific Interest (SSSIs) and National Nature Reserves (NNRs), the latter permitting payments to be made to owners who took part in management agreements.

1952     For selected products, fixed prices were replaced by Minimum Guaranteed Prices for review products (wheat, barley, oats, rye, potatoes, sugar beet, milk, fatstock, eggs and wool) with deficiency payments – calculated on an industry basis. By this date there was a much improved food position (and Balance of Payments position). The policy of general expansion replaced by selective output increases at some cost; this remained the policy to 1956. In 1952 a new set of higher targets was set – 60 per cent of prewar output by 1956 (as opposed to 50 per cent in 1952). As time went on, an output equivalent to 160 per cent of the prewar level was achieved, but not with the commodity balance which was wanted or expected, e.g., over-expansion of milk. In the mid-1950's the idea of a standard quantity was reintroduced and applied to milk from 1954 using powers under the 1947 Act. Note that the greater expansion of food production was to aid the balance of payments and *not* to achieve greater self-sufficiency *per se*.

1952–5     This period saw the gradual decontrol of food marketing, commodity by commodity as food supplies increased. In 1955 meat was de-rationed, marking the end of the functions of the Ministry of Food which was then merged with the Ministry of Agriculture to give MAFF. Its loss-making on food transactions (with a peak of £500m in 1952) was taken over by larger farm subsidies.

1956     By 1956 the return to private enterprise in the food industry required a reshaping of the 1947 Act as the price fixing system had been written around the assumption of State trading. There was also a feeling that the support system was too costly. The amount of support to agriculture by the state (in the form of subsidies, topped up by tariffs, etc.) was half-to three-quarters of aggregate net income of farmers, approximately equal to a 20 – 5 per cent *ad valorem* tariff. Much food could have been imported at the same prices with a substantial saving to the taxpayer. The result was that 1956 witnessed the first non-agreed Annual Price Review settlement, i.e., not agreed by the NFU. NFU complained that the system of annual reviews was too risky, carrying the possibility of price reductions. By way of compromise a White Paper, *Long-term Assurances for Agriculture*, November 1956, was published containing assurance consolidated in the 1957 Act. Also 1956, British Egg Marketing Scheme with its Board; lasted until 1975 when it became the Eggs Authority, wound up in 1986.

1957    The **Agriculture Act** gave assurances that the total value of guarantees would not be reduced by more than 2.5 per cent per year, that guaranteed prices of individual crops or livestock would not be reduced by more than 4 per cent per year (which subsequently limited government's hand in bargaining and resulted in higher prices than necessary, for example with eggs), and that the guaranteed price of livestock products could not be moved by more than 9 per cent in any 3 years. In retrospect the assurances were not of great value because they applied to nominal prices (that is, before inflation had been taken into account), and standard quantities still applied.

Additional features of the 1957 Act were:

1. An increased usage of production grants; these started in a small way in the 1930s but expanded greatly post-1957. At their peak in the 1960s they accounted for nearly half the total value of the guarantee. They allowed guaranteed prices to be nearer world prices, with less distortion of production, lower deficiency payments and reduced risk of fraud. They invoked less criticism from abroad than product price support. They were intended also to promote efficiency and took the place of an active credit policy. Examples (not all in the 1957 Act) were grants for liming, fertilisers, calf-rearing, hill-livestock drainage, TB eradication.
2. Farm Improvement Scheme and Horticultural Improvement Scheme. Introduced, leading to a rapid increase in investment (a form of production grant).
3. Farm amalgamation grants of one-third the cost (excluding the cost of land) of amalgamations were introduced, but with negligible uptake.
4. Setting up of Pig Industry Development Authority (PIDA).

1958    **Agriculture Act.** Changes to rent arbitration procedure altering the guideline for arbitration to that of the market lettings, removing the rent-depressing effect of the 1948 Agricultural Holdings Act's arbitration ruling.

1958    White Paper, '*Assistance for Small Farmers*.' As part of Annual Review procedure of the Small Farmers' Scheme introduced (1959) with grants to small farmers who undertook business development schemes. For the first time used the standard man-day concept. The Scheme helped the able but did not help the low income, poor management group. Note that in the UK, peasant farmers are not protected *per se*, but help is given against low incomes.

## The 1960s

This decade was marked by (a) low world prices for agricultural commodities; (b) therefore an increasing liability for deficiency payments, particularly with domestic supply still expanding. However, (c) the real support to agriculture seems to have been held down *because of* low world food prices.

1962 (and 1968)
**Agriculture and Forestry Associations Act.** Gave Marketing Boards exemption from Restrictive Trade Practices Act (1956).

1963
Grants became available to encourage farmers to retire, also for the costs of amalgamation, e.g., re-siting of buildings.

1964
Annual Review brought in standard quantities for wheat and barley and minimum import prices for these commodities. There was also provision for a variable levy, but this was not used, thanks to kindness of foreign suppliers who raised their prices.

1964
Verdon-Smith Committee of Inquiry into Fatstock and Carcase Meat Marketing and Distribution rejected producers' proposals for a marketing board but proposed a marketing authority similar to PIDA for all meat and livestock.

1965
Home-Grown Cereals Authority set up, with emergency trading power.
**The National Plan.** Identified agriculture as a potential area of expansion for import saving through increased productivity but not by offering farmers higher prices, to the disappointment of the agriculture pressure groups.

1965
**Finance Act.** Introduced Capital Gains Tax at 30 per cent. Subsequently, major concessions were granted, e.g., roll-over relief, exemption on death and concessions on retirement.
White Paper, *The Development of Agriculture*, on the structural problem of agriculture (i.e., the problem of the small farmer) proposed schemes which were incorporated in the 1967 Act.

1967
**Agriculture Act.** (1) Farm Amalgamation and Boundary Adjustment Scheme raised the rate of grant to 50 per cent of costs excluding cost of land. A farmer had to add an 'uncommercial' unit of his own ( < 100 *SMD*) to form a commercial one (600 *SMD* and over). (2) Payment of Outgoers Scheme provided pension or lump sum to farmers of unviable holdings which were amalgamated under (1) above. Farmers' main income source had to be from the farm. (3) Central Council for Agricultural & Horticultural Cooperation. (4) Meat and Livestock Commission (incorporating PIDA). (5) Gave power to create Rural Development Boards in GB, with task of drawing up integrated programmes of rural development covering all aspects of rural land use. Grants to farmers for tourism and recreation provision were envisaged, and Boards were to have reserve powers for land purchase. The only Board to be established (Northern Pennines) was disbanded in 1970.

1968
The Economic Development Committee for Agriculture reported on *Agriculture's Import-Saving Role*. It made the case for import saving, concentrating on the technical possibilities.

1968
**Countryside Act.** Required public bodies to take into account in their activities the desirability of conserving the natural beauty and amenity of the countryside, including its wildlife. This was of relevance to MAFF in its approval of grant aid to buildings, drainage etc. Responsibility for National Parks was transferred from the National Parks Commission to the Countryside Commission, which was empowered to pay farmers in National Parks for amenity and appearance activities.

1969    Farming and Wildlife Advisory Group (FWAG) set up, with participation by voluntary groups, MAFF and NCC. Aim to provide FWAGs at county level to advise farmers and provide a forum for discussion and information.

1970    **The 1970 Agriculture Act.** Reduced covenant period for not selling off parts of amalgamated farms from 40 years to 15 in an attempt to increase the effectiveness of grants for structural change. Also, in the 1970 Act, legislation on statutory smallholdings recognised that smallholdings were no longer a 'step on the ladder', but should be a 'gateway' into the industry from which progression could not be assured. This promoted their reorganisation into larger more viable units.

1971    Variable import levy for beef (July). Although government (Conservative) thinking had since 1969 favoured an EC-type system for the UK, this was the first significant move. Eggs Authority replaced BEMB.

1973    Entry over a five-year transitional period to the European Economic Community and adoption of its **Common Agricultural Policy.** Beyond this year, much UK policy was integrated with the CAP although there has been considerable scope for independent schemes. 'Green currencies' gave UK government much greater control over sterling prices of agricultural commodities than had been envisaged. UK national farm structural policies were replaced by similar ones under EC Directives: on 'Modernisation of farms', 'Cessation of farming and the reallocation of agricultural land for structural improvement' and 'Provision of socio-economic guidance' (Directives 159, 160 and 161 respectively, of 1972).

1974    Countryside Review Committee. Set up by government to investigate all aspects of rural land use. Championed the principles of Consensus, Compromise and Cooperation.

1975    In the **Finance Act** Capital Transfer Tax replaced Estate Duty which had become largely ineffective. Major concessions were soon won for farming.

1975    In the White Paper, '*Food from our own Resources*', the government concluded that a continued expansion of UK agriculture at the rate of the previous decade (2.5 per cent p.a.) would be justified (thereby increasing self-sufficiency) on grounds of comparative advantage (it noted the high cost of imported food with falling value of £ and high world prices) and improved reliability of supplies. The prime advantage was seen to lie in the expansion of dairying (and in beef from more dairy cows) and sugar, and second in cereals and sheepmeat.

1975    UK voted by referendum to stay in the Common Market.

1975    EC directive on 'Mountain and Hill Farming and Farming in Certain Less Favoured Areas' (75/268/EEC) allowed payments to farmers on a livestock headage basis and higher rates of grant aid on capital investment. The Hill Livestock Compensatory Payments are a thinly-disguised direct income support given on social and amenity grounds rather than reasons of agricultural production.

1976    **Agriculture (Miscellaneous Provisions) Act** provided security of tenure for up to three generations, including the present occupier. This soon proved to be a deterrent to the letting of land, landlords preferring to sell vacant farms or to take them 'in hand'.

1976    **Rent (Agriculture) Act.** Removed the automatic right of a farmer to a possession order in respect to a tied cottage. In effect, prevented the eviction of ex-workers unless alternative accomodation was available.

1976    Land Development Tax introduced (on land sold for building etc.).

1979    Northfield Committee Report on ownership and occupation of agricultural land: (a) revealed the relatively small ownership by financial institutions (1.2 per cent); (b) found 90.7 per cent of Great Britain was still held privately; (c) recommended a Land Register or similar inventory of landownership; (d) revealed the understatement of owner-occupation in official statistics.

1979    Farm Animal Welfare Council set up by government.

1979    White Paper *Farming and the Nation* gave some form of assurance (a) that further expansion of production would be welcome; (b) of the continuance of national support measures, e.g. grants, tax reliefs etc. and (c) that the education and advice system would continue. It proposed that 'import prospects and the need for insurance continue to point to the desirability of increased agricultural output in this country; but not so strongly as to justify seeking the maximum output increase, regardless of the cost to the consumer or to the economy at large, or of its impact on the environment'. The White Paper looked for an expansion through productivity improvements of 10-20 per cent over 5 years. It also declared a policy of constraining EEC prices and of exporting UK low cost production.

## The 1980s: a reorientation of policy

The early 1980s have been marked by an increasing awareness of the conflict between 'productivist' agriculture and environmental quality, thrown into prominence by the budgetary problems of the EC and the accumulation of agricultural surpluses. The level of aid to agricultural capital spending has generally been lowered, and incentives to investment through the tax system have been reduced. On the other hand, concessions in general taxes on capital (Capital Transfer Tax and Capital Gains Tax) have lowered their impact.

1981    **Wildlife and Countryside Act.** Prompted by an EEC directive, deals with protection of individual species, habitat protection (especially SSSIs) and access to the countryside. The Act placed a statutory duty on the Nature Conservancy Council to inform owners of designated SSSIs of their existence and of practices which might damage them. Payments can be made to farmers for agreements not to 'improve' such a site. Where a farmer has a grant application turned down because of an NCC objection, the NCC must offer the farmer a management agreement. This arrangement also applies in National Parks. The voluntary principle of conservation policy has come under attack on grounds of injustice, expense and ineffectiveness, and some advocate more land use planning.

1981    Reduction in rates of grant for investment in dairying and pig production.

1982    Capital Gains Tax from now on related to gains after allowing for the effect of inflation, i.e., to real rather than nominal gain.
'Food from Britain' campaign announced to promote the marketing of British food at home and abroad.

1982    **Hops Marketing Act.** Wound up the Hop Marketing Board.

1984    Quotas at farm-level applied to UK milk production as part of a Community-wide attempt to contain production and support costs.

1984    **Agricultural Holdings Act.** Removed the rights of succession for new tenancies but left intact the rights of the families of existing tenant farmers. It aimed to increase the supply of farms to rent.

1984    Official statement of MAFF aims (see Table 14.1).

1984    Extension of areas deemed to suffer from natural handicaps and therefore eligible for special grant aid to 53 per cent of the UK's utilisable agricultural area.
Widespread reduction in rates of aid on capital grant schemes, changes which 'reflect the Government's declared aim of achieving a closer integration of conservation and agricultural policies' (MAFF Press Notice no. 433/1984).

1985    Pending agreement on a new set of EC structural policy measures (the old ones based on the EC Directives of 1972 and 1975 having terminated at 1985 year end), the UK government continued to implement the existing schemes, except the Farm Structure (Payment to Outgoers) Scheme. The Farm Amalgamation Scheme (under the same EC Directive) had already lapsed (1976).

1985    New structural schemes introduced (Oct) as part of the CAP's **New Agricultural Structures Policy.** Payments became available for conservation-linked activities and energy saving as well as items of capital such as buildings and field drainage. (Egg and poultry production does not qualify and special restrictions are placed on pig and dairy investment.) Farms in Less Favoured Areas receive generally higher rates of grant aid and special assistance for tourism and craft expenditure. For the first time young farmers (aged under 40) may qualify for increased aid.

1985    Major European Commission Green Paper on the CAP: *Perspectives for the Common Agricultural Policy.* Of the alternative ways the problem of growing surpluses combined with unsatisfactory incomes can be tackled, the EC clearly advocated a policy of lower product prices coupled, where necessary, with forms of direct income aids. Public debate on the issues raised led to a further document *A Future for Community Agriculture* containing proposals for direct income payments for early retirement, conversion of land to non-farming purposes, maintenance of the rural environment and overt protection of the environment. No general income payments were proposed, a withdrawal from the 'Perspectives' position. These two documents are likely to influence price fixing for 1986 and onwards.

1986    **Agriculture Act.** Designed to implement various changes, including the charging for certain types of advice from the MAFF Advisory Service and the payment to farmers for environment-related activities; special incentive payments made possible for the continuance of traditional farming methods in locations designated as Environmentally Sensitive Areas. Objectives include the maintenance of wildlife and beauty of countryside.

1986    MAFF Advisory Service (ADAS) reorganised as the Farm and Countryside Service and the Research and Development Service.

1986    Eggs Authority wound up.

# Agriculture in the Rural Economy 15

## 15.1  Farming as Part of the Rural System

At various stages throughout this text attention has been drawn to the effect which farming has on the social and natural environment and to its relations with the non-farming sectors of the economy. Similarly, social and economic changes which originate primarily outside agriculture can have major implications for farming. A recent illustration of the influence of external pressures shaping farming change is the British legislation, arising out of public concern with conservation, which aims to protect aspects of the natural environment (see Appendix to Chapter 14). Of course, some of the factors which awoke the conservation interest were brought about by changing patterns of farming, so their creation was not entirely exogenous. On the other hand, the conservation lobby would probably not have become so effective had not improved opportunities for leisure and recreation enabled more people to have time in which to become aware of the issues and to use in promoting their concern; and this increased spare time has come from changes primarily in the non-farming sector of the economy. Even industrial redundancy, early retirement and unemployment may have played some part in reinforcing the environmental movement.

Another example of the interaction between agriculture and the wider economy is the influence on the level of farm-workers' earnings exercised by the presence or absence of alternative employment possibilities in a rural area (see Chapter 7). In the opposite direction, labour-saving farming practices have required far fewer men to be engaged on farms, with implications for the viability of village schools, shops, public transport, and so on.

In this chapter an attempt is made to put agriculture in its context as an activity influenced by its surroundings and also as a means for changing them. This is done in two stages. First, we take a systems approach to explain the nature of the interactions between agriculture and its environment. Second, we describe the position of agriculture within the rural economy and the use of public support of this industry as a way of stimulating the level of economic activity in the countryside.

### 15.1.1   A systems approach

Interactions between sectors are more clearly understood, explained and hopefully better predicted if farming is recognised as being only part of the rural 'system' which is itself only part of a larger world-level system, sometimes termed the 'ecological' system. The notion of a **system** is easier to envisage loosely than to define precisely, although it can be described as an assembly of related parts more closely related to each other than to non-related parts.[1] Human society is a system, one part of the whole. So too is the biological system. Systems can be part of other systems, so that the economic system is a sub-system of the human society system. Each system will therefore have an environment in which it sits. It will also have a boundary, a rather diffuse edge which separates those things which form part of the system from those that do not. For example, if we think of a motor car as a system of related parts – a transportation system – then while an engine is part of that car-system, a bird flying in the sky above the car clearly is not part of it.

Boundaries of systems can be closed or open, indicating whether elements can go in and come out or not. If we think of a club as a system, if outsiders are prevented from joining the club the boundary is thought of as closed. If, however, newcomers are welcomed across an open boundary the nature of the system inside may be altered as they make their influence felt within the club.

The point of adopting this 'systems' approach is that, by taking a broader view, a better understanding of individual sections is obtained. For example, in explaining human behaviour it is very useful to know about the family systems to which individuals belong: a student from an affluent family might be expected to take a different view of getting into debt than another with a poor background. Their parents might bring different pressures to bear about the 'rightness' of borrowing. The important interactions between members of the system are not lost by such a view.

The next question should be: what binds a system together? Alternatively, what are the relationships which give it its structure? Shared beliefs, values or interests may keep some clubs or pressure groups together. Profit making may be the objective which keeps the various parts of a business together. The desire for peace, security and freedom may be what lies behind the system of democracy in Western Europe. Survival of the species is probably what gives much of the biological system its shape.

### 15.1.2 Farming and systems

It is useful to think of human society, itself part of the ecological system, as composed of a number of sub-systems:

A. *the economic system*, being the process of producing, distributing and consuming goods, primarily held together by a network of market relationships
B. *the political system*, within which organisations exist by which conflicts within communities and societies can be regulated;
C. *the socio-cultural system*, the set of values, norms and goals which man has created and by which he lives, held together by a network of social relationships;
D. *the technological system*, comprised of educational and research activities.

Farming is itself a system, or rather a set of systems, for each type of farming and, indeed, each farm could be viewed from the systems angle. Each is a system created to generate income and other objectives for the farmer using available resources within the constraints imposed by biological relationships, the legal system, the price system and so on. However, farming also spreads over all four of the systems described above – a foot in each camp. This point is made in Figure 15.1 where farming, given a dashed boundary, overlaps the other systems. A farmer is both a member of society, sharing some values with other farmers to form a group within society, and a participator in the economic system as a user of scarce resources and a producer of food. Similarly, farming forms part of the political system, with its own pressure groups, and is also part of the technological system. Students of agriculture who are or who will eventually take up farming may consider themselves in the agricultural segment of the technological system. And farming, involving biological activities, has an overlap with the system of the natural environment – but to show this properly in Figure 15.1 would require a third dimension rising up from the page.

Within each of the four systems described above, farming has a boundary (or 'interface') between itself and the rest of society, and this boundary may be more or less open, permeable or impermeable. For example, in the socio-cultural system, the farming community may cut itself off from the rest of society, and it has been frequently observed that farmers do tend to keep to themselves, to marry into other farming families and to remain

## FIGURE 15.1
## Agriculture as a System Within Systems

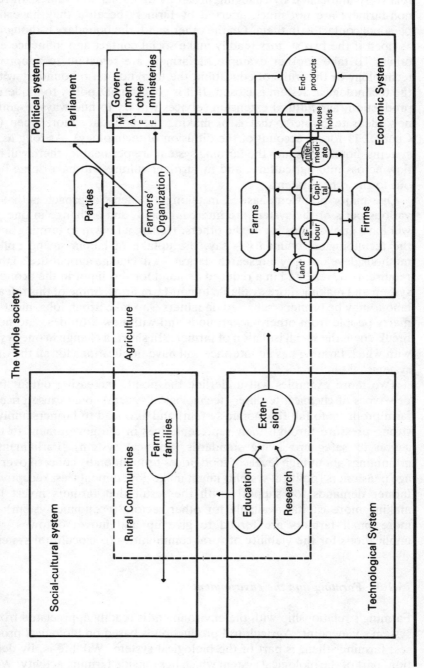

SOURCE   Louwes, S. L. (1977) 'Inhibitors to change in agriculture: is a pluridisciplinary approach needed?' *Eur. Rev. Ag. Econ.* **4**:3, 271–98.

relatively untouched by changing urban living habits and values. In turn, non-farmers are not much affected by farmers because they have little personal contact with them. On the other hand, the boundary is thought of as open if the two sectors readily make social contact and influence each other. To take another example, if farmers are reluctant to adopt new technology or the fruits of education, the farm/non-farm boundary within the technological system is closed. If it is government policy to modernise production, agricultural extension services will try to find ways of getting new ideas to farmers, that is of making the boundary more open (see Chapter 11 for an account of the diffusion of technology). As long as the systems' boundaries into the farming system are permeable, there will be a flow across into agriculture, and in turn agriculture will have some flows out to the other systems.

One major point emphasised in taking a systems approach is that the various parts of the system are interconnected, and a change in one part will have implications for all the others. Hence, a flow in to farming across the technological boundary – say, the uptake of labour-saving milking methods developed by a research station – will change agriculture's labour requirement, reflected in a reduced demand for this input in the economic system and making more available to non-farm firms. Some of this released labour may be farmers' sons and daughters who take urban jobs, meet and marry people from other backgrounds and with new attitudes, and hence break down the social isolation of farmers. In short, a change in one system with which farming has an interface will have implications for all the other systems.

Two more examples will underline the point. Increasing output from new types of chemical fertiliser (technological system) could cause prices of farm products to fall (economic system) and also lead to farmers applying strong pressure through their representatives in the government to take action to safeguard living standards (political system). Parliamentary grumblings about high levels of food prices brought on by concern over old age pensioners (political system) might make governments less receptive to farmer demands for support, with the result that farmers invest less, making more capital available for other sectors (economic system) and more small farmers are forced to give up and move to towns, with implications for the viability of rural communities (sociocultural system).

### 15.1.3 Farming and the environment

Farming's relationship with the environment is readily appreciated from a system's viewpoint. Agricultural production is based on biological processes; farming, then, is part of the biological system. Wildlife is, by definition, part of the biological system which lies outside farming activity. While the simplest nomadic life or gathering wild fruit and berries might hardly change the system, farming as it is usually understood involves harnessing, modifying or taking advantage of the biological system. Farming thus influences the rest of the biological system – wildlife and so on. The countryside is thus in part the result of man's activity on the 'natural environment'. What is not immediately obvious, however, is the extent of that impact.

## 15.2   Systems and Agricultural Policy

We have seen in the previous chapter that agricultural policy has by no means restricted itself to matters closely related to production – the turning of inputs of land, labour, capital and management into agricultural products which are wanted directly by consumers or by other businessmen as intermediate inputs to turn into consumer goods. Some strands of policy are concerned with this – policies of increasing the domestic supply of food or of improving the efficiency of farm production – but some of the most important strands are not.

Currently the major strand is government policy on farmers' incomes. As was seen in the Chapter 12, the concern arises not from any worry that farmers' rewards are inadequate to ensure enough is produced for the country as a whole, since for most major commodities there is oversupply at national or Community level, or both. Rather, the case for support is one based on a social rather than an agricultural rational – that without support living standards would be unacceptably low in some absolute or relative sense, although as we have already pointed out this case is promulgated mainly by regular affirmation of a belief and not by hard evidence, at least not for the UK. There is little doubt that falling farmer incomes would change the relative importance of agriculture as a generator of employment and source of livelihood in the rural economy, and that there would be social implications if this meant more commuting to urban jobs, more unemployment or the inflow of prosperous town dwellers to take the places of former agricultural workers who shift to towns. Government policy might well be justified to arrest these changes, but it would be primarily social in intent, not agricultural. Farm support could be used as a vehicle to achieve social ends, but there may well be alternative, cheaper and more effective ways of pursuing these social goals – such as the provision of non-farm employment in rural areas, subsidies to rural transport services and so on. Both the income and employment issues arise because agriculture has a social dimension – in our present context better expressed as resulting from farming's interface with the social system.

One policy strand which has risen to prominence in the 1980s – the environmental impact of farming and how to contain unwanted damage to the natural environment – is similarly not primarily agricultural. It arises, seemingly, from man's enhanced appreciation of his relations with the other elements in the biological system, especially with the non-farming parts and especially in the awarenesses of non-farmers. Because farming is part of that same system, agricultural activities intrude on man's relationships with the natural environment. Care has to be exercised before putting forward too simple a model of the farming/environment/man interaction since the natural environment is at least in part the product of past farming and land management, notably the appearance of the countryside as dictated by its pattern of hedges, trees, drainage and farm buildings.

Remedying the environmental consequences of farming can take the form of constraining changes in agriculture – by removing the incentives to change, such as high cereal prices which encourage the ploughing of old pasture, or withdrawing grants for draining marshland. Alternatively, the environmental objectives can be approached direct, such as by paying grants for the repair of stone walls or hedge planting, and by direct protective measures such as banning drainage of environmentally valuable marshland and, maybe, the imposition of land use control over long-established grassland, protecting it from the plough. We should pause to consider why such constraints or incentives are necessary. The main reason is that, in the UK and other capitalist market-based economies, farmers are independent business entrepreneurs who make their production decisions on the basis of the prices of their products and the costs of their inputs. When governments intervene in markets by raising product prices or subsidising costs, using the mechanisms described in Chapter 13, farmers will respond to these new price signals. If policy has not been fully thought through, and in practice not every eventuality can be explored, some unforeseen and undesirable outcomes are bound to emerge. The effect of high cereal prices in the UK is a prime example. The remedy is to correct the price signals being sent to farmers towards a more desirable pattern of output and land use. As UK agricultural policy is now in part dictated by decisions taken at Community level, freedom to make correcting changes in light of the developing experience is restricted, as the thwarted attempt by the UK to significantly lower cereal prices in the 1985 EC price-setting exercise exemplified.

However, part of the necessary constraints applied to farmers arises from the nature of the capitalist market economy. As was pointed out in Chapter 13, in an unrestricted free market prices will reflect only the costs of production which the farmer bears directly – fertiliser, fuel, land and labour charges, etc. – his **private costs of production**. He will only take these into account when deciding how much to produce. However, his production may impose considerable costs on the rest of the community in the form of pollution of water-courses, nuisance from stubble burning, loss of amenity when the appearance of the countryside changes through the removal of hedgerows, destruction of wildlife habitat and so on. Some of these **external costs** (also termed 'external diseconomies') are more easily expressed in money terms than others, but they share the characteristic that the price system does not make sure that the farmer bears these wider implications of his actions. Consequently society has to interfere to ensure that these external costs are somehow made to impinge on farmers' decisionmaking. This is done in a variety of ways, such as banning certain highly toxic insecticides, encouraging codes of practice for stubble burning and offering management agreement to farmers in areas of high conservation value to discourage changes in agricultural practice which would lead to their degradation or destruction. These aim at 'internalising' the external costs. Many people concerned with the natural environment believe that this process of internalisation does not go anything like far enough and, as we saw in Chapter 6, would like to see a much tighter control on the freedom of farmers to use their land as they wish.

As well as external costs farmers can provide external benefits to other farmers, to other industries and to the general public. An attractive countryside may be a bonus to the tourist industry and for recreation activities, but farmers do not take such things into account in their farm planning because they do not individually gain by doing so. Again, society will wish to internalise these benefits into private decisionmaking in the interest of the wider community. In the UK this takes the form of financial incentives or free advice for a range of environmentally-desirable practices, encouraged through the Countryside Commission and the Nature Conservancy Council and anticipated to be much expanded through MAFF as the result of the new structural measures to operate from 1986 which enable EC agricultural funds to be used more easily for environmental purposes.

Ironically, it has been the grant aid available for field drainage in the UK, justified partly on the grounds of the beneficial effect which the investment by one farmer has on the productivity of the land of adjacent farmers, which has been accused of perpetrating some of the greatest costs to society in terms of loss of wildlife habitat. This is a clear pointer to the desirability of taking a broad view when designing or monitoring policy intended to act in the national interest.

The parallels between conservation policy and incomes policy are obvious – both are fundamentally non-agricultural in their objectives, up to now both have tended to use agricultural instruments rather than direct approaches, and both have run into problems as a result. Both are undergoing public questioning of their methods of operation in light of their fundamental aims, criticisms to which the designers and administrators of policies are beginning to respond.

## 15.3   The Dangers of a Mono-disciplinary Approach

Farming has been viewed in this book primarily from an economic stand-point. However a wide view of agriculture (or holistic approach) which not only acknowledges the connections which farming has to other parts of the systems to which it belongs, but recognises these connections to be of major interest and concern to the general public, politicians and policy-makers, will not be adequately served by taking an exclusively mono-disciplinary attitude. Economics concerns itself primarily with just one system – the economic system – and thereby can provide some information about the behaviour of farming. Economics is but one of a number of disciplines which look at agriculture from different aspects. A discipline might be described as 'a coherent body of knowledge deriving from a study of an aspect of reality determined by a particular viewpoint'.[2] Hence when we attempt to explain farmers' behaviour, while economics will help provide *part* of the explanation, other disciplines (sociology, social psychology, political science) will enhance the level of understanding, particularly when moving away from the narrowest confines of farm management decisionmaking. And because farming involves both human behaviour and biological processes, natural and physical disciplines need also to be employed. While the concept of an 'externality' in production is an economic one, such as the harmful effects on wildlife of fertiliser contamination of water courses, the existence of pollution depends on the physical relationship between fertiliser application, its retention in the soil and leaching by rainfall.

   The aim of science *as a whole* is to arrive at a satisfactory explanation of reality. As farming forms part of so many systems, a single discipline will run the grave risk of ignoring relevant factors when attempting to explain, interpret, predict and ultimately influence the present pattern of agriculture and changes in it. For example, the attempt to encourage small farmers to leave the industry by financial inducements related to the loss of money income they might incur by quitting farming has been notably unsuccessful. Schemes have failed to recognise adequately the non-monetary rewards from farming which retirement would involve giving up – the 'psychic' income flowing from the independence small farmers enjoy, their enjoyment of participating in technical processes of animal rearing and so on, and from being members of the farming community. Psychology and sociology here had obvious roles to play, but the concentration on solely economic aspects led to an inadequate understanding of the forces involved and an ineffective set of policy measures.

## 15.4  Agriculture and Other Sectors of the Rural Economy

Despite the desirability of taking a broad view which encompasses the biological, economic, political, social and cultural aspects of agriculture, sheer practicality forces the choice of one of these as the major way in which this activity is approached. For the remaining part of this chapter attention will once again be focused on the economic aspects of farming, but without losing sight of the other systems to which it belongs.

Within the economic system, farming competes with other users of resources for land, labour, capital and management. With these it generates part of national income and supplies consumers directly or indirectly with food and other products. We have described the position of agriculture in the national context (the percentage of employment it represents, the proportion of GDP it earns, the percentage of national food supply it produces and so on), but there has been brief reference to the importance of agriculture *in the countryside*, perhaps better described as the rural economy since it is the economic parameters which are here of primary concern. Because government policy is increasingly concerned with issues which embrace many aspects of farming other than food production, and because the countryside is an increasingly complex place, it is necessary to know more about the role of agriculture within its rural context. It is patently obvious that farming is not the only activity taking place in rural areas, if it ever was. Pertinent questions include how important is farming as an employer in the countryside and how effective is the support of agriculture in terms of job-creation? To what extent does the income from farming form a key element in determining the incomes of the whole countryside, and is the maintenance of a prosperous farming sector the best way of ensuring viable rural settlements? How do alternative economic activities, such as forestry, tourism, or rural factories, compare as generators of income or employment and would public funds be more cost effective if used by them rather than by agricultural support?

One starting point is to look at the relative significance of agriculture in the regions of the UK. From Figure 15.2 we can see that, in terms of the share of agricultural product generated in this country, the most important region is the South-East of England. But this region is also top of the list in terms of *all* economic activity. The relevance of agriculture is better indicated by the percentage of regional total economic activity represented by farming; the figures for 1983 show the agricultural sector to be relatively largest in East Anglia (7.2 per cent of GDP) and Northern Ireland (5.1 per cent) and to make its lowest contributions in the South-East and North-West of England (1.0 per cent in both cases). In other respects East Anglia and Northern Ireland are very different; East Anglia generates more than double the amount of agricultural output but has less than a third of the number of farms and, as a result, the average farming income per person

412

## FIGURE 15.2
Percentage Distribution of Agricultural and Gross Domestic Product by
Standard Statistical Region and Agriculture's Share of Regional Product,
1983

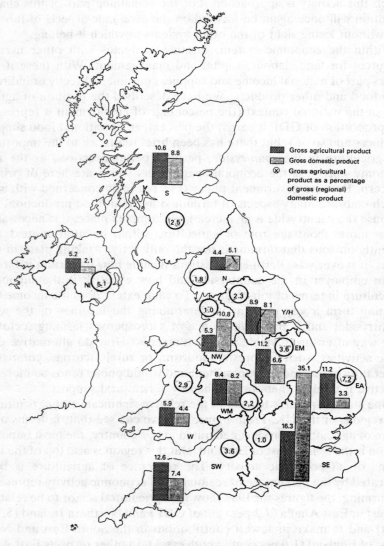

SOURCE   adapted from MAFF (1986) *Farm Incomes in the United Kingdom* (London:
(HMSO).

engaged in agriculture is more than twice the figure for Northern Ireland. Consequently, the part played by agriculture and its suitability to act as a vehicle for stimulating the rural economy is likely to differ. On the other hand, within those regions further down the list in terms of agriculture's importance to their economies, such as the South-West of England and Wales, there will be areas well away from urban influence in which farming assumes a greater economic and social role. Some finer approach than a rundown of regional characteristics is obviously needed to enable 'rural' areas to be identified.

### 15.4.1  What is 'rural'?

While it may seem obvious what is meant by 'rural', problems arise in trying to define the term. The appropriateness of any definition will depend on the purpose for which it is needed. For example, rural land uses might embrace farming and forestry, and a particular area of the country might be described as rural if more than a certain proportion of its surface were in these activities. In this sense, rural implies an arbitrary point within a continuum of land-use proportions. On the other hand, if the overwhelming majority of the people living in the area travelled to urban centres outside the area to work, the area would differ in a very important way from one in which its agriculture and forestry were the dominant employers.[3]

It should be no surprise that there is no single agreed criterion used in official designations of rural areas. One recent definition, used in analysing the results of the 1981 population census, is based on the amount of urban development shown by Ordinance Survey maps. 'Urban area' is identified as urban land extending for 20 ha or more and where, within a set of contiguous urban areas, there is a minimum of 1000 people; rural areas are taken as the residual. On this basis, about 5 million people in England and Wales live in rural areas out of a total of 49 million. Whether this finding means much depends on the aspect of the countryside which is of immediate concern. The problems faced by people living in rural areas are overwhelmingly of a social and economic nature, rather than to do with land use *per se*. If the main interest is in the difficulty of access to urban employment, education or entertainment for people living in rural areas, their degree of remoteness from services associated with towns will depend on the availability of transport and its cost as well as geographical separation. Different members of the rural community will be variously affected: age, income level and ability to drive are obvious factors which will determine how isolated a person may feel. Income levels and their distribution around the average will partly determine whether a problem exists, although in all communities there are likely to be some people who are disadvantaged compared with the rest. One approach has been to calculate an index from a range of factors which individually are linked to rurality. Cloke's 'rurality index'[4] brings together parameters such as population density, population change, occupational structure, distance from an urban centre and migration, using the statistical techniques of principal component analysis to combine them into a single figure. However, such a composite measure may be difficult to understand, and inappropriate when considering specific aspects of rurality, such as the proportion of total personal income of residents earned in agriculture. There is also the problem that between extremely urban areas and extremely rural ones there are intermediate stages which are difficult to rank. The classification into what is rural and what is not thus depends not only on the choice of criterion but also on some normative judgement in drawing the dividing line.

We have seen that UK farming engages a shrinking number of people (Chapter 7), a phenomenon also exibited by all other industrialised countries. While in the past this has also meant a drift of labour to the towns and rural depopulation, this is no longer the case in Britain. Population censuses show that, contrary to popular belief, for about two decades the numbers of people living in rural areas have been on the increase, a reversal of the former long and steady decline. Rather, it is the major conurbations which are showing falls in population (See Table 15.1). This rising rural population (which assumes the census definition of rural) is due primarily to people moving rather than 'natural' factors such as birth rate or death rate changes. The phenomenon is widespread but not uniform, and the most remote rural counties, such as Cumbria and Northumberland, have grown less rapidly.[5] Population loss, where it occurs, is probably localised and related to specific characteristics such as village size and the existence of public facilities; in Norfolk, for example, a country experiencing rapid growth overall, parishes of less than 200 inhabitants have continued to decline.[6]

## 15.5 Agriculture as an Employer and Income Generator

With a generally rising rural population and labour being displaced from farming by technological changes, it is not surprising that agriculture accounts for a small and declining proportion of rural employment, at least among those who are mainly dependent on farming for their livelihood. In contrast with the generally held view, agriculture is only a minority rural employer (rural being defined as in the 1981 Population Census), engaging, – as Table 15.2 shows, – only 14 per cent of the employed residents and exceeded by manufacturing, and by distribution and catering (together). Some rural areas are more dependent on agriculture than others: in rural

**Table 15.1**
**Population Change in Different Categories of Local Government Districts in England and Wales, 1961–1981**

| | 1981 Population ('000s) | Population change | | | |
|---|---|---|---|---|---|
| | | 1961–71 | | 1971–81 | |
| | | ('000s) | (%) | ('000s) | (%) |
| England and Wales | 49 011 | 2 645 | 5.7 | 262 | 0.5 |
| Greater London Boroughs | 6 696 | −540 | −6.8 | −756 | −10.1 |
| Metropolitan Districts | 11 235 | 58 | 0.5 | −546 | −4.6 |
| Non-Metropolitan Districts (all) | 31 080 | 3 127 | 11.8 | 1 564 | 5.3 |
| Large Cities[1] | 2 763 | −41 | −1.4 | −149 | −5.1 |
| Smaller Cities[2] | 1 687 | 37 | 2.2 | −55 | −3.2 |
| Industrial Districts | 6 668 | 459 | 7.6 | 200 | 3.1 |
| Districts that include New Towns | 2 165 | 337 | 21.8 | 283 | 15.1 |
| Resorts and Seaside Retirement Districts | 3 335 | 345 | 12.2 | 156 | 4.9 |
| Other Urban, Mixed Urban Rural, and more accessible Rural Districts | 9 449 | 1 584 | 22.0 | 661 | 7.5 |
| Remoter, largely Rural Districts | 5 013 | 403 | 9.7 | 468 | 10.3 |

NOTES  1. Large Cities are those Districts with over 175 000 population in 1971.
2. Smaller Cities are those 16 Districts with populations of between 70 000 and 175 000 in 1971.
SOURCE  quoted in Whitby, M. *et al*. (1985) 'Rural development Symposium', *J. Agric. Econ*. 36:1, 77–106.

Lincolnshire 24 per cent were employed in agriculture, while in rural Essex the figure was only 10 per cent. While farming as an employer has declined, manufacturing in rural areas has expanded, now markedly exceeding agriculture. Other evidence (though based on a different definition of rurality) suggests that, between 1960 and 1981, manufacturing jobs in small towns and rural areas in Great Britain (local authority districts in which all

**Table 15.2**
**Percentage of Employed Residents Working in Various Industries,
England and Wales, 1981**

|  | Agricul-ture | Energy and Water | Manufac-turing | Construc-tion | Distri-bution and Catering | Transport | Other Services |
|---|---|---|---|---|---|---|---|
| All areas | 2.1 | 3.1 | 27.4 | 6.9 | 19.3 | 6.5 | 34.0 |
| Rural areas | 14.1 | 2.5 | 19.9 | 7.5 | 18.1 | 4.6 | 32.3 |

SOURCE   Office of Population Censuses and Surveys (1984) quoted in Hodge, I. (1984) 'Rural economic development and the environment'. Paper to conference 'UK agriculture and the environment – the economist's approach', Agricultural Economics Society, 23 November 1984.

settlements had a population of less than 35 000) rose by 128 000 when the numbers in regular employment in agriculture fell by 318 000.[7] As with population changes, while rural areas have gained jobs, big losses have occurred in the large cities (see Table 15.3). A rise in service sector employment in rural areas is also apparent. Labour shedding from agriculture, coupled with this industrial relocation and other factors such as retirement migration and commuting, have resulted in agriculture being only a minority direct employer in rural areas.

The proportion of rural personal income generated by agriculture is not easy to ascertain. The more rural counties tend to have lower overall incomes, suggesting that the structure of rural counties is dominated by low-wage industries, of which agriculture is one.[8] We have already seen that hired farm workers receive wages which are near the bottom of the wages league. Furthermore the rewards for similar jobs are apparently somewhat lower in rural than urban areas, reflecting the fewer employment opportunities and transport problems. On the other hand farmers and their spouses, who together make up over half the agricultural labour force (53 per cent in 1983), are represented disproportionately heavily among the higher income receivers in society. However, a substantial minority of farmers gain some of their incomes from non-farming activities, and, even among full-time farmers and part-timers whose major

**Table 15.3**
**Manufacturing Employment Change in Urban and Rural Areas, 1960–78**

|  | 1978 Employment ('000s) | Change 1960–78 ('000s) | (%) |
|---|---|---|---|
| Great Britain | 7110 | −921 | −11.5 |
| London[1] | 769 | −569 | −42.5 |
| Conurbations[2] | 1677 | −605 | −26.5 |
| Free Standing Cities[3] | 1148 | −183 | −13.8 |
| Large Towns[4] | 901 | −20 | −2.2 |
| Small Towns[5] | 1887 | +256 | +15.7 |
| Rural Areas[6] | 728 | +201 | +38.0 |

NOTES    1. Great London.
        2. Great Manchester, Merseyside, Clydeside, West Yorkshire, Tyneside, West Midlands.
        3. Free Standing Cities: defined as other cities with more than 250 000 people.
        4. Large Towns: defined as towns or cities with 10 000–250 000 people.
        5. Small Towns: defined as Districts including at least one town with 35 000–100 000 people.
        6. Rural Areas: defined as Districts in which all settlements have fewer than 35 000 people.
SOURCE    Fothergill, S., Kitson, M. and Monk, S. (1983) *Urban Industrial Decline: the Causes of the Rural–Urban Contrast in Manufacturing Employment Change* (London: HMSO).

self-employed income source is the farm, about one third of income comes from investments or as employees (see Chapter 12). Such complications, allied to insufficient information, means that no precise indication is available of the proportion of personal income in rural areas arising directly from agriculture. However, it would seem unlikely to depart in a major way from the proportion of agricultural employment. Some light is shed by the estimates of gross agricultural product (gross value added) in the gross domestic product of the regions of the UK (Figure 15.2), but these areas contain substantial conurbations and therefore cannot throw up the significance of agriculture in the rural parts of the country. Nevertheless it is worth noting that in the most rural region, East Anglia, agriculture generated only 7 per cent of GDP in 1983.

Increasingly the countryside is the home of non-farmers. The migration *into* the countryside, already noted, has tended to alter its social and occupational balance. Young adults of the professional and managerial classes have been the most mobile and these will have figured disproportionately in the net gain of rural populations, raising the general level of income.[9] Relatively well-off people may retire to the country or have a second home there. However, these immigrants may exacerbate the income distribution problem in rural areas, producing a marked polarisation between the affluent recent incomers (plus the medium and larger farmers) and the generally poorer locals. Unfortunately comprehensive income data are not yet available by which the severity of this problem can be gauged.

In view of the small shares of employment and income accounted for by agriculture in the countryside it is clear that the problems caused by low rural incomes and other aspects of rural deprivation (indicated by poor housing, transport, public services, education and employment opportunities, and so on), where they exist among certain groups in society and in specific areas, are not solely or even mainly agricultural in their causation. And as will be shown below, farming can contribute little to their solution. Rather, the issue is one of rural, as opposed to agricultural, development.

## 15.6 Agriculture as a Stimulator of Economic Activity

Agricultural support is often advocated in the belief that prosperous farming produces a prosperous and viable rural economy (see Chapter 14). At the least, spending by governments on farming is seen to have a preventative role, in that it is assumed that without such support a declining agriculture would result in general economic and social malaise, with falling incomes both in farming and in related sectors, and fewer job opportunities, reduced local services, and so on. But a far more positive role is also sometimes envisaged, with agriculture acting as a catalytic agent in the wider process of rural development.

The objectives of rural development policy are, like those of agricultural policy, not precisely articulated and may be ascertained better from observation of government action rather than by scrutiny of official statements. They may generally be listed thus:[10]

1. the utilisation of resources which would otherwise remain idle but which may be profitably utilised in the economy, especially labour and land
2. the increase in the range of employment opportunities which are open to the labour force
3. the maintenance or development of an acceptable local service base and social environment base for rural residents
4. an improved standard of living and quality of life in assisted areas
5. a reduction of pressures on urban areas.

A pursuit of these objectives will bring benefits both to the rural community and to society at large. A better use of rural resources should, in the longer term, improve living standards generally. More acceptable rural housing for lower-earning families will reduce the demand on the urban stock, and better roads will allow more recreation to take place in the countryside, relieving town facilities to some extent. However the emphasis seems to lie with improving the lot of those who live in rural areas by providing for their needs, as perceived necessary by society at large. Specifically the aims relate to job opportunities, incomes, access to housing and services (health and education especially), social life and mobility. Though farmers may fall into groups for which some of the basic requirements are lacking, it is far more likely that other members of the rural community, with lower incomes and less wealth, will be those for whom assistance will be deemed necessary.

The balance between objectives of rural development policy will vary according to the locality, as too will the potential role of agriculture as an agent for change. Furthermore, politicians will stress different objectives to different audiences; it is understandable that a Minister of Agriculture will tend to emphasise the role of his industry in rural development. However, in lowland areas at least, farming is not generally seen by most observers as an effective instrument by which jobs can be created. Technological change seems to lead persistently to the displacement of farm labour, and a more prosperous farming sector seems more likely to substitute capital for labour than to take on more employees. While it is possible to postulate conditions under which more labour might be used by farmers (such as a rise in the relative cost of capital coupled with incentives for higher output through higher product prices or input subsidies and technical changes that required more labour and less capital) these run counter to most real-world experience.

The argument for supporting agriculture as the lynch-pin of the rural economy is heard most strongly with respect to the Less Favoured Areas: those parts of the country where farmers are deemed to operate under permanent natural handicaps. In the UK these correspond predominantly with the hills and uplands; by 1984 some 9.8m ha were officially classed as being in various handicapped areas, about 53 per cent of the UK's utilisable agricultural area. (The nomenclature was changed to Severely Disadvantaged and Disadvantaged when an extension of the areas was made in 1984, the old LFAs becoming SDAs.) The objectives of the special payments made to farmers in LFAs, according to the relevant EC Directive (75/268), are:

1. to counteract the depopulation caused by declining agricultural incomes and poor working conditions and
2. to ensure the continued conservation of the countryside.

**Table 15.4**
**Summary of Expenditure on Rural Development, 1980**

|  | Radnor (£'000s) | (%) | Eden (£'000s) | (%) |
|---|---|---|---|---|
| Agriculture | 3290 | 81 | 1297 | 54 |
| Forestry* | 0 | 0 | 4 | 0 |
| Manufacturing | 716 | 18 | 1039 | 43 |
| Services | 44 | 1 | 71 | 4 |
| Social development | 36 | 1 | 0 | 0 |
| Environmental enhancement | 2 | 0 | 3 | 0 |
|  | 4086 | 100 | 2413 | 100 |

NOTE    * Only 'special project' expenditure can be directly related to rural development policies.
SOURCE    Hearne, A. (1984) 'Integrated Development in Well Favoured Areas', in Whitby *et al.* (1985) 'Rural development symposium', *J. Ag. Econ.* **36**:1, 77–106.

These objectives, though contained in part of the CAP, are obviously *not* agricultural in essence and contrast strangely with the other productivity-orientated Directives on structural policy which were operative during the 1970s and early 1980s.[11]

Payments made to farmers in disadvantaged areas have been in the form of bounties (production subsidies) on the basis of numbers of livestock kept or on higher rates of grant for buildings or other farm 'improvements'. Spending on these schemes undeniably makes a significant contribution to the total incomes of farmers in the areas concerned. For example, in hill and upland farms surveyed on Exmoor, Dartmoor and Bodmin Moor in 1983–4 the average value of these special beef and sheep production grants was £65 per ha, a sum greater than the average net farm income of £61 per ha.[12] It is also evident from local studies that the majority of development spending coming to these areas from all sources goes directly to farms (see Table 15.4). Yet there is severe criticism of this agricultural spending on the grounds that the LFA Directive has not been an effective or efficient policy and the development which has occurred has not resulted primarily from this Directive but from other less well-funded policies.[13]

The main criticism is that these payments have not approached the objectives set for them; rather, they have been used with the prime intention of raising farmers' incomes, and that the declared social and environmental purposes have been little more than window-dressing. The main beneficiaries have been larger farms;[14] the payments have encouraged the development of larger, less labour-intensive farms at the expense of the smaller ones, resulted in a shedding of labour and the disappearance of the small units. Rather than retain employment in agriculture, the schemes seem to have resulted in its reduction. Overall, the support system has made 'very little contribution to the solution of wider social and economic problems, or to alternative job opportunities'.[15]

### 15.6.1 The multiplier

At the basis of the argument that the wider economy can be stimulated by spending on agricultural support is the notion of the multiplier – that ripple effect which ensures that an injection of spending into one sector of the economy spreads its effect to other sectors, producing an overall rise in income greater than the initial injection. The overall impact is commonly measured in terms of income or in terms of employment created (the income or employment multipliers; multipliers are also encountered in Chapter 4). Hence if the multiplier effect of agricultural support could be shown to be high, such government spending would be an attractive way of stimulating rural development. Advocates of forestry, tourism, manufacturing or service sectors as more suitable vehicles for development spending would need to arm themselves with multiplier coefficients for their favoured industries, augmented by information on the likely effect on the distribution of incomes and employment (**where** jobs are created being of particular importance), on local markets for labour, land and other inputs, the consequences for the national economy of the increased levels of production which would result, the relative impact on the natural environment and so on.

Attempts at measuring the size of the income multiplier have been made in a number of UK studies of development areas (Mid Wales, Eastern Scottish Borders, Isle of Skye), with typical values in the range 1.1 to 1.3,[16] but apparently these have not reached the level of measurement sophistication needed to rank agriculture precisely against other rural activities as an income or employment generator. Nevertheless, most commentators on development seem convinced that farming is an unattractive vehicle for economic stimulation, primarily because of the relatively small numbers of people engaged in it, the inelastic demand characteristics of its products, and technological developments which tend to expand output and depress prices, to the particular detriment of farming in the less productive areas. Rather, attention is given to forestry, industrialisation, tourism and other services. All of these seem to offer better long-term prospects.

## 15.7 Other Ways of Stimulating the Rural Economy

Of the non-agricultural ways of stimulating the rural economy given above, the attractiveness of forestry must be tempered by its apparent sensitivity to productivity improvements such as felling and planting machinery and on the future demand for wood and timber prices, which although assumed to show a modest rise may be undermined by the development of substitutes.[17] However, the fact that a substantial proportion of the UK forestry industry is under public control gives a potentially easy method of influencing the level of rural employment: the Forestry Commission owns and manages nearly half of the UK's 2m ha. of forest and is also responsible for administering another 0.6m ha. of woodland for private owners under 'Dedication Schemes'. Overall, some 21 000 people are directly employed by UK forestry, with another 15 000 indirectly dependent.[18]

The encouragement of industrialisation in economically depressed rural areas has a long history in the UK. The Development and Road Improvements Fund Act of 1909 established a fund, now administered by the Development Commission, which has been used to aid economic development in the countryside, including the establishment of rural industries.[19] Since the mid-1970s the Development Commission has expanded its activities (in England) with the construction of more small factories let to tenants.[20] In global terms the number of jobs created is small (about 8000 in the factories approved up to the end of 1979), even if multiplier effects are included. However they can make a significant impact on employment in the towns or villages in which they are established. Other organisations similarly have encouraged rural industrialisation by providing small business premises – local authorities, the Department of Industry and private developers. Another approach has been to attract industries by providing grants on buildings and equipment, tax relief and similar incentives to areas designated, under regional policy, as requiring special assistance. Most Scottish rural areas, mid- and North Wales, the Pennines and parts of Devon and Cornwall have at times been so classified. Assisted Area status is believed to have increased employment in those Areas by about half a million jobs since 1960.[21]

Service industries are important, providing more than 60 per cent of employment in most rural areas.[22] The rural ratio of service to manufacturing jobs is mostly about 2:1 but in the remoter areas this rises to 4:1.[23] The service sector has grown in relative importance in the UK economy, the fastest growth over the 1960s and 1970s being among financial and business services and professional and scientific services. Nationally, jobs grew by over 70 per cent in these groups between 1960 and 1984, as opposed to 4 per cent for all industries and services. In contrast, distributive trades were static and transport and communications declined. The implication of this change for rural areas is not simple. The bulk of rural tertiary employment is in consumer services (public and private) and retailing – not the fast-growing sectors, but ones which have experienced concentration as shopping patterns have switched away from village shops to town supermarkets and health, education and public utilities have reorganised into larger but fewer centres of supply. Tourism and farm-based recreation, frequently cited as new ways of earning a living in country areas, are obviously more appropriate in some localities than in others. Although the service sector may be the mainstay of rural employment, and despite the overall growth which may benefit individual country towns (through office development or rationalisation of service industries) and certain areas through tourism, it has an uneven local distribution. Improvements in communications technology, such as computers which allow people to work in their own homes and connect to other workers by telephone links, may partly offset this unevenness as the need to travel to work is reduced. The magnitude of this influence is as yet hard to envisage.

## 15.8 Conclusion: Agriculture's Role in UK Rural Development

A contradiction seems to exist over agriculture's role in the development of the rural economy. Most commentators agree that agriculture's importance has been overrated, even in the more remote rural areas. Attempts to support farming as a way of promoting jobs seems to be counter-productive, with most of the income benefit accruing to the larger farm which is encouraged to become more intensive and to expand. The small low-input farmer has not benefited to anything like the same extent and the numbers of small farms have continued to decline. Yet the bulk of financial support to rural areas in general is in the form of aid to farmers; this is also the case in those parts of the country singled out for specially generous assistance. The history of this mismatch of policy objective and instrument would require a longer explanation than is possible here, drawing on the changing objectives of agricultural policy and the political power of various rural interest groups, particularly farmers and landowners.

Public interest in the multifaceted role of agriculture in the countryside has been awakened by the debate over the cost of support which adds to unwanted surpluses and leads to undesirable changes in land use. While the farming lobby still promulgates the notion that a prosperous agriculture is necessary for a viable rural economy, evidence tends not to support this view, although detailed studies of rural employment patterns and income flows require further research. However, sufficient disquiet has been generated that the basic assumption of the wisdom of spending on agriculture as an effective way of promoting rural development is being questioned. Increasing attention is being drawn to ways of redirecting public spending in a more efficient way. The growth of manufacturing and service industries in rural areas indicates one direction in which public spending might accelerate a largely spontaneous change, with benefits in terms of income and jobs, although in recent years it seems that much of this employment growth has been at the expense of losses in city centres. A balance has to be struck between encouraging rural and urban development.

Given that the large-scale abandonment of agricultural support is politically unacceptable and therefore impractical, attention is currently being switched to redirecting farm spending to more development-friendly activities. In particular, the 1984/5 discussion of the new EC structural policy, to be operative from 1986, includes proposals to permit payments to farmers for carrying out nature conservation activities. These, and management agreements negotiated with conservation authorities, are in part job-preserving payments; they also release farmers from the need to secure their income only from the sale of crops or livestock and can produce a more environmentally desirable pattern of land use. Increasing policy interest in farm-based tourism and recreation are other ways of augmenting employment and income without stimulating more agricultural production. Part-time farming has a role to play here, whether combined with on-farm non-farming activities such as self-catering accommodation in converted farm buildings or off-farm employment in small-scale rural industrial development.

According to one commentator[24] the goal to aim for is the creation and maintenance of 'small but mixed patterns of rural communities and away from the weaknesses of the predominantly agricultural communities of the past or the predominantly middle-class commuter and retirement communities of certain rural areas in the present'. It seems clear that public spending on stimulating agricultural production is unlikely to achieve such an aim, but the political will to develop and finance alternative rural development strategies is not yet sufficiently strong to unseat the dominance of agriculture as an absorber of public funds. However, there are moves in that direction.

## Notes

1. Louwes, S. L. (1977) 'Inhibitors to change in agriculture: is a pluridisciplinary approach needed?', *Eur. Rev. Ag. Econ.* **4**:3.
2. Quoted from Louwes, S. L. (1977) op. cit.
3. The idea that there is a simple continuum between rural and urban judged on sociological grounds is challenged by, Pahl (see Pahl, R. E. (1966) 'The rural-urban continuum *Sociologia Ruralis* 6:1, 299–38). Rather, Pahl envisages a complex set of continua superimposed on each other.

4. Cloke, P. J. (1977) 'An index of rurality for England and Wales' *Regional Studies* **11.**
5. Craig, J. (1983) 'Town and country: current changes in the distribution of population in England and Wales', *Journal of the Royal Society of Arts*, **131**:5231.
6. Hodge, I. (1984) 'Rural economic development and the environment' Paper to conference 'UK agriculture and the environment – the economist's approach', Agricultural Economics Society, 23 November 1984.
7. Fothergill, S., Kitson, M. and Monk, S. (1983) *Urban Industrial Decline: the Causes of the Rural–Urban Contrast in Manufacturing Employment Change* (London: HMSO).
8. Hodge, I. and Whitby, M. (1981) *Rural Employment: Trends, Options, Choices* (London: Methuen).
9. Grundy, D. G. 'Rural development in the UK: a solution in search of a problem', in Whitby, M. *et al.* (1985) 'Rural development symposium', *J. Ag. Econ.* 36:1.77–106.
10. As given in Hodge, I. and Whitby, M. op. cit.
11. Tracy, M. (1984) 'Issues of agricultural policy in a historical framework', *J. Ag. Econ.* **35**:3, 307–18.
12. Nixon, B. and Turner, M. M. (1985) *Farm Incomes in South West England 1983/84* (University of Exeter: Agricultural Economics Unit).
13. Hearne, A. 'Integrated rural development in Less Favoured Areas', in Whitby *et al.* op. cit.
14. MacEwen, M. and Sinclair, G. (1983) *New Life for the Hills* (London: Council for National Parks).
15. Ibid., p. 16.
16. Hodge and Whitby, op. cit.
17. Forestry Commission (1977) *The Wood Production Outlook in Britain* (Edinburg: Forestry Commission), and Centre for Agricultural Strategy (1980) *Strategy for the UK Forest Industry*, CAS Report No. 6 (University of Reading).
18. Figures taken from Programme Organising Committee of the Conservation and Development Programme for the UK (1983) *Conservation and Development Programme for the UK* (London: Kogan Page) Ch. 3.
19. Since the mid-1970s the Commission's activities in Wales and Scotland have been taken over by the respective Development Agencies.
20. In designated Development Areas these were constructed by Industrial Estates Corporations and elsewhere by the Council for Small Industries in Rural Areas (CoSIRA).
21. Department of Trade and Industry (1983) *Regional Industrial Policy* (London: HMSO).
22. Phillips, D. and Williams, A. (1984) *Rural Britain: a Social Geography* (Oxford: Blackwell).
23. Gilg, A. (1976) 'Rural employment' in G. E. Cherry (ed.) *Rural planning problems* (London: Leonard Hill).
24. Wibberly, G. P. (1981) 'Strong agricultures but weak rural economies – the undue emphasis on agriculture in European rural development', *Eur. Rev. Ag. Econ.* **8**:2/3, 155–70.

# Selected Further Reading

In a book as broad as this it is impossible to journey far into the various areas mentioned. By way of remedy in this section we list further reading material which should be readily available in libraries or bookshops, or in a few instances, from the source institutions. Many of these titles will have already been cited in the main text. Those references listed in the notes to chapters which are not widely available have not been included here but they too will be valuable to the reader with access to specialist libraries. Those from UK publishers do not have the place of publication indicated as this is not usually a problem in obtaining them. Where a title is appropriate to more than one chapter it is quoted in full on its first appearance; subsequently it is given in abbreviated form.

## Chapter 1:   Introduction and Overview

This deals with the changing nature of agriculture in the economic, social, political and natural environment of the United Kingdom. There is unavoidable repetition of some of this material with later chapters, but alternative books setting out the scene include Newby, H. (1979) *Green and Pleasant Land?* (Penguin; reprinted, with an updating appendix, Gower, 1985); Bowers, J. K. and Cheshire, P. (1983) *Agriculture, the Countryside and Land Use* (Methuen); Blunden, J. and Curry, N. (eds) (1985) *The Changing Countryside* (Open University/Croom Helm).

## Chapter 2:   The Consumer End of the Food Chain

This chapter is concerned with the demand for food and food consumption. The account of the demand for food is based on neoclassical marginal utility theory and consumer preferences. The latter can be further analysed using indifference curves and the theory of revealed preference. Introduction to these concepts can be found in Hill, B. (1980) *An Introduction to Economics for Students of Agriculture* (Pergamon); Lipsey, R. G. (latest edition) *An Introduction to Positive Economics* (Weidenfeld & Nicolson); Ritson, C. (1977) *Agricultural Economics, Principles and Policy* (Crosby Lockwood Staples). The analysis of demand at an intermediate level can be found in Laidler, D. (1974) *Introduction to Microeconomics* (Philip Allen), and the methodology of estimating the demand for food in books on econometrics such as Leser, C. V. (1966) *Econometric Techniques and Problems* (Griffin). A multidisciplinary approach to analysing food demand and consumption which integrates several of these areas is Burk, M. (1968) *Consumption Economics* (New York: Wiley).

References on the history of dietary change include, for the British experience, Burnett, J. (1979) *Plenty and Want: A Social History of Diet in England from 1815 to the Present Day* (Scolar Press), and Barker, T. C. 'Changing patterns of food consumption in the UK' in Yudkin, J. (1978) *Diet of Man: Needs and Wants* (Applied Science Publishers). An official account of food consumption trends in Britain from the research aspect is given in Agricultural Research Council and Medical Research Council (1974) *Food and Nutrition Research* (HMSO). For a general history see Tannahill, R. (1973) *Food in History* (London: Eyre & Methuen).

Recent controversies concerning consumption are dealt with in Cannon, G. (1985) *The Food Scandal* (Century). However, the nature of the debate changes rapidly; useful sources of the latest research findings and surveys of consumption are to be found in the publications of the Consumers' Association (*Which?* magazine), the Health Education Council and the London Food Commission, and the *Caterer and Hotelkeeper, Grocer.* These complement the information which appears in the *National Food Survey*, discussed in this chapter.

# Chapter 3: The Farmer and the Food Industry in the Food Chain

This chapter is concerned with agricultural marketing and the food industry. An introduction to agricultural marketing is given in Barker, J. W. (1981) *Agricultural Marketing* (Oxford University Press). For an American approach see Kohl, R. L. and Uhl, J. N. (1985) *Marketing of Agricultural Products* (New York: Macmillan). Cooperation is discussed in Sargent, M. (1982) *Agricultural Cooperation* (Gower). Marketing Boards have not featured prominently in texts since the 1960s; see especially Warley, T. K. (ed.) (1967) *Agricultural Producers and their Markets* (Blackwell). An exception is Giddings, P. J. (1974) *Marketing Boards and Ministers* (Saxon House), which sees the Boards as political and administrative instruments. Most Boards and Authorities publish annual statistical digests, quarterly reviews and often monthly magazines as well. A useful collection of papers is in Ritson, C. and Warren, R. (eds) (1984) *Agriculture's Marketing Environment* (including a paper by J. Taylor of the Potato Marketing Board).

The food industry is the subject of a regular series of conferences at the University of Reading. Associated with these are Burns, J. A., McInerny, J. P. and Swinbank, A. (eds) (1983) *The Food Industry: Economics and Policies* (Heinemann), and Swinbank, A. and Burns, J. A. (eds) *The EEC and the Food Industries* (University of Reading: Food Study No. 1). Sheffield City Polytechnic also holds conferences on food marketing and papers appear in the journal *Food Marketing*, edited in Sheffield by Brian Beharrell. Food retailing is the subject of many journals and magazines of which three are particularly useful sources for up-to-date commentaries. The Economist Intelligence Bureau publishes *Retail Business* monthly and issues contain regular articles on specific food product markets. A market research orientation is given to the monthly reports of the

Mintel organisation. Less accessible but often of interest are the reports of the Institute of Grocery Distribution, Watford.

# Chapter 4: The Role of International Trade in the Food Chain

The integration of international trade into the food chain and the self-sufficiency issue are dealt with here. The theory of international trade and comparative advantage is described in many introductory economics texts, including those of Hill (1980) and Lipsey mentioned in Chapter 2 above. See also Lindert, P. and Kindleberger, C. (1982) *International Economics* (7th edn., Illinois: Richard Irwin), Sodersten, B. (1980) *International Economics* (2nd ed., Macmillan). Also Williamson, J. (1983) *The Open Economy and the World Economy* (New York: Basic Books) contains a useful short account of the development of international trade since 1945.

Agricultural trade relating to the EC is covered in Buckwell, A., Harvey, D. R., Thomson, K. J. and Parton, K. M. (1982) *The Costs of the Common Agricultural Policy* (Croom Helm). Third World issues are discussed in Matthews, A. (1985) *The Common Agricultural Policy and the Less Developed Countries* (Dublin: Gill & Macmillan). The American view is given in Paarlberg, D. (1980) *Food and Farm Policy: Issues of the 1980s* (Lincoln: University of Nebraska Press), and that from Australia in Bureau of Agricultural Economics (1985) *Agricultural Policies in the EC. Their Origins, Nature and Effects on Production and Trade*, Policy Monograph No. 2 (Canberra). The self-sufficiency issue is dealt with in Ritson, C. (1980) *Self-sufficiency and Food Security* (University of Reading: CAS Paper No. 8). It also forms part of most contemporary polemics on British agriculture, listed under Chapter 14 below.

# Chapter 5: Agriculture in the National Framework

The position of agriculture in the UK and in the EC, and in its world context is described and the longer-term changes outlined in this chapter. The general theme of change is covered in a variety of different ways in the standard textbooks on agricultural economics, such as Hallett, G. (1981) *The Economics of Agricultural Policy* (2nd edn, Blackwell); Hill, B. E. and Ingersent, K. A. (1982) *An Economic Analysis of Agriculture* (2nd edn, Heinemann). A more descriptive approach is taken in the Open University's *The Changing Countryside* (see Chapter 1 above). For the UK, basic statistical information and a review of the state of the agricultural industry appear in the official *Annual Review of Agriculture* (HMSO) and the Central Statistical Office's *Annual Abstract of Statistics* (HMSO). Official and other data sources are brought together in Burrell, A., Hill, B. and Medland, J. (1986) *Statistical Handbook of UK Agriculture* (3rd edn, Ashford, Kent: Wye College). A detailed description of the institutional structure of the industry is in Wormell, P. (1978) *The Anatomy of Agriculture: a Study of Britain's Greatest Industry* (Harrap Kluwer). Other bird's eye views of agriculture, covering a wide range of physical and economic characteristics but concentrating on the former, is Spedding, C. R. W. (ed.)(1983) *Fream's Agriculture* (16th edn, Royal Agricultural Society).

EC agriculture is the subject of many publications of the Commission of the European Communities (Brussels), especially the annual *Agricultural Situation in the Community* and a series of Green Europe reports and newsflashes. A brief description of the recent changes are in Marsh, J. S. and Swanney, P. J. (1980) *Agriculture and the European Community*, and for the longer term in Tracy, M. (1982) *Agriculture in Western Europe: Challenge and Response 1880–1980* Granada. For the widest perspective see Grigg, D. (1982) *The Dynamics of Agricultural Change* (Hutchinson). On agriculture's role in poorer countries, the World Bank produces an annual *World Development Report* which traces the changing importance of agriculture and traces the various difficulties it encounters. Among the many texts on the role of agriculture in the growth of poorer economies see Colman, D. and Nixson, F. (1978) *Economics of Change in Less Developed Countries* (Philip Allan); Mellor, J. (1976) *The New Economics of Growth* (New York: Cornell University Press); Johnston, B. and Kilby, P. (1975) *Agriculture and Structural Transformation* (Oxford University Press); Reynolds, L. (1976) *Agriculture in Development Theory* (New Haven, Conn.: Yale University Press); Eicher, C. K. and Staatz, J. M. (eds) (1984) *Agricultural Development in the Third World* (Baltimore, Md: Johns Hopkins); and Glatak, S. and Ingersent, K. (1984) *Agriculture and Economic Development* (Brighton: Wheatsheaf).

# Chapter 6:   Resources in Agriculture: Land

This chapter considers land not only as a resource for farmers in agricultural production, but examines its multiple use characteristics and its social and environmental dimensions. As a basic description of the physical dimensions of the national land stock and competition for its use, see Best, R. H. (1981) *Land Use and Living Space* (Methuen). Its influence on farming practice is described in Grigg, D. (1984) *An Introduction to Agricultural Geography* (Hutchinson). Many of the polemics of the last decade are heavily slanted towards the effect of agricultural policy on land use, particularly the loss of natural habitats for wildlife. As well as the books cited in Chapter 1 (*The Changing Countryside, Agriculture, the Countryside and Land Use*) see Shoard, M. (1980) *The Theft of the Countryside* Temple Smith; Pye-Smith, C. and Rose, C. (1984) *Crisis and Conservation: Conflict in the British Countryside* (Penguin).

For the planning of land use as it relates to rural areas see Donaldson, J. and Wibberly, G. P. (1977) *Planning and the Rural Environment* (Pergamon); Gilg, A. W. (1979) *Countryside Planning: the First Three Decades* (Methuen); and Blacksell, M. and Gilg, A. (1981) *The Countryside: Planning and Change* (George Allen & Unwin).

The use of land for farm businesses is described in Commission of the EC (1981) *Factors Influencing the Ownership, Tenancy, Mobility and Use of Farmland in the United Kingdom*, Information on Agriculture No. 74 (Brussels). Other quantitative information is collated in the land section of Burrell, A. *et al.* (1986) *Statistical Handbook of UK Agriculture*. The social dimensions of land are explored in Newby (1979) *Green and Pleasant Land?*, cited in Chapter 1, and in Newby, H., Bell, C., Rose, D. and Saunders, p. (1978) *Property, Paternalism and Power: Class and*

430 Selected Further Reading

*Control in Rural England* Hutchinson. A Marxist approach is taken in Massey, D. and Catelano, A. (1978) *Capital and Land: Landownership by Capital in Great Britain* (Arnold).

For the legal aspects of land ownership and use see Gregory, M. and Parrish, M. (1980) *Essential Law for Landowners and Farmers* (Granada). The legal and wider problems associated with land management are covered in Nix, J., Hill, P. and Williams, N. (1986) *Land and Estate Management* (Packard).


## Chapter 7: Resources in Agriculture: Labour

A basic description of the labour force is given in Edwards, A. and Rogers, A. (1974) *Agricultural Resources* (Faber) and changes detailed in Craig, G. M., Jollans, J. L. and Korbey, A. (eds) (1986) *The Case for Agriculture*. CAS Report No. 10 (University of Reading). A sociological view of both farmers and farm-workers is in Newby, H. (1979) *Green and Pleasant Land?* and in the longer study from which that book draws material about the hired labour force, Newby, H. (1977) *The Deferential Worker* (Penguin) and for the farmer Newby, H. *et al.* (1978) *Property, Paternalism and Power: Class and Control in Rural England* (Hutchinson). For a farm-business view of farmers see Harrison, A. (1975) *Farmers and Farm Businesses in England*, Miscellaneous Studies No. 62, (University of Reading: Department of Agricultural Economics). Part-time farming in the UK is comprehensively covered in Gasson, R. (1986) *Other Occupations of Farm Families* (Ashford: Wye College) and in other developed countries in OECD (1978) *Part-time farming in OECD Countries: General Report* (Paris: OECD). The UK hired labour force is described in detail, including earnings, in the annual MAFF publication *Agricultural Labour in England and Wales*.


## Chapter 8: Resources in Agriculture: Capital and Finance

This chapter looks at the amount of capital and finance used by this industry, both at the aggregate and the farm levels. The most comprehensive collection of information at the industry level is in Burrell, A. *et al.* (1986) *Statistical Handbook of UK Agriculture*. Each of the major clearing banks issues brochures on the financial aspects of farming, including the availability of grant aid to investment. At the individual farm level Harrison's *Farmers and Farm Businesses in England* gives a comprehensive picture, although now rather dated. The indebtedness of farms appears in MAFF (1986) *Farm Incomes in the United Kingdom* (HMSO): in its latest form this report of the Farm Management Survey provides a wealth of financial information well in excess of what its title would suggest. The financing of the individual farm business, including the appraisal of investments and approaches to treating capital consumption by making depreciation allowances is covered in Barnard, C. S. and Nix, J. S. (1979) *Farm Planning and Control* (2nd edn, Cambridge University Press) and Warren, M. F. (1982) *Financial Management for Farmers* (Hutchinson). Investment appraisal is considered specifically in Merrett, A. J. and Sykes, A. (1973) *Capital Budgeting and Company Finance* (Longman), and Lumby, S. (1984) *Investment Appraisal* (2nd edn, Van Nostrand Reinhold).

# Chapter 9:   Farming Structure

The structure of the farming industry is described in terms of farm businesses. Again, Harrison's *Farmers and Farm Businesses* is a valuable starting point. For the UK the MAFF *UK Agricultural Statistics* (HMSO) is a detailed information source, and for the EC the Commission's *Agricultural Situation in the Community* contains not only data for each member state but comment too. The relevance of the family farm in the UK is covered in publications from the Centre for Agricultural Strategy, the most important being Tranter, R. B. (ed.) (1983) *Strategies for Family-worked Farms in the United Kingdom* CAS Paper No. 15 (University of Reading). The relationship of economies of scale in agriculture is described in Britton, D. K. and Hill, B. (1975) *Farm Size and Efficiency* (Saxon House). For the US perspective see Ball, A. and Heady, E. O. (1972) *Size, Structure and Future of Farms* (Ames, Iowa: Iowa State University Press), and Vogeler, I. (1981) *The Myth of the Family Farm: Agribusiness Dominance of US Agriculture* (Boulder, Colo.: Western Press).

Taxation and business structure are closely linked. The general picture is described in Stanley, O. (1984) *Taxation of Farmers and Landowners* (Butterworth). The latest edition of Nix, J. S., *Farm Management Pocketbook* (Wye College) contains recent tax information and the National Farmers Union produces a guide for its members.

# Chapter 10:   The Supply of Agricultural Products

This chapter deals with the basis relationships which determine the amount which farmers supply to a market in the short term. Almost all introductory economics texts include sections on production economics, or the theory of the firm. For a farming stance see Hill's *An Introduction to Economics for Students of Agriculture*, and at greater length accounts are given in Hallett, *The Economics of Agricultural Policy*; Hill and Ingersent, *An Economic Analysis of Agriculture*; Ritson *Agricultural Economics: Principles and Policy*; and Upton, M. (1976) *Farm Production Economics and Resource Use* (Oxford University Press). For a long time established American texts in this area have been Heady, E. O. (1952) *The Economics of Agricultural Production and Resource Use* (New York: Prentice Hall), and Bishop, C. C. and Toussaint, W. D. (1958) *Introduction to Agricultural Economic Analysis* (New York: John Wiley), augmented more recently by Doll, J. P. and Orazem, F. (1978) *Production Economics: Theory and Applications* (Columbus, Ohio: Grid).

# Chapter 11:   Change in the Supply of Agricultural Products: Technological Advance

Here the longer-term influence on output of technological change is considered. The books cited for the previous chapter are also relevant here. The spread of new technology is described in Rogers, E. M. (1962) *Diffussion of Innovations* (New York: Free Press), in Part 2 (Jones, G. E., 'Agricultural innovation and farmer decision making') of Open University (1972) *Decision-Making in Britain: Agriculture* (Innovation and Farmer Decision-Making) and in Jones, G. E. (1973) *Rural Life* (Longman). An American perspective is given in Summers, G. F. (ed) (1983) *Technology and Social Change in Rural Areas* (Boulder, Colorado: Westview Press).

## Chapter 12:   Incomes and Agriculture

It is unusual for the incomes of UK farmers to receive special attention in spite of their role in agricultural adjustment and relevance to agricultural policy. A seminal work in this area is Bellerby, J. R. (1956) *Agriculture and Industry, Relative Income* (Macmillan). An account of the income issue is in Open University, *The Changing Countryside*, and Howarth, R. W. (1985) *Farming for Farmers?* (Institute of Economic Affairs). The 1986 edition of MAFF *Farm Incomes in the United Kingdom* is a radically improved version of an annual series and contains estimates of the income position of both the industry and individual farms, drawing on a variety of sources.

## Chapter 13:   The Price System and Agriculture

This chapter deals with the price characteristics of agricultural products and the mechanisms by which governments attempt to manipulate the markets. The basic workings of the free markets are described in standard introductory texts. The failure of the system to deal adequately with externalities and to provide public goods and to prevent public bads is covered in Ritson, *Agricultural Economics: Principles and Policy*. Texts dealing more specifically with such issues include Mishan, E. J. (1982) *An Introduction to Political Economy* (Hutchinson); Mishan, E. J. (1981) *Introduction to Positive Economics* (Oxford University Press); Nath, S. K. (1973) *A Perspective of Welfare Economics* (Macmillan); and Dorfman, R. and Dorfman, N. S. (1977) *Economics of the Environment* (Norton).

The devices which governments use to intervene in agricultural product markets are further described in Hill and Ingersent, *Economic Analysis of Agriculture*; and in Ritson, *Agricultural Economics: Principles and Policies*. These texts also deal with the mechanisms used by the Common Agricultural Policy. The CAP systems are more fully explained in Harris, S., Swinbank, A. and Wilkinson, G. (1983) *The Food and Farm Policies of the European Community* (John Wiley), and, less fully, in Marsh and Swanney, *Agriculture and the European Community*. An up-to-date review of the different systems applied to the range of commodities subject to CAP support is given in AgraEurope *CAP Monitor*, a subscription information system held by some libraries, and issues of the EC's own *Green Europe* deal with crop products and livestock products. The farm management publications from Universities and Colleges, of which the best-known is Nix's *Farm Management Pocketbook* from Wye College, contain information on the market interventions as they affect the farmer.

# Chapter 14:  Public Policy Towards Agriculture

In addition to the texts given in the paragraph above, the fundamental question of why governments of virtually all industrialised countries have policies to shape their farming industries is dealt with by James, P. J. (1971) *Agricultural Policy in Wealthy Countries* (Angus & Robertson); Hallett, *The Economics of Agricultural Policy*, and Bowler, I. (1979) *Government and Agriculture: a Spatial Perspective* (Longman). The shape of the CAP is best described in Harris, S. *et al.*, *The Food and Farm Policies of the European Community*, although a more readily understood, shorter version appears in Hill, B. E. (1984) *The Common Agricultural Policy: Past, Present and Future* (Methuen). Legislation on which UK policy is based is contained in Rogers, A. *et al.* (1985) *The Countryside Handbook* (Croom Helm/Open University). An assessment of the burden to the UK of participating in the CAP is given in Buckwell, A. *et al.*, *The Costs of the Common Agricultural Policy*. The official description is available from the information office of the European Communities as *The Agricultural Policy of the European Community* (Office of the Official Publications of the European Communities). Various aspects of the CAP are described in *Green Europe*, a series from the same source. Agricultural policy in the USA is described in Paarlberg, *Food and Farm Policy: Issues of the 1980s*, and Rausser, G. C. and Farrell, K. R. (1985) *Alternative Agricultural and Food Policies and the 1985 Farm Bill* (University of California: Gianni Foundation of Agricultural Economics).

Political pressures on the direction of policy are described in Self, P. and Storing, H. J. (1962) *The State and the Farmer* (George Allen & Unwin), and Wilson, G. W. (1977) *Special Interests and Policymaking* (John Wiley). The functioning of the political machine is also related in Wormell *Anatomy of Agriculture*. All are unfortunately rather dated. More recent and concerned with the EC generally is Neville-Rolfe, E. (1984) *The Politics of Agriculture in the European Community* (Policy Studies Institute). The rise of the environmental pressure groups and their influence on policy is described in O'Riordan, T. (1981) *Environmentalism* (2nd edn, Pion), and Lowe, P. and Goyder, J. (1983) *Environmental Groups in Politics* (George Allen & Unwin). The institutions concerned with the promotion of environmental issues as they effect farming are described in Pye-Smith and Rose, *Crisis and Conservation: Conflict in the British Countryside*, and Bury, P. (1985) *Agriculture and Countryside Conservation* (Wye College).

In the late 1970s and 1980s agricultural policy has attracted a stream of critical books, mostly overstating their case. The best is probably Newby, *Green and Pleasant Land?* and the most ferocious Shoard, *The Theft of the Countryside*. Those by Bowers and Cheshire, *Agriculture, the Countryside and Land Use*, and Howarth, *Farming for Farmers*, take a mainly economic standpoint but are flawed. The two books by Body, R. (1982) *Agriculture: the Triumph and the Shame* and (1984) *Farming in the Clouds* are primarily political polemics: the former should be read in conjunction with *Agriculture: the Triumph and the Shame – And Independent Assessment* (Wye College: Centre of European Agricultural Studies, and University of Reading: Centre for Agricultural Strategy).

## Chapter 15:    Agriculture in the Rural Economy

The first part of this chapter considers the relationship of the agricultural system with other systems, natural, economic, social, political. The systems approach is described in Spedding, C. R. W. (1982) *An Introduction to Agricultural Systems* (Applied Science Publishers), and Haines, M. (1982) *An Introduction to Farming Systems* (Longman). On a larger scale is Grigg, D. B. (1974) *The Agricultural Systems of the World* (Cambridge University Press), and Dalton, G. E. (1975) *Study of Agricultural Systems* (Applied Science Publishers).

The latter part of the chapter examines the role of agriculture in the economic system of rural areas and, particularly, the part it can play in stimulating the economic activity in them. An overall picture is given in the Open University's *The Changing Countryside* and in Phillips, D. and Williams, A. (1984) *Rural Britain: a Social Geography* (Blackwell). Planning in the countryside is dealt with in the titles given in Chapter 6 above. The process of rural development, including using various approaches to stimulating rural areas, is considered in Whitby, M. C. and Willis, K. G. (1978) *Rural Resource Development – An Economic Approach* (2nd edn, Methuen), and with specific regard to employment in Hodge, I. and Whitby, M. (1981) *Rural Employment: Trends, Options, Choices* (Methuen). Agriculture's integration with the rest of the economy, including input-output analysis and a consideration of multipliers, is dealt with in Craig, Jollans and Korbey, *The Case for Agriculture: An Independent Assessment*. The rise in importance of non-agricultural activities and the change in the nature of farming are both covered in Healey, M. J. and Ibery, B. W. (1985) *The Industrialisation of the Countryside* (Norwich: Geo. Books).

# Index

How do we know where genes are?

Using information drawn from the practicals, discuss some of the ways in which gene loci can be allocated to specific positions.

1000 words